METHODS IN MOLECULAR BIOLOGY™

Series Editor
John M. Walker
School of Life Sciences
University of Hertfordshire
Hatfield, Hertfordshire, AL10 9AB, UK

For other titles published in this series, go to
www.springer.com/series/7651

Yeast Functional Genomics and Proteomics

Methods and Protocols

Edited by

Igor Stagljar

Department of Biochemistry & Department of Molecular Genetics, Terrence Donnelly Centre for Cellular and Biomolecular Research, University of Toronto, Toronto, ON, Canada

Editor
Igor Stagljar, Ph.D.
Department of Biochemistry & Department of Molecular Genetics
Terrence Donnelly Centre for Cellular and Biomolecular Research
University of Toronto
Toronto, ON
Canada

QH
470
.S23
Y43
2009

ISSN: 1064-3745 e-ISSN: 1940-6029
ISBN: 978-1-934115-71-8 e-ISBN: 978-1-59745-540-4
DOI: 10.1007/978-1-59745-540-4
Springer Dordrecht Heidelberg London New York

Library of Congress Control Number: 2008944036

© Humana Press, a part of Springer Science+Business Media, LLC 2009
All rights reserved. This work may not be translated or copied in whole or in part without the written permission of the publisher (Humana Press, c/o Springer Science+Business Media, LLC, 233 Spring Street, New York, NY 10013, USA), except for brief excerpts in connection with reviews or scholarly analysis. Use in connection with any form of information storage and retrieval, electronic adaptation, computer software, or by similar or dissimilar methodology now known or hereafter developed is forbidden.
The use in this publication of trade names, trademarks, service marks, and similar terms, even if they are not identified as such, is not to be taken as an expression of opinion as to whether or not they are subject to proprietary rights.
While the advice and information in this book are believed to be true and accurate at the date of going to press, neither the authors nor the editors nor the publisher can accept any legal responsibility for any errors or omissions that may be made. The publisher makes no warranty, express or implied, with respect to the material contained herein.

Cover illustration: Taken from Chapter 5, Figure 6B.

Printed on acid-free paper

Springer is part of Springer Science+Business Media (www.springer.com)

Dedication

This book is dedicated to my parents Sonja and Mirko Stagljar for their great guidance and support in my life

Preface

Often defined as "the unicellular human" and "everybody's favorite fungus," the baker's yeast *Saccharomyces cerevisiae* has long been considered one of the most highly studied model organisms in the study of basic cellular processes. Along with this notion, yeast-based functional genomics and proteomics technologies, developed over the past decade, have contributed greatly to our understanding of bacterial, yeast, fly, worm, and human gene functions. More than 1,000 different papers and hundreds of reviews dealing with functional genomics and proteomics in yeast have appeared, but no comprehensive yeast-based functional genomics and proteomics textbook has yet been written and published. This book aims to be the standard textbook in the field of yeast-based functional genomics and proteomics and should serve as a stand-alone protocols handbook suitable for daily use in research laboratories. It includes recent advanced protocols in addition to major basic yeast-based functional genomics and proteomics techniques. In this way, both yeast researchers and those who wish to use yeast as a model system for functional genomics and proteomics will find this book useful.

Chapter "Comparative Genome Hybridization on Tiling Microarrays to Detect Aneuploidies in Yeast" serves as an introduction in how to use DNA microarrays to detect copy number variations in yeast. Chapter "Identification of Transcription Factor Targets by Phenotypic Activation and Microarray Expression Profiling in Yeast" describes in detail a methodology showing how overexpression of all yeast transcription factors combined with DNA microarray expression profiling and data analysis can be used to identify DNA-binding sequences for transcription factors. Chapter "SGAM: An Array-Based Approach for High-Resolution Genetic Mapping in *Saccharomyces cerevisiae*" contains state-of-the-art protocols for one of the best-known yeast functional genomics techniques, the synthetic genetic array (SGA) analysis, and focuses on a specific SGA application for high-resolution genetic mapping, referred to as SGA mapping (SGAM). Chapter "Reporter-Based Synthetic Genetic Array Analysis: A Functional Genomics Approach for Investigating the Cell Cycle in *Saccharomyces cerevisiae*" describes a modification of the SGA, termed reporter-based SGA (R-SGA) analysis, and its application in studying the expression of all yeast genes under a particular condition. Chapter "The Fidgety Yeast: Focus on High-Resolution Live Yeast Cell Microscopy" gives an excellent insight into experimental strategies for live yeast cell imaging, geared towards imaging-based large-scale screens, whereas Chapter "A Genomic Approach to Yeast Chronological Aging" describes a novel functional genomics approach for quantitatively measuring the yeast chronological life span. Chapters "Chemogenomic Approaches to Elucidation of Gene Function and Genetic Pathways" and "Identification of Inhibitors of Chromatin Modifying Enzymes Using the Yeast Phenotypic Screens" contain series of protocols that were essentially invented to study drug action in yeast and thus set up a foundation for yeast-based chemical genomics approaches. Chapter "Exploiting Yeast Genetics to Inform Therapeutic Strategies for Huntington's Disease" shows a perfect example of how yeast functional genomics approaches can efficiently be used to study a devastating human neurodegenerative disorder, Huntington's disease. Chapter "Global Proteomic Analysis of *Saccharomyces cerevisiae*

Identifies Molecular Pathways of Histone Modifications" describes a proteomics method, the global proteomic analysis in *S. cerevisiae* (GPS), for the global analysis of the molecular machinery required for proper histone modifications. Chapters "Systematic Characterization of the Protein Interaction Network and Protein Complexes in *Saccharomyces cerevisiae* Using Tandem Affinity Purification and Mass Spectrometry," "Protein Microarrays," "Analysis of Protein–Protein Interactions Using Array-Based Yeast Two-Hybrid Screens," and "Analysis of Membrane Protein Complexes Using the Split-Ubiquitin Membrane Yeast Two-Hybrid System" contain a collection of protocols for studying protein complexes and protein–protein interactions such as tandem affinity purification (TAP) linked to mass spectrometry, protein microarrays, the array-based yeast two-hybrid approach, and membrane yeast two-hybrid (MYTH) system. Protocols described in the last chapter aim to describe how computational analyses help us to understand the yeast proteome.

Finally, I wish to take this opportunity to thank all authors for their great commitment, cooperation, and contributions that made my first editing job easier. I also wish to express my sincere thanks to Dr. John M. Walker for providing guidance on how to generate this book.

Toronto, ON   *Igor Stagljar*

Contents

Preface... vii
Contributors.. xi

1 Comparative Genome Hybridization on Tiling Microarrays
   to Detect Aneuploidies in Yeast... 1
   Barry Dion and Grant W. Brown

2 Identification of Transcription Factor Targets by Phenotypic
   Activation and Microarray Expression Profiling in Yeast 19
   Gordon Chua

3 SGAM: An Array-Based Approach for High-Resolution
   Genetic Mapping in *Saccharomyces cerevisiae*.............................. 37
   Michael Costanzo and Charles Boone

4 Reporter-Based Synthetic Genetic Array Analysis:
   A Functional Genomics Approach for Investigating the Cell Cycle
   in *Saccharomyces cerevisiae*.. 55
   **Holly E. Sassi, Nazareth Bastajian, Pinay Kainth,
   and Brenda J. Andrews**

5 The Fidgety Yeast: Focus on High-Resolution Live Yeast Cell Microscopy 75
   Heimo Wolinski, Klaus Natter, and Sepp D. Kohlwein

6 A Genomic Approach to Yeast Chronological Aging............................ 101
   Christopher R. Burtner, Christopher J. Murakami, and Matt Kaeberlein

7 Chemogenomic Approaches to Elucidation of Gene Function
   and Genetic Pathways... 115
   Sarah E. Pierce, Ronald W. Davis, Corey Nislow, and Guri Giaever

8 Identification of Inhibitors of Chromatin Modifying Enzymes
   Using the Yeast Phenotypic Screens 145
   Benjamin Newcomb and Antonio Bedalov

9 Exploiting Yeast Genetics to Inform Therapeutic Strategies
   for Huntington's Disease 161
   Flaviano Giorgini and Paul J. Muchowski

10 Global Proteomic Analysis of *Saccharomyces cerevisiae*
   Identifies Molecular Pathways of Histone Modifications 175
   Jessica Jackson and Ali Shilatifard

11 Systematic Characterization of the Protein Interaction Network
   and Protein Complexes in *Saccharomyces cerevisiae* Using Tandem
   Affinity Purification and Mass Spectrometry 187
   **Mohan Babu, Nevan J. Krogan, Donald E. Awrey, Andrew Emili,
   and Jack F. Greenblatt**

12 Protein Microarrays 209
   Joseph Fasolo and Michael Snyder

13 Analysis of Protein–Protein Interactions Using Array-Based
   Yeast Two-Hybrid Screens 223
   Seesandra V. Rajagopala and Peter Uetz

14 Analysis of Membrane Protein Complexes Using the Split-Ubiquitin
   Membrane Yeast Two-Hybrid (MYTH) System 247
   *Saranya Kittanakom, Matthew Chuk, Victoria Wong, Jamie Snyder,
   Dawn Edmonds, Apostolos Lydakis, Zhaolei Zhang, Daniel Auerbach,
   and Igor Stagljar*

15 Computational Analysis of the Yeast Proteome: Understanding and
   Exploiting Functional Specificity in Genomic Data 273
   *Curtis Huttenhower, Chad L. Myers, Matthew A. Hibbs,
   and Olga G. Troyanskaya*

Index 295

Contributors

BRENDA J. ANDREWS • *Department of Molecular Genetics, Banting and Best Department of Medical Research, Terrence Donnelly Centre for Cellular and Biomolecular Research, University of Toronto, Toronto, ON, Canada*

DANIEL AUERBACH • *DUALSYSTEMS Biotech Inc., Zurich, Switzerland*

DONAL E. AWREY • *Campbell Family Institute for Breast Cancer Research Therapeutics, Toronto, ON, Canada*

MOHAN BABU • *Banting and Best Department of Medical Research, University of Toronto, Terrence Donnelly Center for Cellular and Biomolecular Research, Toronto, ON, Canada*

NAZARETH BASTAJIAN • *Department of Medical Genetics and Microbiology, Banting and Best Department of Medical Research, University of Toronto, Toronto, ON, Canada*

ANTONIO BEDALOV • *Clinical Research Division, Fred Hutchinson Cancer Research Center, Seattle, WA, USA*

CHARLIE BOONE • *Department of Molecular Genetics, Banting and Best Department of Medical Research, Terrence Donnelly Centre for Cellular and Biomolecular Research, University of Toronto, Toronto, ON, Canada*

GRANT W. BROWN • *Department of Biochemistry and Terrence Donnelly Centre for Cellular and Biomolecular Research, University of Toronto, Toronto, ON, Canada*

CHRISTOPHER R. BURTNER • *Department of Biochemistry, University of Washington, Seattle, WA, USA*

GORDON CHUA • *Institute for Biocomplexity and Informatics, Department of Biological Sciences, University of Calgary, Calgary, AB, Canada*

MATTHEW CHUK • *Department of Biochemistry & Department of Molecular Genetics, Terrence Donnelly Centre for Cellular and Biomolecular Research, University of Toronto, Toronto, ON, Canada*

MICHAEL COSTANZO • *Banting and Best Department of Medical Research, University of Toronto, Toronto, ON, Canada*

RON W. DAVIS • *Stanford Genome Technology Center, Stanford University, Palo Alto, CA, USA*

BARRY DION • *Department of Biochemistry, Terrence Donnelly Centre for Cellular and Biomolecular Research, University of Toronto, Toronto, ON, Canada*

DAWN EDMONDS • *Department of Biochemistry & Department of Molecular Genetics, Terrence Donnelly Centre for Cellular and Biomolecular Research, University of Toronto, Toronto, ON, Canada*

ANDREW EMILI • *Department of Molecular Genetics, Banting and Best Department of Medical Research, Terrence Donnelly Centre for Cellular and Biomolecular Research, University of Toronto, Toronto, ON, Canada*

JOSEPH FASOLO • *Department of Molecular, Cellular, and Developmental Biology, Yale University, New Haven, CT, USA*
GURI GIAEVER • *Department of Pharmaceutical Sciences, Terrence Donnelly Centre for Cellular and Biomolecular Research, University of Toronto, Toronto, ON, Canada*
FLAVIANO GIORGINI • *Department of Genetics, University of Leicester, Leicester, UK*
JACK F. GREENBLATT • *Department of Molecular Genetics, Banting and Best Department of Medical Research, Terrence Donnelly Centre for Cellular and Biomolecular Research, University of Toronto, Toronto, ON, Canada*
MATTHEW A. HIBBS • *Lewis-Sigler Institute for Integrative Genomics, Carl Icahn Laboratory, Princeton University, Princeton, NJ, USA; Department of Computer Science, Princeton University, Princeton, NJ, USA*
CURTIS HUTTENHOWER • *Lewis-Sigler Institute for Integrative Genomics, Carl Icahn Laboratory, Princeton University, Princeton, NJ, USA; Department of Computer Science, Princeton University, Princeton, NJ, USA*
JESSICA JACKSON • *Department of Genetics, Washington University School of Medicine, St. Louis, MO, USA*
MATT KAEBERLEIN • *Department of Pathology, University of Washington, Seattle, WA, USA*
PINAY KAINTH • *Department of Medical Genetics and Microbiology, Banting and Best Department of Medical Research, University of Toronto, Toronto, ON, Canada*
SARANYA KITTANAKOM • *Department of Biochemistry & Department of Molecular Genetics, Terrence Donnelly Centre for Cellular and Biomolecular Research, University of Toronto, Toronto, ON, Canada*
SEPP D. KOHLWEIN • *Institute of Molecular Biosciences, University of Graz, Graz, Austria*
NEVAN J. KROGAN • *Department of Cellular and Molecular Pharmacology, University of California, San Francisco, CA, USA*
APOSTOLOS LYDAKIS • *Department of Molecular Genetics, Terrence Donnelly Centre for Cellular and Biomolecular Research, University of Toronto, Toronto, ON, Canada*
PAUL J. MUCHOWSKI • *Departments of Biochemistry and Biophysics, and Neurology, Gladstone Institute of Neurological Disease, University of California, San Francisco, San Francisco, CA, USA*
CHRISTOPHER J. MURAKAMI • *Department of Pathology, University of Washington, Seattle, WA, USA*
CHAD L. MYERS • *Department of Computer Science, University of Minnesota, Minneapolis, MN, USA*
KLAUS NATTER • *Institute of Molecular Biosciences, University of Graz, Graz, Austria*
BENJAMIN NEWCOMB • *Clinical Research Division, Fred Hutchinson Cancer Research Center, Seattle, WA, USA*
COREY NISLOW • *Department of Molecular Genetics, Banting and Best Department of Medical Research, Terrence Donnelly Centre for Cellular and Biomolecular Research, University of Toronto, Toronto, ON, Canada*
SARAH E. PIERCE • *Stanford Genome Technology Center, Stanford University, Palo Alto, CA, USA*

SEESANDRA V RAJAGOPALA • *J. Craig Venter Institute, Rockville, MD, USA*
HOLLY E. SASSI • *Department of Medical Genetics and Microbiology, Banting and Best Department of Medical Research, University of Toronto, Toronto, ON, Canada*
ALI SHILATIFARD • *Stowers Institute for Medical Research, Kansas City, MO, USA*
JAMIE SNYDER • *Department of Biochemistry & Department of Molecular Genetics, Terrence Donnelly Centre for Cellular and Biomolecular Research, University of Toronto, Toronto, ON, Canada*
MICHAEL SNYDER • *Department of Molecular, Cellular, and Developmental Biology, Yale University, New Haven, CT, USA*
IGOR STAGLJAR • *Department of Biochemistry & Department of Molecular Genetics, Terrence Donnelly Centre for Cellular and Biomolecular Research, University of Toronto, Toronto, ON, Canada*
OLGA G. TROYANSKAYA • *Lewis-Sigler Institute for Integrative Genomics, Carl Icahn Laboratory, Princeton University, Princeton, NJ, USA; Department of Computer Science, Princeton University, Princeton, NJ, USA*
PETER UETZ • *J. Craig Venter Institute, Rockville, MD, USA*
HEIMO WOLINSKI • *Institute of Molecular Biosciences, University of Graz, Graz, Austria*
VICTORIA WONG • *Department of Biochemistry & Department of Molecular Genetics, Terrence Donnelly Centre for Cellular and Biomolecular Research, University of Toronto, Toronto, ON, Canada*
ZHAOLEI ZHANG • *Department of Molecular Genetics, Banting and Best Department of Medical Research, Terrence Donnelly Centre for Cellular and Biomolecular Research, University of Toronto, Toronto, ON, Canada*

Chapter 1

Comparative Genome Hybridization on Tiling Microarrays to Detect Aneuploidies in Yeast

Barry Dion and Grant W. Brown

Summary

Chromosomal aberrations resulting in aneuploidies have been implicated in the development of most cancers and numerous other genetic disorders. Aneuploidies are a key feature of genomic instability, so classification of these copy number changes will be important in understanding how rearrangements arise and how ongoing instability is maintained. Traditional methods for detecting copy number changes have relatively poor resolution, making accurate detection of breakpoints impossible. The advent of microarray technology and its advance over the years has improved the ability to detect aneuploidies with greater accuracy. Mammalian comparative genome hybridization on microarrays (array-CGH) has been applied to the study of many carcinomas, identifying common copy number changes in key regions including known oncogenes. However, the large size of mammalian genomes has made it impractical to perform whole genome CGH at high resolution. Yeast has been established as a useful model for studying pathways relevant to oncogenesis, particularly those that maintain the integrity of the genome. Given the smaller size of the yeast genome, oligonucleotide tiling arrays have been developed that allow for nucleotide resolution of the whole genome on a single chip. Here we describe in detail how to use these arrays to detect copy number variations in yeast. This method will be useful in many different studies, but particularly in monitoring and cataloguing the changes resulting from genetic instability.

Key words: Comparative genome hybridization, Tiling microarrays, Aneuploidy, Array-CGH, Whole genome amplification

1. Introduction

Chromosomal abnormalities such as amplifications, deletions, and translocations resulting in DNA copy alterations have been implicated in the development of most cancers and a number of other human genetic disorders. Specifically, amplifications of oncogenes, deletions of tumor suppressors, and oncogenic fusions have all been identified in human cancers *(1)* and can contribute to tumorigenesis. Other chromosomal aberrations, such as duplications and deletions can result in genetic diseases such as Down syndrome and Cri du Chat syndrome *(2, 3)*. Given the obvious importance of genomic instability in the development of many pathologies it is clear that the methods to detect and map aneuploidies are of great utility in determining the mechanisms by which these aneuploidies arise.

In the past, comparative genome hybridization (CGH) has been applied to metaphase chromosome spreads to assess DNA copy number changes in mammalian cells. However, this method was limited by its low resolution, between 10 and 20 Mb *(4, 5)*. In its modern form, CGH is performed on microarrays which involves the labeling of reference and sample DNA with different fluorochromes. The DNA is then competitively hybridized to DNA probes representing the whole or partial genome of interest. The resultant signal intensity ratio between the two samples corresponds to the copy number imbalance. Array-CGH, first performed in 1997, improved the resolution level to 75–130 kb, compared to previous CGH methods *(6)*. Advances in array technologies since then have allowed a vast improvement in the resolution capabilities of array-CGH. Since CGH requires a net change in DNA content to detect aneuploidies, it is unable to detect reciprocal translocations. Despite this limitation, array-CGH is seeing increased use as a means of comparing genomes for the purpose of identifying alterations resulting in copy number changes.

In mammalian cells, array-CGH has been effective in detecting aneuploidies that are characteristic of many human cancers and genetic diseases. For example, studies of fallopian tube carcinoma have revealed a large number of copy number changes *(7, 8)*. The improved resolution of array-CGH has been applied to further identify and refine regions of copy number alterations quantitatively while mapping the aberrations directly to the human genome sequence at a resolution of ~1.4 Mb *(9)*. Mapping of these genetic changes revealed recurrent amplifications in a number of known oncogenes *(9)*. Colorectal cancers have also been surveyed by array-CGH at 1–2 Mb resolution, identifying high frequency losses and gains previously identified by metaphase CGH, as well as additional recurrent aberrations that were not previously identified *(10)*. More recently array-CGH studies in breast cancers have identified common copy number changes to a resolution of 100 kb on chromosome 8 *(11)*. These are only a few of the many type of cancers for which array-CGH has been able to identify changes in DNA copy number which may play an important role in tumorigenesis. Array-CGH has allowed researchers and clinicians to further define the characteristics of various cancers, identifying previously unidentified aberrations, to complement those alterations already known. This knowledge can be used to further our understanding of the cause and progression of cancers, in addition to providing clinicians with an efficient method for diagnosis.

Despite the improvement in CGH using microarrays, mammalian array-CGH is still hindered by relatively poor resolution. To date only sub-megabase resolution microarrays have been developed that span the whole human genome *(12)*. Although breakpoint mapping information obtained from low resolution arrays can be used to create specific tiling oligonucleotide arrays that allow for CGH at high resolution *(13)*, the large size of mammalian genomes remains a challenge in high resolution array-CGH.

The relatively small genome of *Saccharomyces cerevisiae* allows array-CGH to be performed at a much higher resolution compared to mammalian studies to date. State of the art tiling microarrays span the entire genome on a single chip at 4 bp resolution *(14, 15)*. A variety of studies have employed the use of these yeast whole-genome tiling arrays to monitor copy number changes. Adaptive rearrangements resulting from nutrient limitation during experimental evolutions have been mapped at single gene resolution *(16)*. Array-CGH has also been used to monitor the efficiency of DNA replication origins and the timing of the initiation of DNA replication *(17)*. CGH on microarrays has also been used to compare different yeast strains and species, demonstrating its usefulness in species determination and differentiation of strains within a species *(18)*. Clearly array-CGH can be applied to a wide variety of questions addressing patterns and rates of changes during genome evolution *(19)*.

S. cerevisiae has seen extensive use as a model in the analysis of genomic instability, motivating and informing further studies in mammalian systems. Examples of this include the discovery of cell cycle checkpoints *(20, 21)*, checkpoint mediators such as Rad9 and Mrc1 *(22, 23)*, and the gross chromosomal rearrangement studies of the Kolodner lab, which helped to define the genetic basis for genomic instability *(24, 25)*. Further research is needed to advance our limited understanding of how genome rearrangements arise, the pathways that suppress them, and whether defects in these pathways result in ongoing instability that is present in many cancers *(24)*. Understanding genome rearrangement is key to understanding the causes and effects of genomic instability and requires high resolution techniques to identify aberrations and to map their breakpoints accurately. Array technology has not yet produced affordable methods by which to study genomic aberrations at high resolution in mammalian cells. The availability of these high resolution technologies for yeast provides a convenient model system by which to elucidate the molecular mechanisms behind genome instability and uncontrolled cell growth. Given that many cancers arise as a result of a loss of genome stability, these studies in yeast can be used to direct studies in mammalian cells more efficiently, providing further mechanistic insight into the causes of cancers.

In this chapter we outline a procedure to detect aneuploidies (changes in DNA copy number), such as amplifications and deletions, at high resolution in yeast. The basic protocol includes DNA isolation, amplification (where necessary), DNA fragmentation and labeling, array hybridization, and data analysis (**Fig. 1**). The array used for detection of these aneuploidies is a single oligonucleotide tiling array containing 6.5 million probes interrogating both strands of *S. cerevisiae* genomic sequence with 25-mer probes tiled at eight nucleotide intervals *(26)*. This array allows for high resolution mapping of breakpoints, and with more sophisticated data analysis can allow for the detection of single nucleotide polymorphisms *(14)*.

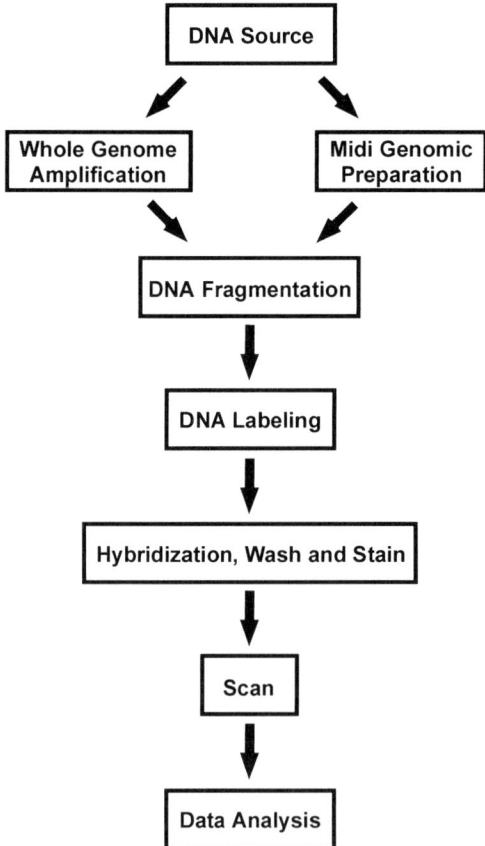

Fig. 1. Schematic of the array-CGH method for detection of aneuploidies.

2. Materials

2.1. DNA Isolation

2.1.1. Yeast Genomic DNA Preparation

1. Genomic DNA Midi-prep Kit (Qiagen Genomic-tip 100/G Cat# 10243).
2. Genomic DNA buffer set (Qiagen Cat# 19060).
3. YPD (see **Subheading 2.7**).
4. 125-ml Erlenmeyer flasks.
5. RNase A (see **Subheading 2.8**).
6. Zymolyase.
7. Proteinase K (see **Subheading 2.8**).
8. 100% Isopropanol.
9. 70% Ethanol.
10. 50-ml Conical centrifuge tubes.
11. 15-ml Conical centrifuge tubes.
12. Polycarbonate round bottom centrifuge tubes.
13. 1.5-ml microfuge tubes.
14. 10 mM Tris–HCl (see **Subheading 2.8**).

2.1.2. Yeast Genomic DNA Mini-Prep

1. YeaStar genomic DNA kit (Zymo Research Cat# D2002).
2. Culture tubes.
3. YPD (see **Subheading 2.7**).
4. 1.5-ml microfuge tubes.
5. 10 mM Tris–HCl (see **Subheading 2.8**).
6. Chloroform.

2.1.3. DNA Quantification

1. Quant-iT dsDNA HS assay kit (Invitrogen Cat# Q32851).
2. Qubit assay tubes (Invitrogen Cat# Q33856).
3. Qubit fluorometer (Invitrogen Cat# Q32857).
4. DNA to be quantified.

2.2. Whole Genome Amplification

1. GenomePlex complete whole genome amplification kit (Sigma Cat# WGA2).
2. Thin-walled PCR tubes.
3. DNA to be amplified.

2.3. Purification of Amplified DNA

1. QIAquick PCR purification kit (Qiagen Cat# 28104).
2. Microfuge tubes.
3. Amplified DNA.

2.4. DNA Fragmentation

1. 10× One-phor-all buffer PLUS (GE Healthcare Amersham Cat# 27-091-02).
2. Cobalt chloride ($CoCl_2$) (*see* **Subheading 2.8**).
3. Deoxyribonuclease I, amplification grade (Invitrogen Cat# 18068-015).
4. Thin-walled PCR tubes.
5. Agarose.
6. 20× Lithium boric (LB) acid ultralow-conductive medium for DNA electrophoresis (Faster Better Media LLC Cat# LB20).
7. 5× LB loading medium (Faster Better Media LLC Cat# LB5N).
8. SYBR Green.
9. TAE (*see* **Subheading 2.8**).
10. 10 bp DNA ladder molecular weight standards (Invitrogen Cat# 10821-015).
11. Sterile ddH_2O.
12. DNA to be fragmented.

2.5. DNA Labeling

1. Biotin-N^6-ddATP (Enzo Life Sciences Cat# 42809).
2. Terminal deoxynucleotidyl transferase (MBI Fermentas Cat# EP0161).
3. Fragmented DNA.
4. Thin-walled PCR tubes.

2.6. Array Hybridization and Washing

1. Hybridization buffer (*see* **Subheading 2.8**).
2. Sterile dH_2O (filtered).
3. Bovine serum albumin (BSA) (*see* **Subheading 2.8**).
4. Herring testes carrier DNA, denatured (Clonetech Cat# 50277).
5. b213 control oligonucleotide:
Biotin-CTG AAC GGT AGC ATC TTG AC 3′
6. MES stain buffer (*see* **Subheading 2.8**).
7. Streptavidin, R-phycoerythrin conjugate (SAPE) (Invitrogen Cat# S866).
8. Biotinylated anti-streptavidin antibody (Vector Laboratories Cat# BA-500).
9. Normal goat IgG (*see* **Subheading 2.8**) (Sigma Cat# I 5256).
10. 20× SSPE (UltraPure) (Invitrogen Cat# 15591).
11. Wash A: Non-stringent wash buffer (*see* **Subheading 2.8**).
12. Wash B: Stringent wash buffer (*see* **Subheading 2.8**).
13. Microtube tough-spots (Mandel Cat# US-9185-050X).

14. Compressed gas duster.
15. Lens paper.
16. 95% Ethanol.
17. Microfuge tubes.
18. Biotin labeled DNA.
19. *S. cerevisiae* Tiling Array (Affymetrix Cat# 520055).

2.7. Media

1. *YPD (1 l liquid):* 10 g yeast extract, 20 g bio-tryptone, 20 g d-glucose. Resuspend all ingredients in 1 l ddH$_2$O, dissolve by stirring. Autoclave at 121°C, 15 psi for 30 min.

2.8. Solutions

1. 10 mM Tris–HCl: Dissolve 1.211 g Tris base into 800 ml ddH$_2$O. Stir while adjusting pH to 8.5 with concentrated HCl. Adjust volume to 1 l with ddH$_2$O and autoclave at 121°C, 15 psi for 30 min.
2. 50× TAE *(27)*: Dissolve 242 g Tris base in 650 ml ddH$_2$O. Add 57.1 ml of glacial acetic acid and 100 ml 0.5 M EDTA (pH 8.0). Stir solutions with magnetic stir and adjust volume to 1 l with ddH$_2$O. The working solution is 1× (40 mM Tris-acetate, 1 mM EDTA).
3. 12× MES stock: Dissolve 7.74 g MES free acid monohydrate and 21.26 g MES sodium salt in 80 ml ddH$_2$O. Stir with magnetic stirrer and adjust pH to between 6.5 and 6.7. Filter through a 0.22-µm polyethersulfone filter. Store at 4°C shielded from light. If solution turns yellow, discard and make a new batch.
4. 2× Hybridization buffer: Mix 8.3 ml 12× MES stock, 17.7 ml 5 M NaCl, 4 ml 0.5 M EDTA, 0.1 ml 10% Tween 20, 19.9 ml ddH$_2$O together. Filter through a 0.22-µm polyethersulfone filter. Store at 4°C shielded from light. If solution turns yellow, discard and make a new batch.
5. 1× Hybridization buffer: Dilute 2× hybridization buffer with ddH$_2$O by half to a final concentration of 100 mM MES, 1 M NaCl, 0.05% Tween 20. Filter through a 0.22-µm polyethersulfone filter. Store at 4°C shielded from light. If solution turns yellow discard, and make a new batch.
6. Wash A: Non-stringent wash buffer: Mix 300 ml 20× SSPE, and 1 ml 10% Tween 20 with 699 ml ddH$_2$O.
7. Wash B: Stringent wash buffer: Mix 83.3 ml 12× MES stock, 5.2 ml 5 M NaCl and 1 ml 10% Tween 20 with 910.5 ml ddH$_2$O. Protect from light with aluminum foil and store at 4°C. If solution turns yellow discard and make a new batch.
8. 30 mM CoCl$_2$: Dissolve 35.69 mg cobalt chloride hexahydrate in 5 ml ddH$_2$O. Filter through a 0.22-µm polyethersulfone filter and store at room temperature.

9. 20 mg/ml BSA: Dissolve 200 mg of bovine albumin (fraction V) into 10 ml of ddH$_2$O by vortexing. Filter through a 0.22 μm polyethersulfone sterilizing filter and store at −20°C.
10. 30 mg/ml normal goat IgG: Reconstitute 30 mg normal goat IgG with 1 ml 150 mM NaCl solution. Aliquot into smaller portions and store at −20°C. Do not freeze and thaw repeatedly.
11. 5 M NaCl *(27)*: Dissolve 292.2 g of sodium chloride in 800 ml ddH$_2$O. Adjust volume to 1 l with ddH$_2$O and autoclave at 121°C, 15 psi for 30 min.
12. 10% Tween 20: Mix 10 ml Tween 20 with 90 ml ddH$_2$O, stir with magnetic stir bar. Filter through a 0.22-μm polyethersulfone filter and store at 4°C.
13. 0.5 M EDTA (pH 8.0) *(27)*: Dissolve 186.1 g of disodium ethylenediaminetetra-acetate·2H$_2$O with 800 ml ddH$_2$O. Stir with magnetic stirrer while adjusting pH to 8.0 with NaOH. Adjust volume to 1 l with ddH$_2$O and autoclave at 121°C, 15 psi for 30 min.
14. 20 mg/ml Proteinase K: Dissolve 250 mg of proteinase K (activity > 30 U/mg) into 2.5 ml ddH$_2$O by vortexing gently. Aliquot into 100 μl fractions and store at −20°C.
15. 100 mg/ml RNase A: Dissolve 100 mg of DNase-free ribonuclease A (>60 K U/mg) in 1 ml ddH$_2$O. Vortex and store at −20°C.
16. 3 M sodium acetate (pH 5.2) *(27)*: Dissolve 102 g of sodium acetate·3H$_2$O in 200 ml ddH$_2$O. Adjust pH to 5.2 with glacial acetic acid. Adjust volume to 250 ml with ddH$_2$O. Autoclave at 121°C, 15 psi for 30 min to sterilize.

3. Methods

3.1. Yeast Genomic DNA Midi-Prep

1. Inoculate 35 ml of YPD with a single yeast colony.
2. Grow overnight, shaking (180 rpm) at 30°C.
3. Measure density of cells in a spectrophotometer at 600 nm.
4. Harvest 7.0×10^9 cells (OD$_{600}$ value of 1 equals 2×10^7 cells/ml) at $2,100 \times g$ for 5 min.
5. Extract genomic DNA using Qiagen Genomic-tip 100/G kit (Cat# 10243) and Qiagen Genomic DNA Buffer set (Cat# 19060) (or with equivalent product) with the following modifications:
   (a) Incubate with zymolyase at 30°C for 1 h.
   (b) Incubate with proteinase K at 50°C for 1 h.

(c) Prewarm elution buffer QF to 50°C.

(d) Pellet precipitated DNA by centrifugation at $23,500 \times g$ for 15 min for both the isopropanol and 70% ethanol steps.

(e) Resuspend pelleted DNA in 400 µl of 10 mM Tris–HCl, thoroughly washing the sides of the centrifuge tube.

(f) Transfer to 1.5-ml microfuge tube and dissolve overnight or at 55°C for 2 h.

6. Measure concentration of genomic DNA (*see* **Subheading 3.8**).

7. Concentrate genomic DNA if concentration is less than 0.403 µg/µl (*see* **Subheading 3.10**).

3.2. Yeast Genomic DNA Mini-Prep

1. Inoculate 5 ml of YPD with single colony (*see* **Note 1**).
2. Grow overnight at 30°C with rotation.
3. Extract genomic DNA using YeaStar Genomic DNA kit (Cat# D2002) Protocol I or with equivalent product with the following modifications:
   (a) Spin 5×10^7 cells at $2,700 \times g$ for 2 min (*see* **Note 1**).
   (b) Incubate with R-zymolyase (provided in kit) for 60 min and vortex for 1 min.
   (c) Vortex YD lysis buffer with sample for 1 min at medium-low setting.
   (d) Centrifuge at $17,900 \times g$ for 2 min.
   (e) Elute in 60 µl of 10 mM Tris–HCl and centrifuge for 30 s.
4. Measure the concentration of genomic DNA (*see* **Subheading 3.8**).
5. Concentrate genomic DNA if concentration is less than 5 ng/µl (*see* **Subheading 3.10**).

3.3. Whole Genome Amplification

1. Amplify genomic DNA prepared with genomic DNA mini-prep kit (*see* **Subheading 3.2**) using GenomePlex complete whole genome amplification kit (Cat# WGA2) with the following modifications:
   (a) Prepare a 5 ng/µl DNA solution from extracted DNA.
   (b) Add 10 µl of 5 ng/µl DNA solution to 1 µl of 10× fragmentation buffer.
2. Measure amplified DNA concentration (*see* **Subheading 3.8**) (*see* **Note 2**).
3. Purify final product before use (*see* **Subheading 3.9**) (*see* **Note 3**).

3.4. DNA Fragmentation

1. Prepare DNase master mix: 7.4 µl ddH$_2$O, 1 µl 10× one-phor-all buffer, 0.6 µl 30 mM CoCl$_2$ and 1 µl 1 U/µl DNase I (Invitrogen, amplification grade). Mix solution with a pipette.

2. Prepare 0.403 µg/µl DNA (from **Subheadings 2.1** or **2.3**) solution with ddH$_2$O to a total volume of 37.25 µg/µl in a thin-walled PCR tube. Mix solution with a pipette.

3. Add to each DNA sample: 4.5 µl 10× one-phor-all buffer, 2.25 µl 30 mM CoCl$_2$ and 1.5 µl of the DNase master mix. Mix solution with a pipette.

4. Use a thermocycler to incubate for 4 min at 37°C, then 95°C for 10 min, then decrease the temperature to 4°C (*see* **Note 4**).

5. Check digestion on a thin (<5 mm) 0.1× lithium boric (LB) acid/ 2% agarose gel. Immerse gel in 0.1× LB running buffer such that there is only a thin (~1 mm) layer covering the gel.

6. Load gel with 0.5 µl (500 ng) of 10 bp ladder and 1 µl of each DNA sample plus 1 µl 5× LB loading medium and 3 µl ddH$_2$O and run at 250 V for 27 min.

7. Stain for 20 min in 1× SYBR green in 1× TAE (*see* **Note 5**).

8. Visualize with UV light. Smear of fragmented DNA should appear centered at approximately 25 bp.

9. If the smear is centered at a larger size, repeat **steps 4–8** after adding an additional 1.5 µl of the DNase master mix to each sample (*see* **Note 6**).

10. Once desired fragment size is obtained, store at –20°C until required for labeling.

3.5. DNA Labeling

1. Add 1 µl biotin-N^6-ddATP (1 nmol/µl) and 1.54 µl TdT (20 U/µl) to the fragmented DNA sample.

2. Incubate for 1 h at 37°C and cool to 4°C.

3. Store at 4°C or use immediately. Do not freeze after labeling.

3.6. Array Hybridization

1. Preheat hybridization oven to 45°C.

2. Fill each *S. cerevisiae* genome tiling array chip with 1× hybridization buffer and incubate for at least 10 min in the hybridization oven at 45°C, spinning at 60 rpm.

3. Prepare chip hybridization master mix. For each chip add 150 µl 2× hybridization buffer, 94 µl ddH$_2$O, 7.5 µl 20 mg/ml BSA, 5.6 µl 50 nM b213 control oligonucleotide and 3 µl 10 mg/ml herring testes carrier DNA.

4. Add the labeled DNA sample (45 µl) to 255 µl of chip hybridization master mix.

5. Heat the hybridization mix at 95–100°C for 10 min.

6. Cool on ice for 5 min.

7. Remove 1× hybridization buffer from each chip.

8. Fill each chip with the DNA hybridization mix (*see* **Note 7**). Cover each gasket with a tough-spot to prevent leakage.

9. Place chips in oven and hybridize for 20 h at 45°C, spinning at 60 rpm.
10. After hybridization remove the hybridization solution and save in its corresponding tube at 4°C (*see* **Note 8**).
11. Fill each chip with 1× hybridization buffer.

3.7. Array Washing and Scanning

1. Prepare SAPE solution. For each chip mix 600 μl 2× MES stain buffer, 120 μl 20 mg/ml BSA, 12 μl 1 mg/ml streptavidin phycoerythrin (SAPE) and 468 μl ddH$_2$O. Mix well and divide into two aliquots of 600 μl in 1.5-ml microfuge tubes. Wrap in aluminum foil until ready for use.
2. Prepare antibody solution. For each chip mix 300 μl 2× MES stain buffer, 60 μl 20 mg/ml BSA, 2 μl 30 mg/ml normal goat IgG, 3.6 μl 500 μg/ml biotinylated anti-streptavidin antibody and 234 μl ddH$_2$O in a microfuge tube.
3. Using the GeneChip Operation System (GCOS) from Affymetrix, prime the Gene Chip Fluidics Station 450. Run the Prime_450 protocol. Ensure that the Wash A and Wash B intake tubes are in their corresponding solutions.
4. After priming remove the microfuge tubes and replace with the prepared solutions: SAPE, Antibody, and SAPE solutions in positions 1, 2, and 3, respectively.
5. Run the EukGE-WS2v4_450 fluidics protocol to wash and stain the arrays.
6. When complete, ensure that no bubbles are present in the chip. If any bubbles are present re-engage wash block and allow the fluidics station to run wash A through the chip. Repeat until no air bubbles are present.
7. Remove chip and shut down fluidics station using the SHUTDOWN_450 protocol.
8. Place tough-spots on the gaskets of each chip to prevent any leakage in the scanner.
9. Ensure that the front glass plate on the chip is clean and streak free using lens paper and 95% ethanol if the gas duster is unable to remove any dirt/dust.
10. Scan chips in the GeneChip Scanner 3000
11. Once complete, store chips at 4°C until alignment has been confirmed and analysis is underway, in case a rescan or rehybridization is required.

3.8. DNA Quantification (See Note 9)

1. Measure DNA using the Quant-iT dsDNA HS Assay Kit (Cat# Q32851) and the Qubit fluorometer (Cat# Q32857) from Invitrogen. No modifications to the protocol outlined by the manufacturer need to be made.

3.9. Amplified DNA Purification (See Note 3)

1. Purify DNA from whole genome amplification using QiaQuick PCR purification kit. Elute with appropriate volume of EB.

3.10. DNA Precipitation (Modified from (27))

1. Determine the volume of the DNA solution.
2. Add 0.1 volumes of 3 M sodium acetate (pH 5.2) to the DNA solution. Mix well with a pipette.
3. Add 2.5 volumes of ice cold 95% ethanol and mix well with a pipette.
4. Incubate at –20°C for 30 min.
5. Centrifuge at 0°C for 15 min at $17,900 \times g$.
6. Carefully remove the supernatant taking care not to disturb the pellet (which may be invisible).
7. Wash with 70% ethanol.
8. Centrifuge at 4°C for 5 min at $17,900 \times g$.
9. Carefully remove the supernatant taking care not to disturb the pellet (which may be invisible).
10. Open the tube and allow the ethanol to evaporate. Do not exceed 15 min.
11. Resuspend in desired volume of 10 mM Tris–HCl.

3.11. Data Analysis

1. Obtain the .CEL file that is created by the GCOS software (Affymetrix) after automatic alignment of the tiling arrays using the control oligonucleotide (*see* **Note 10**).
2. Obtain the map file corresponding to the particular array. This file relates array position to chromosome position.
3. Download Tiling Analysis Software (TAS) Version 1.1 and the corresponding manual from the Affymetrix website at: http://www.affymetrix.com/support/developer/downloads/TilingArrayTools/index.affx.
4. Use the two-sample comparison option in TAS to detect aneuploidies by giving appropriate treatment and reference (control). CEL files as input. Normalize data using quantile normalization plus scaling. Normalize experiments together. Once treatment, control, and genomic map have been given as input, save to create a.TAG file.
5. Use the .TAG file to analyze intensities according to the TAS manual. Use the following starting parameters modified from the defaults:
   (a) Export
       Save signal values only.
   (c) Scale
       Log2.
   (e) Probe analysis
       Bandwidth: 40.
       Test type: Two sided.
       Intensities: PM only.

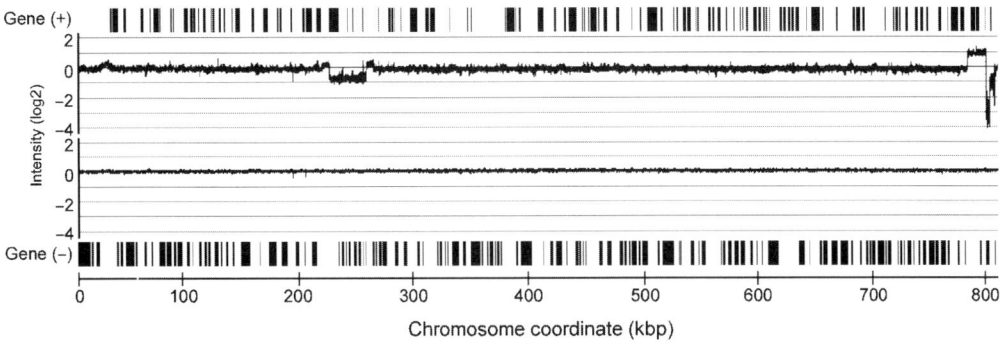

Fig. 2. Data output from the Integrated Genome Browser. The integrated genome browser (IGB) was used to visualize the ratio of signal intensities between experiment and control samples on a log2 scale, plotted against chromosomal coordinate. The full length of chromosome II from two diploid strains is shown, with an experimental strain with aneuploidies on top, and a wild-type strain on the bottom. A single copy deletion is present, as indicated by an average log2 ratio of −1, between 200 and 300 kb. A single copy amplification and a two copy deletion are visible, corresponding to average log2 ratios of ~0.58 and<−1 respectively, at 800 kb. The locations of open reading frames along the x-axis for both the Watson and Crick strands are represented by (+) and (−).

(d) Interval analysis

Threshold: 6.64.

Max. gap: 80.

Min. gap: 40.

Less than threshold.

Modifications of parameters may be necessary after initial analysis to improve signal to noise ratio.

6. Download Integrated Genome Browser (IGB) and User's guide from the Affymetrix website at http://www.affymetrix.com/support/developer/tools/download_igb.affx.

7. Use the files created for each chip after intensity analysis (.BAR) according to the IGB user's guide to visualize the changes in DNA copy number plotted along each chromosomal axis (*see* **Note 11**). An example of this visualization is shown in **Fig. 2**. The log2 experiment to control ratio of signal intensities is plotted against the chromosomal location of each probe. Results from a strain that displayed aneuploidies (**Fig. 2**, top) and a strain that did not (**Fig. 2**, bottom) are shown.

4. Notes

1. Whole genome amplification (WGA) makes it possible to convert a small amount of DNA, on the scale of nanograms, to microgram quantities, allowing for array-CGH to be performed. There are a number of experimental conditions that could limit the available amount of DNA or the

amount of culture available from which to prepare DNA. WGA is presented here as a way to overcome these limitations. When extracting DNA from culture for WGA using the YeaStar genomic DNA kit it is not necessary to grow an overnight culture from a single colony. In fact a smaller number of cells (as low as 1×10^6 cells) can be harvested directly from glycerol stocks, which is useful in experiments that are time- or generation-sensitive. In general, hybridization to tiling microarrays requires 15 μg of DNA. We find typically that DNA prepared from 1.0×10^7 cells with the YeaStar genomic DNA kit yields approximately 1 μg of DNA. We recommend that trial genomic preparations with known number of cells be performed to determine the exact number of cells required. We have been able to produce enough amplified DNA for array-CGH with 50 ng of genomic DNA input. An example of IGB histograms comparing an unamplified and an amplified DNA sample from the same diploid strain is shown in **Fig. 3**. All aneuploidies detected in the unamplified sample were apparent in the amplified sample, although the background noise was somewhat higher in the amplified sample.

2. In general, we find DNA quantification by fluorometry to be more reliable than measuring absorbance at 260 nm (*see* **Note 9**), although there is some potential for underestimation of yield with amplified samples which might contain significant amounts of single-stranded DNA.

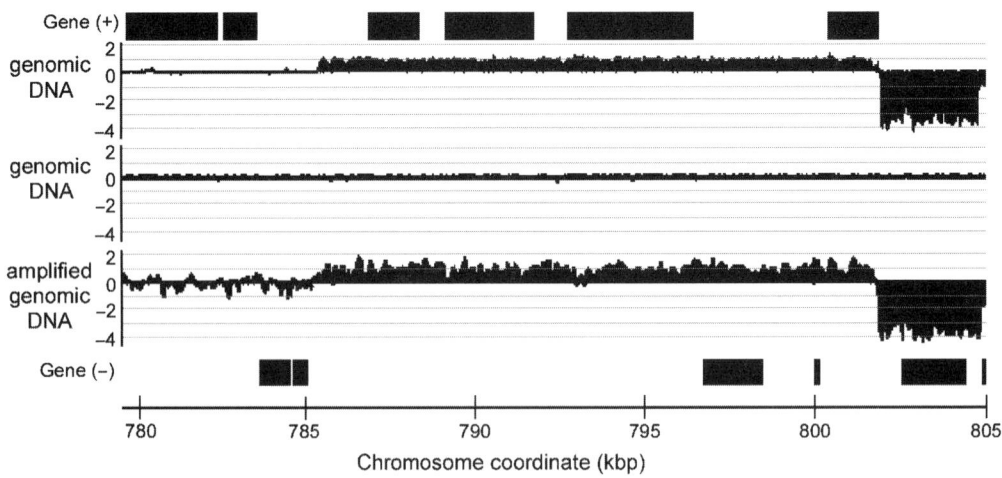

Fig.3. Comparison between amplified and unamplified genomic DNA preparations. The IGB histogram of aneuploidies detected on chromosome II in the 780–805 kbp region is shown. The unamplified genomic DNA sample is shown on top, and the sample amplified by the WGA method is shown on the bottom. No significant difference in log2 ratios were apparent. An unamplified wild type genomic DNA sample is shown in the middle and lacks any genomic aberrations.

3. Most commercially available purification kits for DNA are compatible with the fragment size produced by WGA, however, they often have a maximum DNA binding capacity of 10 μg. Therefore it is necessary to determine the concentration of DNA and divide the sample accordingly such that no more than 10 μg is loaded on each column. If necessary, ethanol precipitation can be used to concentrate pooled samples (*see* **Subheading 3.10**).

4. DNA fragmentation is an extremely sensitive procedure and requires careful experimentation, as over digestion can occur quite easily. We recommend that digestion times be reduced by half during the first experiment. When fragmenting amplified DNA times should also be reduced in half as the DNA is already in relatively small fragments compared to unamplified genomic DNA.

5. SYBR green is used to stain the gel as it is more sensitive than detection with ethidium bromide. We recommend using fresh 1× SYBR green stain solution for each gel in order to detect the DNA without having to run more than 1 μl of sample on the gel.

6. Typically this DNA fragmentation method requires two rounds of digestion at 4 min each. There may be cases where more than two rounds with the DNase are required. After each round of digestion we recommend assessing the smear pattern and adjusting the length of the 37°C incubation time.

7. Ideally, after loading the hybridization buffer in the chip, there should be no air bubbles visible in the glass window. However, small bubbles are well tolerated given the array size and the movement generated by the relatively quick rotation during the pre- and overnight hybridization steps.

8. Saving the hybridization mix for each sample will prove useful if problems occur downstream of its preparation. It may be necessary to rehybridize the sample if problems occur with the fluidics, scanning, or alignment of the arrays.

9. There are numerous methods to measure DNA. In this method fluorometry is used with a fluorescent dye, which fluoresces when bound to double stranded DNA. This method was chosen in order to preserve the sample by using only small amounts during quantification.

10. In some cases a .CEL file is not produced after the scanning of a chip. This is most likely due to the failure of the software to automatically align the subgrids, which help to define the probe position. Depending on the severity it is usually possible to manually align the few misaligned subgrids. Procedures for manual alignment can be found in the GCOS

Handbook. In more severe cases, where too many subgrids are misaligned or the main grid fails to align, it may be necessary to rehybridize with the saved hybridization mix, after spiking with additional 2.5–3.5 µl of b213 control oligonucleotide.

11. Data is presented in IGB as a histogram of the log2 ratio of treated sample to reference (control) sample, versus the yeast chromosomal coordinates of each microarray probe (*see* **Fig. 2**). Through visual inspection it is easy to determine the quality of the signal-to-noise ratio and whether software parameters will need to be adjusted in the previous steps. Increase in DNA copy number by one or two copies will appear as a spike in the log2 ratios of 0.58 and 1 respectively. Likewise a loss of one copy of a DNA sequence will appear as a decrease in the log2 ratio of –1 in a diploid cell. Complete loss of a DNA sequence will be quite obvious as a large decrease, exceeding –1. Aneuploidies will be obvious regions where the log2 ratios are either above or below the origin. Zooming in on a particular region and assessing where a particular amplification or deletion begins and ends can determine the breakpoints of a given copy number variation. The actual structure of the chromosomal aberration is, however, not apparent from this analysis. Analysis of data is dependent primarily on visual inspection, however, a number of programs have been developed to detect the approximate aneuploidies and even single nucleotide polymorphisms (SNPs). Chromosomal Aberration Region Miner (ChARM) has been developed to accurately to detect changes in the DNA copy number and define their breakpoints *(28)*. Programs for arrays of similar resolution have been written to detect SNPs, allowing the accurate detection (>90%) of aberrations at nucleotide resolution *(14)*.

Acknowledgments

We thank Corey Nislow for experimental advice and generous use of his Affymetrix equipment; Maitreya Dunham for strains, experimental work and advice; William Lee for experimental advice; and Marinella Gebbia and Malene Urbanus for their technical assistance. Research in the Brown Lab is supported by the Canadian Cancer Society and the Canadian Institutes of Health Research.

References

1. Cahill, D. P., Lengauer, C., Yu, J., Riggins, G. J., Willson, J. K., Markowitz, S. D., Kinzler, K. W., and Vogelstein, B. (1998). Mutations of mitotic checkpoint genes in human cancers, *Nature* **392**, 300–303.
2. Lejeune, J., Gautier, M., and Turpin, R. (1959). [Study of somatic chromosomes from 9 mongoloid children.], *C R Hebd Seances Acad Sci* **248**, 1721–1722.
3. Niebuhr, E. (1978). The Cri du Chat syndrome: epidemiology, cytogenetics, and clinical features, *Hum Genet* **44**, 227–275.
4. Albertson, D. G. (2003). Profiling breast cancer by array CGH, *Breast Cancer Res Treat* **78**, 289–298.
5. Kallioniemi, A., Kallioniemi, O. P., Sudar, D., Rutovitz, D., Gray, J. W., Waldman, F., and Pinkel, D. (1992). Comparative genomic hybridization for molecular cytogenetic analysis of solid tumors, *Science* **258**, 818–821.
6. Solinas-Toldo, S., Lampel, S., Stilgenbauer, S., Nickolenko, J., Benner, A., Dohner, H., Cremer, T., and Lichter, P. (1997). Matrix-based comparative genomic hybridization: biochips to screen for genomic imbalances, *Genes Chromosomes Cancer* **20**, 399–407.
7. Heselmeyer, K., Hellstrom, A. C., Blegen, H., Schrock, E., Silfversward, C., Shah, K., Auer, G., and Ried, T. (1998). Primary carcinoma of the fallopian tube: comparative genomic hybridization reveals high genetic instability and a specific, recurring pattern of chromosomal aberrations, *Int J Gynecol Pathol* **17**, 245–254.
8. Pere, H., Tapper, J., Seppala, M., Knuutila, S., and Butzow, R. (1998). Genomic alterations in fallopian tube carcinoma: comparison to serous uterine and ovarian carcinomas reveals similarity suggesting likeness in molecular pathogenesis, *Cancer Res* **58**, 4274–4276.
9. Snijders, A. M., Nowee, M. E., Fridlyand, J., Piek, J. M., Dorsman, J. C., Jain, A. N., Pinkel, D., van Diest, P. J., Verheijen, R. H., and Albertson, D. G. (2003). Genome-wide-array-based comparative genomic hybridization reveals genetic homogeneity and frequent copy number increases encompassing CCNE1 in fallopian tube carcinoma, *Oncogene* **22**, 4281–4286.
10. Nakao, K., Mehta, K. R., Fridlyand, J., Moore, D. H., Jain, A. N., Lafuente, A., Wiencke, J. W., Terdiman, J. P., and Waldman, F. M. (2004). High-resolution analysis of DNA copy number alterations in colorectal cancer by array-based comparative genomic hybridization, *Carcinogenesis* **25**, 1345–1357.
11. Rodriguez, V., Chen, Y., Elkahloun, A., Dutra, A., Pak, E., and Chandrasekharappa, S. (2007). Chromosome 8 BAC array comparative genomic hybridization and expression analysis identify amplification and overexpression of TRMT12 in breast cancer, *Genes Chromosomes Cancer* **46**, 694–707.
12. Ishkanian, A. S., Malloff, C. A., Watson, S. K., DeLeeuw, R. J., Chi, B., Coe, B. P., Snijders, A., Albertson, D. G., Pinkel, D., Marra, M. A., Ling, V., MacAulay, C., and Lam, W. L. (2004). A tiling resolution DNA microarray with complete coverage of the human genome, *Nat Genet* **36**, 299–303.
13. Selzer, R. R., Richmond, T. A., Pofahl, N. J., Green, R. D., Eis, P. S., Nair, P., Brothman, A. R., and Stallings, R. L. (2005). Analysis of chromosome breakpoints in neuroblastoma at sub-kilobase resolution using fine-tiling oligonucleotide array CGH, *Genes Chromosomes Cancer* **44**, 305–319.
14. Gresham, D., Ruderfer, D. M., Pratt, S. C., Schacherer, J., Dunham, M. J., Botstein, D., and Kruglyak, L. (2006). Genome-wide detection of polymorphisms at nucleotide resolution with a single DNA microarray, *Science* **311**, 1932–1936.
15. Juneau, K., Palm, C., Miranda, M., and Davis, R. W. (2007). High-density yeast-tiling array reveals previously undiscovered introns and extensive regulation of meiotic splicing, *Proc Natl Acad Sci U S A* **104**, 1522–1527.
16. Dunham, M. J., Badrane, H., Ferea, T., Adams, J., Brown, P. O., Rosenzweig, F., and Botstein, D. (2002). Characteristic genome rearrangements in experimental evolution of Saccharomyces cerevisiae, *Proc Natl Acad Sci U S A* **99**, 16144–16149.
17. Green, B. M., Morreale, R. J., Ozaydin, B., Derisi, J. L., and Li, J. J. (2006). Genome-wide mapping of DNA synthesis in *Saccharomyces cerevisiae* reveals that mechanisms preventing reinitiation of DNA replication are not redundant, *Mol Biol Cell* **17**, 2401–2414.
18. Watanabe, T., Murata, Y., Oka, S., and Iwahashi, H. (2004). A new approach to species determination for yeast strains: DNA microarray-based comparative genomic hybridization using a yeast DNA microarray with 6000 genes, *Yeast* **21**, 351–365.
19. Shiu, S. H., and Borevitz, J. O. (2006). The next generation of microarray research: applications in evolutionary and ecological genomics, *Heredity* **100**, 141–149.

20. Hartwell, L. H., and Weinert, T. A. (1989). Checkpoints: controls that ensure the order of cell cycle events, *Science* **246**, 629–634.
21. Weinert, T. A., and Hartwell, L. H. (1988). The RAD9 gene controls the cell cycle response to DNA damage in Saccharomyces cerevisiae, *Science* **241**, 317–322.
22. Alcasabas, A. A., Osborn, A. J., Bachant, J., Hu, F., Werler, P. J., Bousset, K., Furuya, K., Diffley, J. F., Carr, A. M., and Elledge, S. J. (2001). Mrc1 transduces signals of DNA replication stress to activate Rad53, *Nat Cell Biol* **3**, 958–965.
23. Sun, Z., Hsiao, J., Fay, D. S., and Stern, D. F. (1998). Rad53 FHA domain associated with phosphorylated Rad9 in the DNA damage checkpoint, *Science* **281**, 272–274.
24. Kolodner, R. D., Putnam, C. D., and Myung, K. (2002). Maintenance of genome stability in Saccharomyces cerevisiae, *Science* **297**, 552–557.
25. Putnam, C. D., Pennaneach, V., and Kolodner, R. D. (2005). *Saccharomyces cerevisiae* as a model system to define the chromosomal instability phenotype, *Mol Cell Biol* **25**, 7226–7238.
26. David, L., Huber, W., Granovskaia, M., Toedling, J., Palm, C. J., Bofkin, L., Jones, T., Davis, R. W., and Steinmetz, L. M. (2006). A high-resolution map of transcription in the yeast genome, *Proc Natl Acad Sci U S A* **103**, 5320–5325.
27. Sambrook, J., Fritsch, E. F., and Maniatis, T. (1989). Molecular Cloning A Laboratory Manual, 2 edn., Cold Spring Harbor Laboratory Press, Cold Spring Harbor, New York.
28. Myers, C. L., Dunham, M. J., Kung, S. Y., and Troyanskaya, O. G. (2004). Accurate detection of aneuploidies in array CGH and gene expression microarray data, *Bioinformatics* **20**, 3533–3543.

Chapter 2

Identification of Transcription Factor Targets by Phenotypic Activation and Microarray Expression Profiling in Yeast

Gordon Chua

Summary

A major obstacle to identify physiological transcriptional targets is that the conditions that induce the majority of yeast transcription factors (TFs) are unknown. Microarray analyses of deletion mutants indicate that most TFs are inactive under standard growth conditions. To overcome this, we screened an ordered array of yeast open reading frames (ORFs) to identify TFs that confer reduced fitness upon overexpression, suggesting that overexpression results in an activated state (phenotypic activation). Approximately one-third of all yeast TFs exhibited this phenotype. Here, we describe in detail our methodology to characterize these TF overexpression strains including microarray expression profiling, data analysis, and motif searching. Our analyses show that in many cases, the differentially regulated genes correspond to physiological functions and known targets of well-characterized TFs. The expected binding sites of several TFs were also identified in the promoters of these genes. Moreover, novel DNA-binding sequences and putative targets were identified for less-characterized TFs. These results demonstrate that phenotypic activation is an effective approach to rapidly characterize TFs on a large scale, which should also be feasible in other organisms.

Key words: Transcription factor, Overexpression, Phenotypic activation, Yeast, Microarray, FunSpec, RankMotif

1. Introduction

A primary objective to understand the workings of an organism is to obtain a complete mapping of the transcriptional-regulatory network, defined here as interactions between DNA-binding transcription factors (TFs) and the cognate sequences they bind in order to control expression of target genes. One straightforward approach to systematically characterize TF activities would be to microarray-profile all the TF deletion strains in an organism, where differentially downregulated or upregulated genes in the mutants would potentially represent target genes of transcriptional activators and repressors, respectively. However, we discovered that a large majority of yeast TF mutants (~85%) exhibit profiles indistinguishable from microarray noise when grown typically in the laboratory, suggesting that most TFs are not active under these conditions, or that there exists a high degree of functional overlap among TFs *(1)*. To overcome these obstacles, we sought for alternate approaches to systematically characterize TFs and identify their target genes. One such approach stemmed from the construction of an overexpression array of 5,280 yeast strains, each containing a unique gene under control of the strong, inducible *GAL1/10* promoter *(2)*. Phenotypic characterization of the overexpression array revealed that 769 genes (15% of the genome) were detrimental to normal growth when ectopically expressed *(2)*. Interestingly, these toxic overexpressors were most enriched for TFs (32.6% of all TFs), representing a greater than twofold enrichment for this functional class of proteins compared to the rest of the genome *(2)*. On the basis of these observations, we hypothesized that the reduced growth rate is attributed to induction of TF activity by ectopic expression (hence the term "phenotypic activation"; *see* **Note 1**) *(3)*.

Differential gene expression caused by phenotypic activation of TFs is globally detected by DNA microarrays to elucidate the biological function of TFs and identify putative gene targets (**Fig. 1**). Strains containing either a 2-μm vector with a *GAL1/10*-driven TF gene or an empty vector are grown concurrently and induced for 3 h after a raffinose to galactose shift. Total RNA is extracted from these strains using hot phenol, and poly-A mRNA is subsequently isolated using oligo-dT cellulose. To couple Cy3 and Cy5 dyes to the samples, the mRNA is initially labeled with aminoallyl-dUTPs during reverse transcription. Cy3/Cy5-coupled cDNA samples are hybridized in dye reversal onto a spotted microarray containing 60-mer oligonucleotide probes specific for all coding genes of *Saccharomyces cerevisiae* *(4)*.

The microarray data is subjected to an analysis pipeline designed to filter out microarray noise and elucidate TF function and putative gene targets. We determined that Pearson correlations >0.3 between experimental replicates are indicative that the differentially expressed genes are more likely due to TF overexpression rather than microarray noise *(5)*. The differentially regulated genes are also examined for significant enrichment in Gene Ontology functional categories using FunSpec to identify putative function and targets of TFs *(6)*. Finally, the promoter regions of differentially expressed genes are subjected

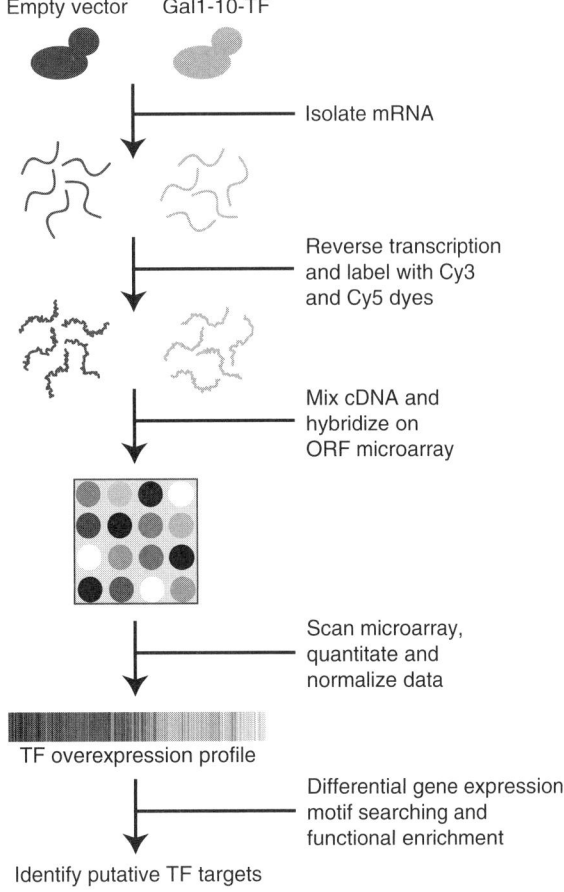

Fig. 1. Flow chart describing the procedure involved in the microarray profiling of transcription factor overexpression strains and data analysis to identify putative targets.

to a motif-finding algorithm called RankMotif to identify putative TF-binding sites *(3)*.

2. Materials

2.1. Cell Culture

1. Synthetic uracil dropout medium with 2% raffinose (SR-Ura).
2. 40% Galactose.
3. 250-ml baffled culture flasks.
4. 50-ml Falcon tubes.
5. Liquid nitrogen.
6. Spectrophotometer (microplate reader).

2.2. Total RNA Isolation

1. AE buffer: 50 mM sodium acetate, pH 5.2, 10 mM ethylenediamine tetraacetic acid (EDTA), 1% sodium dodecyl sulfate (SDS), diethylpyrocarbonate-treated (DEPC) H_2O.
2. Unbuffered liquefied acid phenol (pH = 4.5–5.5) (this is extremely toxic and corrosive and care should be taken not to receive exposure). Store at 4°C.
3. Phenol: chloroform: isoamyl alcohol (25:24:1) (this is extremely toxic and corrosive and care should be taken not to receive exposure). Store at 4°C.
4. Chloroform: isoamyl alcohol (24:1). Store at 4°C.
5. 3 M sodium acetate, pH 5.2, DEPC H_2O.
6. 95–100% Ethanol.
7. 70% Ethanol.
8. Isopropanol.
9. 0.1% DEPC H_2O.
10. Acid-washed 425–600 μm glass beads (Sigma).
11. Plasticware: 10-ml disposal pipettes, 15-ml Falcon tubes.
12. Multi-tube vortexer (VWR).

2.3. Poly-A mRNA Isolation

1. 2× Loading buffer: 40 mM Tris–HCl, pH 7.6; 1 M NaCl; 2 mM EDTA; 0.2% sodium lauryl sarosine (SLS); DEPC water.
2. Middle wash buffer: 20 mM Tris–HCl, pH 7.6; 150 mM NaCl; 1 mM EDTA; 0.1% SLS; DEPC water.
3. Elution buffer: 10 mM Tris–HCl, pH 7.6; 0.1 mM EDTA; DEPC water.
4. Oligo-dT cellulose (SIGMA): This can be recycled by washing the resin with 0.1 N NaOH, which will elute and/or degrade any attached RNA. Store at 4°C.
5. Poly-prep columns (Bio-Rad).
6. 95–100% Ethanol.
7. 0.1 N NaOH.
8. DEPC H_2O.
9. Linear acrylamide (Ambion). Store at −20°C.
10. 3 M sodium acetate, pH 5.2, DEPC water.
11. Plasticware: 10-ml disposal pipettes, 6-ml and 50-ml Falcon tubes.
12. Nutator.

2.4. Reverse Transcription and Aminoallyl Labeling of mRNA

1. PCR purification columns and associated reagents (Qiagen).
2. T18-VN primer (1 μg/μl).

3. Superscript II reverse transcriptase with 5× buffer and 0.1 M dithiothreitol (Invitrogen).
4. 10 mM dNTPs.
5. 1 mM aminoallyl-dUTP (Sigma): Dissolve 1 mg of AA-dUTP in 191 μl DEPC water. Store at −20°C.
6. 1 N NaOH.
7. 0.5 M EDTA.
8. 1 M Tris–HCl, pH 7.6.
9. 80% Ethanol.
10. DEPC H_2O.
11. 42 and 65°C water baths.
12. Speed-Vac.

2.5. Cy3/Cy5 Dye Coupling of cDNA

1. Anhydrous dimethyl sulfoxide (DMSO, Sigma).
2. 2× Bicarb buffer: (one bicarbonate capsule (Sigma), 25 ml water, 125 μl 37% HCl). Note: this buffer should be made fresh and is good only for 1 week.
3. Cy3 and Cy5 monofunctional dye packs (GE Healthcare).
4. 4 M hydroxylamine.
5. PCR-purification columns and associated reagents (Qiagen).

2.6. Microarray Hybridization

1. Quackenbush prehybridization solution: 5× SSC, 1% SDS, and 1% BSA. Store in 50-ml aliquots at −20°C.
2. Hybridization solution: 5× SSC, 25% formamide, 0.1% SDS. Store in 1-ml aliquots at −20°C.
3. Wash 1 solution: 1× SSC, 0.2% SDS.
4. Wash 2 solution: 0.1× SSC, 0.2% SDS.
5. Wash 3 solution: 0.1× SSC.
6. Microscope slide mailers.
7. Lifter cover slips.
8. Curved forceps.
9. UV crosslinker.
10. Speed-Vac.
11. Rack for microscope slides.
12. Hybridization oven (42°C).
13. Hybridization chambers.
14. 250-ml glass dishes.
15. Compressed air (e.g., Dust-off).

2.7. Acquisition, Processing, and Analysis of Microarray Data

1. Laser microarray scanner with integrated software for quantitation and normalization of data (e.g., Axon GenePix 4200A scanner with GenePix Pro/Acuity software – Molecular Devices).

3. Methods

3.1. Cell Culture

1. For each experiment, prepare two overnight cultures of isogenic strains, one containing a 2-μm vector with a *GAL1/10*-driven TF gene and the other containing an empty vector (pEGH, *URA3+*). Inoculate several medium-size colonies (~10) from an agar plate into 100 ml SR-Ura medium in a 250-ml baffled culture flask. In addition, set up a second diluted culture (1/5–1/10×) to ensure that one of the overnight cultures will be in the appropriate cell density range the following day. Shake cultures at 30°C overnight.

2. For each strain, select an overnight culture in mid-log phase ($<5 \times 10^6$ cells/ml) and inoculate a new 95-ml culture in SR-Ura medium at an initial cell density of 7×10^5–9×10^5 cells/ml.

3. Add 5 ml 40% galactose (2%) into each culture and shake at 30°C for 3 h (*see* **Note 2**).

4. Just prior to 3 h in galactose, determine the optical density at 600 nm (OD_{600}) of each culture. The cell densities should range between 2×10^6 and 3×10^6 cells/ml. Pair up the empty vector and TF overexpression cultures with the closest cell densities.

5. Harvest cultures in 50-ml Falcon tubes by centrifugation at $2,000 \times g$ for 2 min, pour out the supernatant, and freeze cell pellets in liquid nitrogen.

6. Store cell pellets at –80°C.

3.2. Total RNA Isolation

1. Preparation prior to isolation procedure: Preheat AE buffer and acid phenol in a 65°C water bath. For each sample, label three 15-ml Falcon tubes.

2. Remove frozen cell pellets from the freezer (two 50-ml Falcon tubes/100 ml culture).

3. Combine the samples by adding 4 ml AE buffer (65°C) to one of the frozen cell pellets, loosening the frozen pellet by shaking gently or pipetting and pouring the contents into the other Falcon tube. Add 4 ml of acid phenol (65°C) and

~200 μl acid-washed glass beads. **Steps 2** and **3** should be performed quickly to prevent RNA degradation.

4. Vortex the samples in a multitube vortexer for 30 s and place in a 65°C water bath for 4 min. Repeat four times and then incubate 10 min on ice.

5. Centrifuge at 2,000 × g for 5 min at 4°C and transfer the aqueous layer into a 15-ml Falcon tube containing 2 ml cold phenol: chloroform: isoamyl alcohol (25:24:1). Vortex three times for 30 s with 1 min intervals in between.

6. Centrifuge at 2,000 × g for 5 min at 4°C and transfer the aqueous layer into a 15-ml Falcon tube containing 2 ml chloroform: isoamyl alcohol (24:1). Vortex once for 15 s.

7. Centrifuge at 2,000 × g for 5 min at 4°C and transfer the aqueous layer into an empty 15-ml Falcon tube. Add 1/10 volume of 3 M sodium acetate (pH 5.2) (300–400 μl), vortex for 1 min, and then add an equal volume of isopropanol (3–4 ml). Mix well by gently inverting the tubes, and incubate for 10 min at room temperature or –20°C overnight.

8. Centrifuge at 3,200 × g for 30 min at 4°C, remove the supernatant and wash the pellet with 1 ml 70% ethanol. Centrifuge at 3,200 × g for 5 min at 4°C, remove the supernatant, and air-dry the pellet for 15–20 min.

9. Dissolve the RNA pellet in 1 ml of DEPC water for each sample by pipetting. The samples may be heated at 65°C for 5–10 min to fully dissolve the RNA pellet.

10. Determine the quantity of RNA at OD_{260}. Typical total RNA yield for a 100-ml yeast culture ranges from 500 to 2,000 μg. Store total RNA in –80°C, or continue with poly-A mRNA isolation.

3.3. Poly-A mRNA Isolation

1. Preparation prior to isolation procedure: Remove total RNAs out from freezer and thaw on ice. Save a 10-μl aliquot of each RNA sample for an EPPS/formaldehyde gel to check the integrity of RNA (optional). For each sample, label one poly-prep column, two 6-ml Falcon tubes, and a 1.5-ml microcentrifuge tube. Preheat the elution buffer at 65°C.

2. Preparing columns: Place columns on a wash rack (e.g., Microcentrifuge Tube Rack, ResMer Resin, VWR) with a reservoir underneath (e.g., the lid of a P1000 tip box). Weigh out 0.7 g oligo-dT cellulose (SIGMA) and place in a 50-ml Falcon tube. This amount of oligo-dT cellulose is sufficient for 12–14 columns. Wash the oligo-dT cellulose three times with 50 ml DEPC water and once with 50 ml 0.1 N NaOH. Resuspend the oligo-dT cellulose in 50 ml 0.1 N NaOH.

Mix on Nutator for 10–15 min. Mix the oligo-dT cellulose slurry by inverting the Falcon tube, and quickly add 4 ml slurry into each column using a 10-ml plastic pipette. After the slurry has settled, use the remaining slurry to equalize the volume of oligo-dT cellulose among the columns. Wash the columns by adding 4 ml DEPC water, and then with 2 ml 1× column loading buffer after the water has flowed through the column.

3. Heat the samples to 65°C for 5 min in the water bath. Cool quickly by chilling on ice for 3 min.
4. While samples are cooling, add an equal volume (1 ml) of 2× column loading buffer.
5. Place each column into a 6-ml Falcon tube while inserting a P2 pipette tip between the column and the tube to allow venting and flow through of the sample. Stand the Falcon tubes with columns in a rack.
6. Carefully pour the samples into the columns. The sample should flow through in 2–5 min. If it takes much longer than this, pipette gently to resuspend the resin and let it resettle – the column should then run a bit faster.
7. Reload the columns by pouring the sample from the Falcon tube back into the top of each column, and reinsert the column into the Falcon tube as in **step 5**.
8. Reload the columns again for the third time (repeat **step 7**).
9. After the samples have run through the columns, store them in the –20°C freezer. They can be retrieved later if the mRNA yield is very low.
10. Place the columns into the wash rack. Wash twice with 2 ml 1× column wash buffer.
11. Wash once with 400 μl middle wash buffer.
12. Place the column into a new 6-ml Falcon tube with a P2 pipette tip as described before.
13. Elute three times with 330 μl 65°C prewarmed elution buffer.

A second column run is required to further purify the poly-A mRNA samples.

14. Place columns on the wash rack and wash once with 4 ml DEPC water and then 4 ml 1× column wash buffer.
15. Repeat **steps 3–11**.
16. Place each column into a 1.5-ml microcentrifuge tube with a P2 pipette tip as described before.
17. Elute twice with 250 μl 65°C prewarmed elution buffer.

18. Add 50 μl 3 M sodium acetate (pH 5.2) and 6 μl linear acrylamide into each sample. Vortex the samples for 30 s. Add 1.1 ml of 95% EtOH and vortex the microcentrifuge tube on its side for 30 s. Precipitate mRNA in –20°C overnight.

19. Centrifuge mRNA samples at 16,000 × g at 4°C for 30 min. Remove supernatant by pipetting, pulse-centrifuge, and remove the remaining supernatant. Do not wash the samples. Place Kimwipe over the tubes and air-dry samples at room temperature for 30 min. Resuspend in 20 μl DEPC water by pipetting.

20. To remove traces of oligo-dT cellulose that may interfere with subsequent steps, heat the samples at 65°C for 5–10 min, centrifuge at 16,000 × g for 1 min, and carefully transfer the supernatant into a new microcentrifuge tube.

21. Determine the quantity of RNA at OD_{260}. Typical total poly-A mRNA yield ranges from 0.5 to >20 μg. Store mRNA in –80°C.

22. Recycling the oligo-dT cellulose and column: Cap the bottom of the columns and add 2 ml of 0.1 N NaOH to each column. Resuspend the oligo-dT cellulose in each column by pipetting, and transfer contents into a 50-ml Falcon tube. Add 1 ml of 0.1 N NaOH to the column, resuspend, and transfer remaining oligo-dT cellulose into the Falcon tube. It is important to remove as much residual oligo-dT cellulose from the column to prevent blockage in future mRNA isolations. Wash oligo-dT cellulose three times with DEPC water. After the final wash, add 50 ml DEPC water and store at 4°C. Rinse the columns twice with DEPC water before storage.

3.4. Reverse Transcription and Aminoallyl Labeling of cDNA

1. Each experiment will be performed twice with fluor reversal (control sample–Cy3 vs. experimental sample–Cy5, and experimental sample–Cy3 vs. control sample–Cy5). As a result, each mRNA sample will be reverse-transcribed twice. Aliquot equal amounts (1–2 μg) of control and experimental mRNA samples into microcentrifuge tubes. Dry down samples in a Speed-Vac (medium-heat setting to prevent RNA hydrolysis).

2. Dissolve each sample in 10.5 μl DEPC water and 1 μl T18VN primer (1 μg/μl).

3. Denature the samples at 65°C for 5 min.

4. Incubate at 42°C for 5 min to anneal the T18VN primer.

5. While the samples are incubating at 42°C, prepare a master reaction mixture containing 4 μl 5× RT buffer, 2 μl 0.1 M DTT, 1 μl 10 mM dNTPs, 1 μl 1 mM aminoallyl-dUTP, and

0.5 μl superscript II enzyme per sample. Add 8.5 μl of reaction mixture to each sample.

6. Incubate 50 min at 42°C.
7. Add 10 μl of a 1:1 mixture 0.5 M EDTA: 1 N NaOH to each of the samples.
8. Incubate at 65°C for 20 min to hydrolyze the RNA. During this incubation, set up and label a QIAGEN purification column and microcentrifuge tube for each sample, and preheat some water at 65°C.
9. Add 10 μl 1 M Tris–HCl pH 7.6 to each sample.
10. To purify the cDNA samples and to remove Tris to prevent undesired coupling of the monofunctional NHS-ester Cy-dyes to free amine groups in solution, add 60 μl water to each sample to bring the reaction volume to 100 μl.
11. Add 500 μl Qiagen buffer PB to each sample, mix well by pipetting, and apply to the PCR purification columns.
12. Centrifuge the columns at $3,800 \times g$ for 1 min at room temperature and discard the flow-through.
13. Add 600 μl 80% ethanol to each column, centrifuge at $3,800 \times g$ for 1 min, and discard the flow-through.
14. Repeat the ethanol washes two more times.
15. Centrifuge at $16,000 \times g$ for 1 min to dry the column.
16. Elute each sample from the column into a microcentrifuge tube twice with 40 μl 65°C water.
17. Dry the cDNA samples in Speed-Vac (high-heat setting).
18. The cDNA pellet can now be frozen at –80°C for at least several weeks prior to coupling.

3.5. Cy3/Cy5 Dye Coupling of cDNA

1. Resuspend each of the dried cDNA samples in 3.5 μl water.
2. Arrange each pair of cDNA samples of which one will be labeled with Cy3 and the other with Cy5.
3. Preparation of Cy3 and Cy5 dyes: Resuspend each Cy3 and Cy5 dye pack in 15 μl anhydrous DMSO by pipetting up and down 30–50 times, and vortexing for 1 min, prior to pulse centrifugation. Immediately recap and reseal the DMSO. Each dye pack can label 12–14 cDNA samples.
4. Dye coupling: Add 30 μl 2× Bicarb buffer to the Cy3 dye. Quickly aliquot 3.5 μl to each of the Cy3 cDNA samples. Repeat for Cy5 samples.
5. Vortex cDNA samples for 1 min, pulse-centrifuge, and incubate for 30 min at room temperature in the dark.
6. Repeat the previous step.

Identification of Transcription Factor Targets by Phenotypic Activation   29

7. Add 3.5 μl 4 M hydroxylamine to each cDNA sample to quench the reaction. Incubate for 15 min at room temperature in the dark. While the reaction is quenching, set up and label the Qiagen PCR purification columns and microcentrifuge tubes.

8. Purify labeled cDNA away from the dyes: Arrange the tubes in Cy3/Cy5 pairs – they will be combined in this step.

9. For all cDNA sample pairs: Add 70 μl water to the Cy3 cDNA sample. Add 500 μl buffer PB (Qiagen) to the Cy3 cDNA sample, mix by pipetting, combine with the Cy5 cDNA sample, and apply the mix to a Qiagen PCR purification column.

10. Centrifuge at $2,600 \times g$ for 1 min and discard the flow-through.

11. Apply 700 μl buffer PE (Qiagen), centrifuge at $2,600 \times g$ for 1 min, and discard the flow-through.

12. Repeat the previous step.

13. Centrifuge at $16,000 \times g$ for 1 min to dry the column.

14. Elute each cDNA sample from the column into a microcentrifuge tube twice with 30 μl elution buffer (Qiagen). The cDNA samples should be slightly purple in color. The efficiency of Cy3/Cy5 coupling to the cDNA samples can be determined at wavelengths 532 and 635 nm, respectively, with a spectrophotometer.

15. Dry the cDNA samples in a Speed-Vac (high-heat setting). Proceed to prehybridize the microarray slides.

3.6. Microarray Hybridization and Washing

1. Hybridization volumes, reagents, chambers, and conditions vary according to specific expression microarray platforms. The following protocol is used for expression microarrays manufactured by spotting 60-mer oligonucleotides (Open Biosystems) onto polylysine-treated glass microscope slides.

2. Microarray slide prehybridization: Thaw Quackenbush prehybridization buffer at 65°C and then incubate at 42°C.

3. UV-crosslink the microarray slides.

4. Place microarrays into slide mailers, fill with prehybridization buffer, and incubate at 42°C for a minimum of 45 min.

5. At the end of prehybridization, wash the microarrays 4–5 times with water by filling the slide mailer with water, rocking it, and pouring out the water.

6. Dry the microarrays by centrifuging them in slide racks in a benchtop centrifuge containing a swinging-bucket rotor and microplate carriers at $200 \times g$ for 5 min. Store the microarrays

in a slide box at 42°C until ready to hybridize the Cy3/Cy5-coupled cDNA samples.

7. Hybridizations: When the coupled cDNA samples are dry and the microarray slides are ready, resuspend each sample in 40 μl hybridization buffer. Add oligo spike-ins specific for control probes on the microarray if required.

8. Just prior to hybridization, heat the samples at 65°C for 3 min to denature, and then incubate at 42°C until ready for hybridization.

9. Place a microarray slide (array side up) at the bottom part of the chamber.

10. Pipette a Cy3/Cy5-coupled cDNA sample onto the microarray, taking care not to touch it with the pipette tip. Avoid the formation of bubbles since they will prevent uniform hybridization on the microarray.

11. Place one edge of the lifter cover slip on one side of the microarray slide with the lifter side down, and carefully lower the other edge onto the sample using a pair of curved forceps. Gently center the lifter cover slip over the microarray with forceps.

12. Fill the humidifying well at the end of the hybridization chamber with 4× SSC.

13. Seal the upper and lower portions of the hybridization chamber with screws/clips and incubate at 42°C overnight (or at least 6 h).

14. Record the label on the microarray and description of the samples in a hybridization sheet.

15. Repeat **steps 9–14** for the next sample.

16. Microarray slide washes: After hybridization is complete, fill three 250-ml glass dishes with Wash 1, Wash 2, and Wash 3.

17. Open a hybridization chamber, and pick up a microarray slide with forceps. Submerge the microarray slide in Wash 1. The lifter cover slip should fall off almost immediately when the microarray slide is submerged. Be careful to avoid scratching the array.

18. Dip the microarray slide up and down 20 times in Wash 1.

19. Dip the microarray slide up and down 20 times in Wash 2.

20. Dip the microarray slide up and down 20 times in Wash 3.

21. Quickly dry the microarray slide (array side first) by immediately blowing off the wash solution with a dust gun or some other compressed gas, going from one side to the other in about 5 s. Store the microarray slide in a slide box.

22. Repeat **steps 17–21** for the next microarray slide.

3.7. Acquisition, Processing, and Analysis of Microarray Data

1. Scan microarrays, quantitate images, and normalize data by Lowess smoothing with the laser scanner and integrated software according to the manufacturer's instructions (7). Each pair of fluor-reversal experiments is combined by averaging the ratios of normalized median intensity of TF overexpression vs. the empty vector control for each spot on the microarray.

2. Examine the microarray data for specific upregulation of the TF mRNA in the corresponding TF overexpression experiment to confirm the induction of the TF gene in the presence of galactose and the correct orientation of the microarray samples (**Fig. 2**).

3. Calculate the Pearson correlation between each pair of fluor-reversal experiments. A Pearson correlation of >0.3 between replicates is indicative that the differential gene expression in an experiment is more likely caused by TF overexpression rather than microarray noise (5). This correlation threshold may be different for other microarray platforms (Agilent, Affymetrix, Nimblegen, etc.).

4. Input subsets of upregulated and downregulated genes separately at various thresholds of differential expression into FunSpec (http://funspec.med.utoronto.ca/) to identify significant functional enrichment of Gene Ontology categories (*see* **Note 3**). The segregation of upregulated and downregulated genes is required to characterize transcriptional activators and repressors, respectively. The significant functional

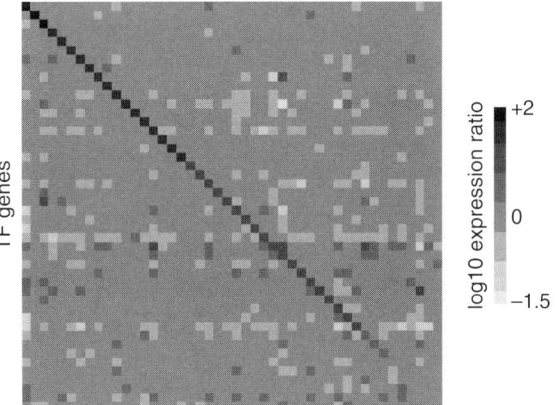

Fig. 2. Induction of transcription factor genes in 46 overexpression strains exhibiting phenotypic activation. The majority of transcription factor genes are highly induced in the corresponding transcription factor overexpression strains from microarray profiling (*diagonal*). The induction of transcription factor genes along the diagonal are highest and lowest in the upper left and lower right, respectively.

categories appearing in the list of differentially regulated genes should provide information on the TF's putative function and targets since both are expected to be involved in a common biological process (**Fig. 3**).

5. Subject the microarray data of each TF overexpression experiment to motif searching using RankMotif (*see* **Note 4**). This probabilistic-inference algorithm is applied to the promoter regions of differentially expressed genes to identify prevalent 8-mer sequences that represent putative TF-binding sites (**Fig. 3**). Motif predictions can be validated using gel mobility shift assays involving the DNA-binding domain of the TF and two tandem copies of the predicted motif *(3)*.

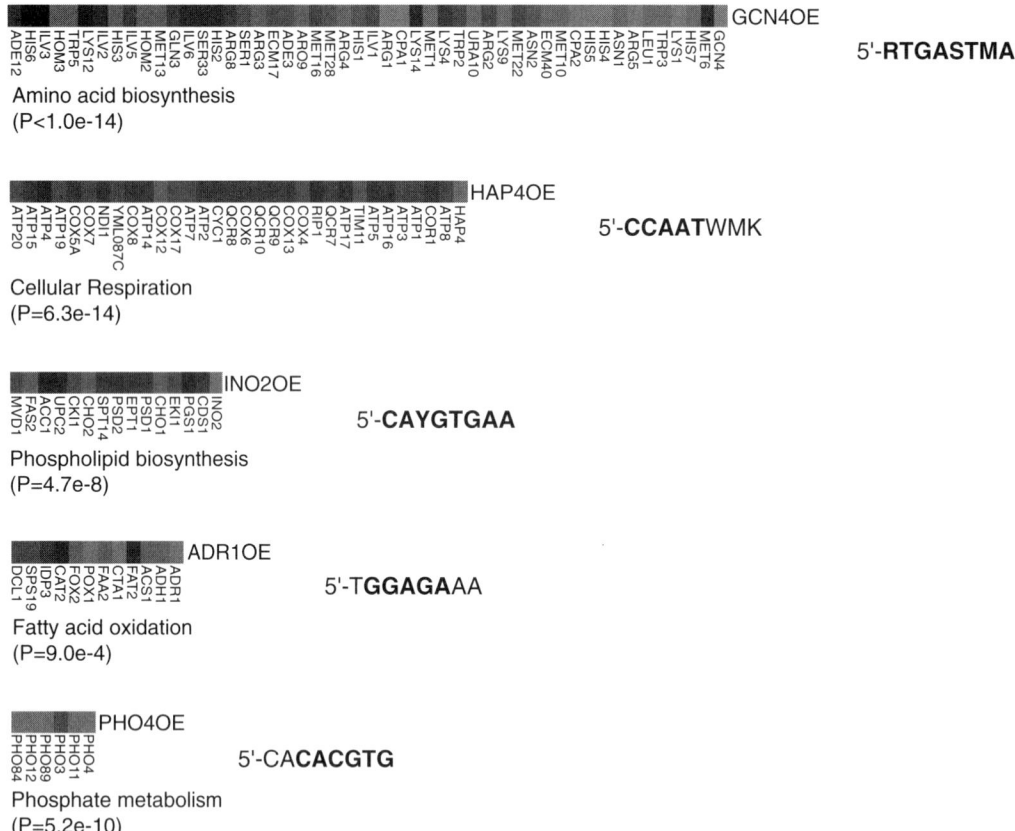

Fig. 3. Putative targets identified for several transcription factors by phenotypic activation and microarray expression profiling. Genes upregulated in response to transcription factor overexpression were searched for (1) significant functional enrichment of gene ontology categories (*P* values) using FunSpec and; (2) overrepresented 8-mer sequences in promoter regions using RankMotif (S = GC; W = AT; R = AG; Y = CT; K = GT; M = AC). Nucleotides of predicted motifs that match known motifs are shown in *bold*.

4. Notes

1. Microarray expression profiles recovered from 23 well-characterized TFs that failed to exhibit an overexpression-related growth defect were all muted, indicating that toxic overexpression is a strong indicator of TF activation *(3)*. It is possible that some of the nontoxic TF overexpressors simply contained epitope-tagged proteins that were nonfunctional. There is currently another overexpression array available which uses a different fusion protein tagging format from that used in this study, and this array is also designed to allow rapid swapping of various epitope tags *(8)*. These assorted versions of the overexpression array can then be screened for novel occurrences of phenotypic activation. In addition, many TFs in the helix–loop–helix, basic–leucine–zipper, and Gal4 classes are obligate dimers and some proportion of these would be obligate heterodimers, which may require overexpression of both subunits in order to obtain hyperactivation of the TF *(9)*. Therefore, one possible strategy to tease out the target genes of these TFs is to identify associations of distinct TFs by protein–protein interaction techniques followed by co-overexpression to phenotypically activate them.

2. Despite obtaining meaningful microarray expression profiles from the phenotypic activation of TFs, identifying and validating their target genes and binding motifs remain challenging. A large part of the difficulty lies in eliminating or filtering out the differential gene expression caused by the induction of secondary pathways resulting from the toxic effects of TF overexpression. One possible approach to overcome this obstacle is to microarray expression-profile earlier and later time points of TF induction. This would conceivably better discriminate primary targets from secondary ones and lead to better predictions by motif-finding algorithms. We had selected a single 3 h induction time point for each TF gene under control of the *GAL1/10* promoter because time course studies initially on several TFs had indicated that this induction period was optimal with not much advantage gained from earlier or later times *(3, 10)*. This is likely not the case for all TFs.

3. One complication in identifying significant enrichment of Gene Ontology categories and motif sequences from microarray expression data is that the optimum arbitrary threshold to define genes that are differentially regulated is difficult to determine. The optimum cut-off selection of differentially regulated genes will certainly vary among expression profiles of different TFs. As a result, several analyses of distinct subsets of differentially expressed genes based on various arbitrary

thresholds are required. One approach that circumvents this problem is the use of threshold-independent statistical methods such as the Wilcoxon–Mann–Whitney (WMW) test to identify significant enrichment of various categories. In this method, the expression ratios of all genes from a particular microarray experiment are initially ranked from the highest to the lowest. The WMW test determines whether the differences in the median expression ratio ranks between genes belonging to a given category and those that are not are statistically significant. This approach has been successfully applied to microarray data to determine functional enrichment, motif finding, and comparison of various genome-wide datasets *(3, 11, 12)*.

4. The probabilistic-inference algorithm RankMotif has proven more robust for microarray expression data with considerable secondary effects than other motif-finding algorithms such as BioProspector *(3, 13)*. For example, motif searches by BioProspector frequently outputs the stress–response element 5′-CCCCT-3′ found in the promoters of numerous upregulated genes as a secondary response to TF overexpression. To overcome this problem, RankMotif is designed to search for a motif that is specific for a particular experiment as well as an additional motif that is overrepresented in multiple experiments. RankMotif is available upon request to Quaid Morris (University of Toronto).

Acknowledgments

This author would like to thank Professor Timothy R. Hughes for his support, advice, and critical reading of this manuscript. This work was supported by the Charles H. Best Postdoctoral Fellowship to G.C. and the Natural Sciences and Engineering Research Council to T.R.H.

References

1. Chua, G., Robinson, M. D., Morris, Q., and Hughes, T. R. (2004). Transcriptional networks: reverse-engineering gene regulation on a global scale. *Curr. Opin. Microbiol.* 7, 638–646.
2. Sopko, R., Huang, D., Preston, N., Chua, G., Papp, B., Kafadar, K., Snyder, M., Oliver, S. G., Cyert, M., Hughes, T. R., Boone, C., and Andrews, B. (2006). Mapping pathways and phenotypes by systematic gene overexpression. *Mol. Cell* 21, 319–330.
3. Chua, G., Morris, Q. D., Sopko, R., Robinson, M. D., Ryan, O., Chan, E. T., Frey, B. J., Andrews, B. J., Boone, C., and Hughes T. R. (2006). Identifying transcription factor functions and targets by phenotypic activation. *Proc. Natl. Acad. Sci. USA* **103**, 12045–12050.

4. Hughes, T. R., Marton, M. J., Jones, A. R., Roberts, C. J., Stoughton, R., Armour, C. D., Bennett, H. A., Coffey, E., Dai, H., He, Y. D., Kidd, M. J., King, A. M., Meyer, M. R., Slade, D., Lum, P. Y., Stepaniants, S. B., Shoemaker, D. D., Gachotte, D., Chakraburtty, K., Simon, J., Bard, M., and Friend, S. H. (2000). Functional discovery via a compendium of expression profiles. *Cell* **102**, 109–126.

5. Grigull, J., Mnaimneh, S., Pootoolal, J., Robinson, M. D., and Hughes, T. R. (2004). Genome-wide analysis of mRNA stability using transcription inhibitors and microarrays reveals posttranscriptional control of ribosome biogenesis factors. *Mol. Cell. Biol.* **24**, 5534–5547.

6. Robinson, M. D., Grigull, J., Mohammad, N., and Hughes, T. R. (2002). FunSpec: a web-based cluster interpreter for yeast. *BMC Bioinformatics* **3**, 35.

7. Mnaimneh, S., Davierwala, A. P., Haynes, J., Moffat, J., Peng, W.-T., Zhang, W., Yang, X., Pootoolal, J., Chua, G., Lopez, A., Trochesset, M., Morse, D., Krogan, N. J., Hiley, S. L., Li, Z., Morris, Q., Grigull, J., Mitsakakis, N., Roberts, C. J., Greenblatt, J. F., Boone, C., Kaiser, C. A., Andrews, B. J., and Hughes, T. R. (2004.) Exploration of essential gene functions via titratable promoter alleles. *Cell* **118**, 31–44.

8. Gelperin, D. M., White, M. A., Wilkinson, M. L., Kon, Y., Kung, L. A., Wise, K. J., Lopez-Hoyo, N., Jiang, L., Piccirillo, S., Yu, H., Gerstein, M., Dumont, M. E., Phizicky, E. M., Synder, M., and Grayhach, E. J. (2005). Biochemical and genetic analysis of the yeast proteome with a movable ORF collection. *Genes Dev.* **19**, 2816–2826.

9. Luscombe, N. M., Austin, S. E., Berman, H. M., and Thornton, J. M. (2000). An overview of the structures of protein–DNA complexes. *Genome Biol.* **1**, REVIEWS001.

10. Roberts, C. J., Nelson, B., Marton, M. J., Stoughton, R., Meyer, M. R., Bennett, H. A., He, Y. D., Dai, H., Walker, W. L., Hughes, T. R., Tyers, M., Boone, C., and Friend, S. H. (2000). Signaling and circuitry of multiple MAPK pathways revealed by a matrix of global gene expression profiles. *Science* **287**, 873–880.

11. Berger, M. F., Philippakis, A. A., Qureshi, A. M., He, F. S., Estep, P. W. III, and Bulyk, M. L. (2006). Compact, universal DNA microarrays to comprehensively determine transcription factor binding site specificities. *Nat. Biotechnol.* **24**, 1429–1435.

12. Chen, X., Hughes, T. R., and Morris, Q. (2007). RankMotif ++: a motif-search algorithm that accounts for relative ranks of K-mers in binding transcription factors. *Bioinformatics* **23**, i72–i79.

13. Liu, X., Brutlag, D. L., and Liu, J. S. (2001). BioProspector: discovering conserved DNA motifs in upstream regulatory regions of co-expressed genes. *Pac. Symp. Biocomput.* **6**, 127–138.

Chapter 3

SGAM: An Array-Based Approach for High-Resolution Genetic Mapping in *Saccharomyces cerevisiae*

Michael Costanzo and Charles Boone

Summary

The development of genome-scale resources and high-throughput methodologies has enabled systematic assessment of gene function *in vivo*. Synthetic genetic array (SGA) analysis automates yeast genetic manipulation, permitting diverse analysis of ~5,000 viable deletion mutants in *Saccharomyces cerevisiae*. SGA methodology has enabled genome-wide synthetic lethal screening and construction of a large-scale genetic interaction network for yeast. Genetic networks often reveal new components of specific pathways and functional relationships between genes whose products buffer one another or impinge on a common essential pathway. Because SGA analysis can be used to manipulate any genetic element linked to a selectable marker, it is a highly versatile approach that can be adapted for a variety of different genetic screens, including synthetic lethality, dosage suppression, and dosage lethality. This chapter focuses on a specific SGA application for high-resolution genetic mapping, referred to as SGA mapping (SGAM), which enables the identification of suppressor mutations and thus provides a powerful means for interrogating gene function and pathway order.

Key words: Yeast, Genetics, Genetic mapping, Synthetic lethal, SGA, SGAM, Deletion mutant, Double mutant, Suppression

1. Introduction

Genetic interaction analysis is an important approach for assessing gene function *in vivo*, including the mapping of genes whose products form biological pathways as well as the ordering of gene products within a pathway. Synthetic genetic interactions describe a phenomenon in which the action of one gene is modified by another or several other genes resulting in an altered or unexpected phenotype. Interactions of this kind are usually identified when a second mutation (e.g., loss of function) or increased gene dosage (e.g., gain of function) suppresses or enhances the original mutant phenotype *(1)*. This type of screening approach has been used extensively in yeast, worms, flies, mice, and other model organisms *(1)*.

Synthetic lethality refers to a specific example of a genetic interaction in which mutation of a single gene, while having little or no effect on the organism, results in cell death when combined with an otherwise viable mutation of a second gene *(1)*. Typically, synthetic lethal interactions indicate a functional relationship whereby two genes impinge on a common and essential biological process *(1)*. Systematic studies indicate that only 19% of yeast genes are required for viability *(2, 3)*. The fact that 81% of the predicted genes are not required for life (under standard laboratory conditions) highlights the robustness of biological circuits and suggests that, in addition to providing a global view of functional relationships between genes and pathways, large-scale surveys of genetic interactions will provide insights into mechanisms governing cellular buffering *(4–6)*.

Synthetic genetic array (SGA) analysis is a high-throughput (HTP), array-based method for the comprehensive identification of genetic interactions in yeast *(7, 8)*. This approach automates yeast genetics enabling systematic construction of double mutant strains. Synthetic lethal interactions are mapped by comparing single mutant and double mutant fitness on the basis of yeast colony growth. The first application of SGA technology *(8)* involved synthetic lethal screening with a complete arrayed set of viable gene deletion mutants *(2, 3)*. A marked query mutation was crossed into the set of ~5,000 viable haploid gene deletion strains, such that a series of robotic arraying procedures enabled the selection of haploid double mutant meiotic progeny, which were examined for growth defects *(7, 8)*. In a large-scale application of SGA analysis, we generated a genetic interaction network consisting of ~4,000 synthetic lethal interactions among ~1,000 genes *(7)*.

In contrast to synthetic lethality or genetic enhancement, genetic suppression describes a scenario where the double mutant exhibits a less severe phenotype than either single mutant. Genetic suppression is also a very effective method for uncovering functional relationships including the identification of genes functioning in the same complex or pathway *(1)*. Selecting for spontaneous mutations that suppress the fitness defect of a specific query mutation can identify extragenic suppressors relatively easily, but they are often difficult to identify. Although initially applied primarily for synthetic lethality screens, SGA analysis also permits a new method, dubbed SGAM (SGA mapping), for high-resolution mapping of genetic suppressor mutations *(9)*. SGAM takes advantage of the fact that the location of each deletion allele is known precisely and thus each SGA screen involves a genome-wide set of two-factor crosses. This method has been used successfully to map locations of several extragenic suppressors of yeast deletion alleles associated with severe fitness defects *(9–11)*. Notably, SGAM was recently used to identify a dominant suppressor of a disease-associated gene in *Saccharomyces cerevisiae(11)*. In this chapter, we describe the steps required to perform SGAM and and how to perform the analysis to identify extragenic suppressors of yeast deletion mutants.

2. Materials

2.1. Media and Stock Solutions

1. G418 (Geneticin, Invitrogen): Dissolve in water at 200 mg/mL, filter-sterilize, and store aliquots at 4 °C.

2. clonNAT (nourseothricin, Werner BioAgents, Jena, Germany): Dissolve 100 mg/mL in water at 100 mg/mL, filter-sterilize, and store in aliquots at 4 °C.

3. Canavanine (L-canavanine sulfate salt; Sigma): Dissolve 100 mg/mL in water, filter-sterilize, and store in aliquots at 4 °C.

4. Thialysine (*S*-[2-aminoethyl]-L-cysteine hydrochloride; Sigma): Dissolve 100 mg/mL in water, filter-sterilize, and store aliquots at 4 °C.

5. Amino acid supplement powder mixture for synthetic media (complete): 3 g adenine (Sigma), 2 g uracil (ICN), 2 g inositol, 0.2 g *p*-aminobenzoic acid (Acros Organics), 2 g alanine, 2 g arginine, 2 g asparagine, 2 g aspartic acid, 2 g cysteine, 2 g glutamic acid, 2 g glutamine, 2 g glycine, 2 g histidine, 2 g isoleucine, 10 g leucine, 2 g lysine, 2 g methionine, 2 g phenylalanine, 2 g proline, 2 g serine, 2 g threonine, 2 g tryptophan, 2 g tyrosine, and 2 g valine (Fisher). Dropout (DO) powder mixture is a combination of the aforementioned ingredients minus the appropriate supplement. 2 g of the DO powder mixture is used per liter of medium (*see* **Note 1**).

6. Amino acid supplement for sporulation medium: 2 g histidine, 10 g leucine, 2 g lysine, and 2 g uracil; 0.1 g of the amino acid supplement powder mixture is used per liter of sporulation medium (*see* **Note 1**).

7. β-Glucuronidase (Sigma): Prepare 0.5% solution in water and store at 4 °C.

8. Glucose (Dextrose, Fisher): Prepare 40% solution, autoclave, and store at room temperature.

9. YEPD: Add 120 mg adenine (Sigma), 10 g yeast extract, 20 g peptone, and 20 g bacto agar (BD Difco) to 950 mL water in

a 2-L flask. After autoclaving, add 50 mL of 40% glucose solution, mix thoroughly, cool to approximately 65 °C, and pour plates.

10. YEPD + G418: Cool YEPD medium to approximately 65 °C, add 1 mL of G418 stock solution (200 mg/L), mix thoroughly, and pour plates.

11. YEPD + clonNAT: Cool YEPD medium to approximately 65 °C, add 1 mL of clonNAT stock solution (100 mg/L), mix thoroughly, and pour plates.

12. YEPD + G418/clonNAT: Cool YEPD medium to approximately 65 °C, add 1 mL of G418 stock solution (200 mg/L) and 1 mL clonNAT stock solution (100 mg/L), mix thoroughly, and pour plates.

13. Enriched sporulation: Add 10 g potassium acetate (Fisher), 1 g yeast extract, 0.5 g glucose, 0.1 g amino acid supplement powder mixture for sporulation, and 20 g bacto agar to 1 L water in a 2-L flask. After autoclaving, cool medium to approximately 65 °C, add 250 μL G418 stock solution (50 mg/L), mix thoroughly, and pour plates.

14. (SD/MSG) − His/Arg/Lys + canavanine/thialysine/G418: Add 1.7 g yeast nitrogen base without amino acids or ammonium sulfate (BD Difco), 1 g L-glutamic acid sodium salt hydrate (MSG; Sigma), 2 g amino acid supplement powder mixture lacking histidine, arginine, and lysine (DO − His/Arg/Lys), and 100 mL water in a 250-mL flask. Add 20 g bacto agar to 850 mL water in a 2-L flask. Autoclave separately. Combine the autoclaved solutions, add 50 mL 40% glucose, cool the medium to approximately 65 °C, add 0.5 mL canavanine (50 mg/L), 0.5 mL thialysine (50 mg/L) and 1 mL G418 (200 mg/L) stock solutions, mix thoroughly, and pour plates (*see* **Note 2**).

15. (SD/MSG) − His/Arg/Lys + canavanine/thialysine/clonNAT: Add 1.7 g yeast nitrogen base without the appropriate amino acids or ammonium sulfate, 1 g MSG, 2 g amino acid supplement powder mixture (DO − His/Arg/Lys), and 100 mL water in a 250-mL flask. Add 20 g bacto agar to 850 mL water in a 2-L flask. Autoclave separately. Combine the autoclaved solutions, add 50 mL 40% glucose, cool the medium to approximately 65 °C, add 0.5 mL canavanine (50 mg/L), 0.5 mL thialysine (50 mg/L), and 1 mL clonNAT (100 mg/L) stock solutions, mix thoroughly, and pour plates.

16. (SD/MSG) − His/Arg/Lys + canavanine/thialysine/G418/clonNAT: Add 1.7 g yeast nitrogen base without amino acids or ammonium sulfate, 1 g MSG, 2 g amino acid supplement powder mixture (DO − His/Arg/Lys), and 100 mL water

in a 250-mL flask. Add 20 g bacto agar to 850 mL water in a 2-L flask. Autoclave separately. Combine the autoclaved solutions, add 50 mL 40% glucose, cool the medium to approximately 65 °C, add 0.5 mL canavanine (50 mg/L), 0.5 mL thialysine (50 mg/L), 1 mL G418 (200 mg/L), and 1 mL clonNAT (100 mg/L) stock solutions, mix thoroughly, and pour plates.

17. (SD/MSG) complete: Add 1.7 g yeast nitrogen base without amino acids or ammonium sulfate, 1 g MSG, 2 g amino acid supplement powder mixture (complete), and 100 mL water in a 250-mL flask. Add 20 g bacto agar to 850 mL water in a 2-L flask. Autoclave separately. Combine the autoclaved solutions, add 50 mL 40% glucose, mix thoroughly, cool the medium to approximately 65 °C, and pour plates.

18. SD – His/Arg/Lys + canavanine/thialysine: Add 6.7 g yeast nitrogen base without amino acids (BD Difco), 2 g amino acid supplement powder mixture (DO – His/Arg/Lys), and 100 mL water in a 250-mL flask. Add 20 g bacto agar to 850 mL water in a 2-L flask. Autoclave separately. Combine the autoclaved solutions, add 50 mL 40% glucose, cool the medium to approximately 65 °C, add 0.5 mL canavanine (50 mg/L) and 0.5 mL thialysine (50 mg/L) stock solutions, mix thoroughly, and pour plates (*see* **Note 3**).

19. SD – Leu/Arg/Lys + canavanine/thialysine: Add 6.7 g yeast nitrogen base w/o amino acids, 2 g amino acid supplement powder mixture (DO – Leu/Arg/Lys), and 100 mL water in a 250-mL flask. Add 20 g bacto agar to 850 mL water in a 2-L flask. Autoclave separately. Combine autoclaved solutions, add 50 mL 40% glucose, cool the medium to approximately 65 °C, add 0.5 mL canavanine (50 mg/L) and 0.5 mL thialysine (50 mg/L) stock solutions, mix thoroughly, and pour plates.

2.2. Plates and Accessories

1. We use the BioMatrix robot (S & P Robotics Inc., Toronto, ON) and OmniTrays (Nunc, cat. no. 242811) for all replica pinning procedures (*see* **Note 4**).

2.3. Manual Pin Tools

The following manual pin tools can be purchased from V & P Scientific, Inc. (San Diego, CA).

1. 96 floating pin E-clip style manual replicator (cat. no. VP408FH).

2. 384 floating pin E-clip style manual replicator (cat. no. VP384F).

3. Extra floating pins (FP): 1.58 mm diameter with chamfered tip (*see* **Note 5**).

4. Registration accessories: Library Copier (cat. no. VP381), Colony Copier (cat. no. VP380).

5. Pin cleaning accessories: Plastic bleach or water reservoirs (cat. no. VP421), pyrex alcohol reservoir with lid (cat. no. VP420), pin cleaning brush (cat. no. VP425) (*see* **Note 6**).

2.4. Robotic Pinning Systems

1. BioMatrix colony arrayer robot (S&P Robotics, Toronto, ON).
2. Singer RoTor benchtop robot (Singer Instruments, Somerset, UK) (*see* **Note 4**).

2.5. Strains and Plasmids

1. Starting strains for SGA were constructed in the following genetic backgrounds: Y7092 (*MAT*α *can1Δ::STE2pr*-Sp_*his5 lyp1Δ ura3Δ0 leu2Δ0 his3Δ1 met15Δ0*), Y8205 (*MAT*α *can1Δ::STE2pr*-Sp_*his5 lyp1Δ::STE3pr-LEU2 ura3Δ0 leu2Δ0 his3Δ1*).
2. p4339 (pCRII-TOPO::*NatMX4*).
3. Y8835 (*MAT*α *can1Δ::STE2pr*-Sp_*his5 lyp1Δ ura3Δ::NatMX4 leu2Δ0 his3Δ1 met15Δ0*) is the wild-type control strain for the *NatMX4*-marked query strains.
4. The collection of *MAT*a deletion strains can be purchased from Invitrogen (http://www.invitrogen.com). Deletion strains are stamped onto 96-well agar plates, American Type Culture Collection (http://www.atcc.org/cydac/cydac.cfm) as stamped 96-well agar plates, EUROSCARF (http://www.uni-frankfurt.de/fb15/mikro/euroscarf/index.html) as stamped 96-well agar plates, and Open Biosystems (http://www.open-biosystems.com/yeast_collections.php) as stamped 96-well agar plates or frozen stocks in 96-well plates.

3. Methods

3.1. SGA Query Strain Construction

3.1.1. Nonessential Genes: PCR-Mediated Gene Deletion

1. Synthesize two gene-deletion primers, each containing 55 bp of sequence at the 5′ end that is specific to the region upstream or downstream of the gene of interest (*Gene X*), excluding the start and stop codons, and 22 bp of sequence at the 3′ end that is specific for the amplification of the *NatMX4* cassette *(12)*. The *MX4* cassette amplification sequences include the forward amplification primer (5′-ACATGGAGGCCCAGAATACC-CT-3′) and the reverse amplification primer (5′-CAGTATAG-CGACCAGCATTCAC-3′).
2. Amplify the *NatMX4* cassette flanked with 55 bp target sequences from p4339 with the gene-deletion primers designed in **Step 1**.
3. Transform the polymerase chain reaction (PCR) product into the SGA starting strain, Y7092. Select transformants on YEPD + clonNAT medium.
4. Verify correct targeting of the deletion cassette by PCR.

3.1.2. Nonessential Genes: Switching Method

1. Obtain the deletion strain of interest (xxxΔ::KanMX4) from the MATa deletion collection and mate with Y8205 and isolate diploid zygotes by micromanipulation.

2. Transform the resulting diploid with EcoRI-digested p4339, which switches the gene deletion marker from KanMX4 to NatMX4. Select transformants on YEPD + clonNAT medium.

3. Transfer the resultant diploids to enriched sporulation medium and incubate at 22 °C for 5 days.

4. Resuspend a small amount of spores in sterile water and plate on SD – Leu/Arg/Lys + canavanine/thialysine to select MATα meiotic progeny. Incubate at 30 °C for approximately 2 days.

5. Replica plate to YEPD + clonNAT to identify the MATα meiotic progeny that carry the query deletion marked with NatMX4 (xxxΔ::NatMX4) (see **Note 7**).

3.1.3. Isolating Extragenic Suppressors of Slow-Growing Deletion Mutants

We have used SGAM to identify suppressors of slow-growing deletion mutants *(9–11)*. Fitness is a convenient phenotype for suppressor analyses because it can be easily assessed by measurement of colony size (on solid growth medium) or growth rate (in liquid growth medium).

1. Construct a heterozygous gene deletion mutant using a wild-type diploid strain and the PCR-mediated gene disruption strategy described above to replace the gene of interest with the NatMX4 selection cassette (see **Subheading 3.1.1**).

2. Sporulate the heterozygous deletion mutant and perform tetrad analysis to ensure that progeny from a single meiotic event yield two unmarked spores with wild-type growth rates and two NatMX4-marked spores with an observable slow-growth phenotype.

3. To generate suppressors of the slow-growth phenotype, inoculate MATa NatMX4-marked deletion strain into rich media and incubate at 30 °C with shaking for up to 10 days.

4. Streak the culture onto solid, rich YEPD medium and isolate putative suppressor strains by selecting single colonies exhibiting wild-type colony size.

5. Construct an SGA query strain using the newly isolated suppressor strain as described above (see **Subheading 3.1.2**)

3.2. Sterilization Procedure for the Pin Tools

3.2.1. Manual Pin Tools

1. Set up the wash reservoirs as follows: three trays of sterile water of increasing volume (30, 50, and 70 mL), one tray of 40 mL 10% bleach, and one tray of 90 mL 95% ethanol (see **Note 8**).

2. Allow the replicator to sit in the 30-mL water reservoir for approximately 1 min to remove the cells from the pins.

3. Place the replicator in 10% bleach for approximately 20 s.

4. Transfer the replicator to the 50-mL water reservoir and then to the 70-mL water reservoir to rinse the bleach off the pins.

5. Transfer the replicator to 95% ethanol.

6. Let excess ethanol drip off the pins, then flame.

7. Allow replicator to cool.

3.2.2. Robotic Pin Tools (BioMatrix Colony Arrayer System)

Use the following procedure to clean and sterilize the replicator pins prior to use of the robot.

1. Fill the sonicator bath with 390 mL sterile distilled water.

2. Clean the replicator pins in the sonicator bath for 5 min.

3. Remove the water and fill the sonicator bath with 390 mL 70% ethanol.

4. Sterilize the replicator in the sonicator bath for 20 s per cycle, and repeat the cycle twice.

5. Let the replicator sit in a tray of 100 mL of 95% ethanol for 5 s.

6. Allow the replicator to dry over a fan for 20 s.

Use the following procedure to sterilize the pins at the end of each replica pinning step.

1. Set up the wash reservoirs as follows: Program the water bath to automatically fill with sterile distilled water from the bottle supply source, manually fill the brush station with 320 mL sterile distilled water, fill the sonicator with 390 mL 70% ethanol, and basin with 100 mL 95% ethanol.

2. Let the replicator sit in the water bath for 10 s per cycle, and repeat cycle four more times to remove residual cells from the replicator pins.

3. Clean the replicator pins further at the brush station for three cycles.

4. Sterilize the replicator in the 70% ethanol sonicator bath for 20 s per cycle, and repeat twice.

5. Let the replicator sit in the 95% ethanol reservoir for 5 s.

6. Allow the replicator to dry over the fan for 20 s.

3.3. SGA Procedure

Figure 1 illustrates the selection steps in SGA analysis.
  Query strain and DMA (Double Mutant Array).

1. Grow the query strain in a 5-mL YEPD overnight culture.

2. Pour the query strain culture onto a YEPD plate, and use the replicator to transfer the liquid culture onto two fresh YEPD plates, generating a source of newly grown query cells for mating to the DMA in the density of 1,536 colonies (*see* **Note 9**). Allow the cells to grow at 30 °C for 2 days (*see* **Note 10**).

3. Replicate the DMA to fresh YEPD + G418 media. Allow the cells to grow at 30 °C for 1 day (*see* **Note 11**).

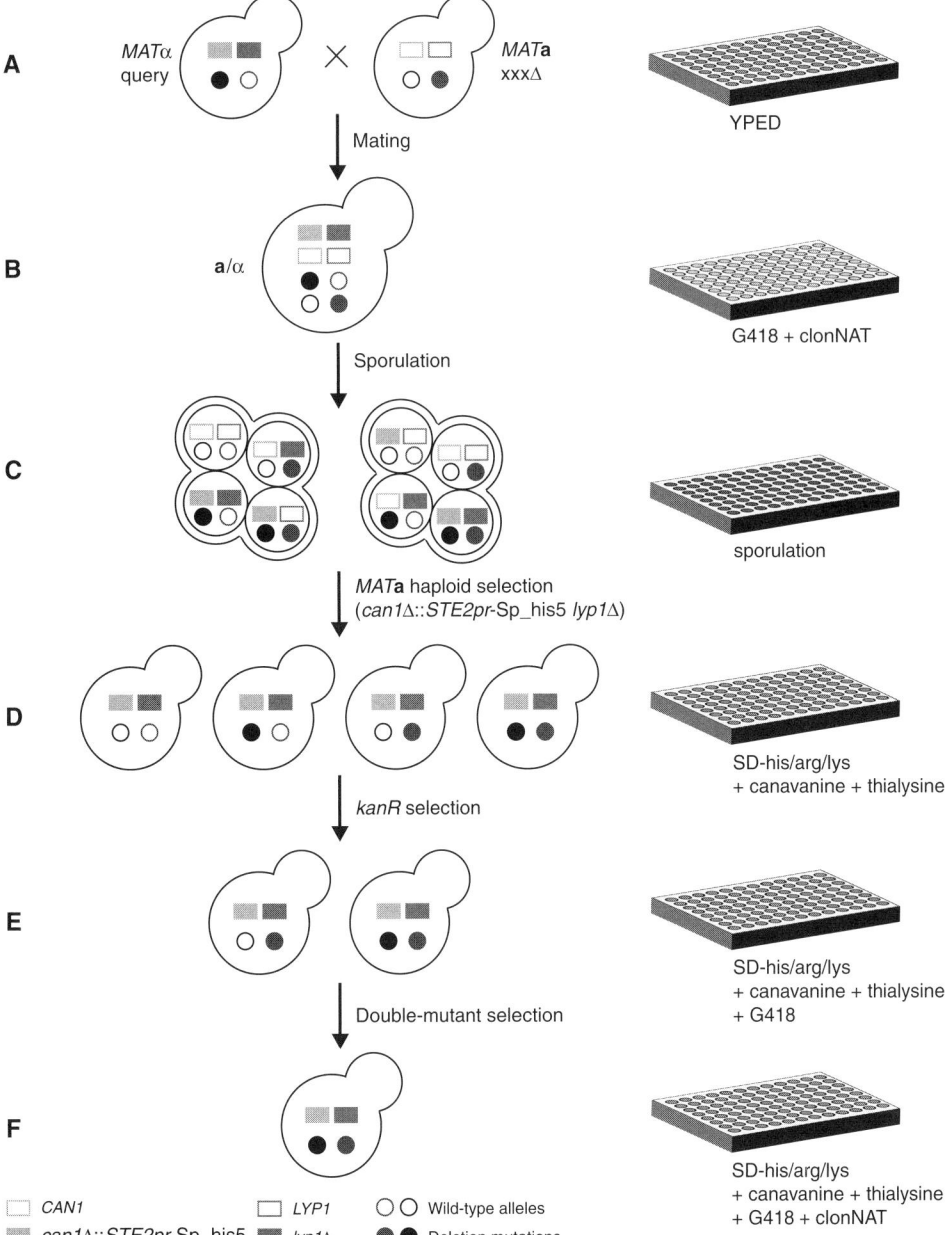

Fig. 1. Synthetic genetic array (SGA) methodology. (**a**) A *MAT*α strain carries a query mutation linked to a dominant selectable marker (*filled black circle*), such as the nourseothricin-resistance marker, *NatMX4,* and the SGA reporter, *can1Δ::STE2pr–Sp_his5* (in which *STE2pr–Sp_his5* is integrated into the genome such that it deletes the open reading frame (ORF) of the *CAN1* gene, which normally confers sensitivity to canavanine). The query strain also lacks the *LYP1* gene. Deletion of *LYP1* confers resistance to thialysine. This query strain is crossed to an ordered array of *MAT*a deletion mutants (*xxxΔ*). In each of these deletion strains, a single gene is disrupted by the insertion of a dominant selectable marker, such as the kanamycin-resistance (*KanMX4*) module (the disrupted gene is represented as a *filled red circle*). (**b**) The resulting heterozygous diploids are transferred to a medium with reduced carbon and nitrogen to induce sporulation and form haploid meiotic spore progeny. (**c**) Spores are transferred to a synthetic medium that lacks histidine, which allows selective germination of *MAT*a meiotic progeny owing to expression of the SGA reporter, *can1Δ::STE2pr–Sp_his5*. To improve this selection, canavanine and thialysine, which select *can1Δ* and *lyp1Δ* while killing *CAN1* and *LYP1* cells, respectively, are included in the selection medium. (**d**) The *MAT*a meiotic progeny are transferred to a medium that contains kanamycin which selects single mutants equivalent to the original array mutants and double mutants. (**e**, **f**) An array of double mutants is selected on a medium that contains both nourseothricin and kanamycin. Adapted from *(1)* (*see Color Plates*).

Mating the query strain with the DMA.

4. Pin the 1536-format query strain onto a fresh YEPD plate.
5. Pin the DMA on top of the query cells.
6. Incubate the mating plates at room temperature for 1 day.

MATa/α diploid selection and sporulation.

7. Pin the resulting *MATa/α* zygotes onto YEPD + G418/clonNAT plates.
8. Incubate the diploid selection plates at 30 °C for 2 days.
9. Pin diploid cells onto enriched sporulation medium.
10. Incubate the sporulation plates at 22 °C for 5 days (*see* **Note 12**).

MATa meiotic progeny selection.

11. Pin spores onto SD – His/Arg/Lys + canavanine/thialysine plates.
12. Incubate the haploid selection plates at 30 °C for 2 days.

MATa-KanMX4 meiotic progeny selection.

13. Pin the *MATa* meiotic progeny onto (SD/MSG) – His/Arg/Lys + canavanine/thialysine/G418 plates.
14. Incubate the *KanMX4*-selection plates at 30 °C for 2 days.

MATa-KanMX4-NatMX4 meiotic progeny selection.

15. Pin the *MATa* meiotic progeny onto (SD/MSG) – His/Arg/Lys + canavanine/thialysine/G418/clonNAT plates.
16. Incubate the *KanMX4/NatMX4* selection plates at 30 °C for 1–2 days.
17. Score double mutants for fitness defects (*see* **Note 13**).

3.4. SGAM analysis

SGAM depends on the viable gene deletion mutant array which comprises a systematic series of physically defined genetic markers covering almost every centimorgan of the yeast genome *(2, 3, 9)*. Given an SGA query strain carrying a marked query mutation that compromises cellular fitness, such as a *NatMX4*-marked gene deletion mutation, and a spontaneous suppressor allele, which suppresses the fitness defect of the query mutation, SGAM can be used to map the location of the suppressor. Since meiotic recombination is necessary to generate haploid double mutants in SGAM, a set of gene deletions that are linked and located on either side of the query allele on the same chromosome (termed a linkage group) form double mutants at a reduced frequency and appear synthetic lethal/sick with the query mutation. If a suppressor allele is required for normal growth of cells carrying the query allele, then it will also be associated with a linkage group, which maps the position of the suppressor genetically. This concept is illustrated in **Fig. 2** *(9)*. In this example, deletion of a query gene (queryΔ::*NatMX4*; n, *green box*) leads to a lethal phenotype in an otherwise wild-type background. However, a sup-

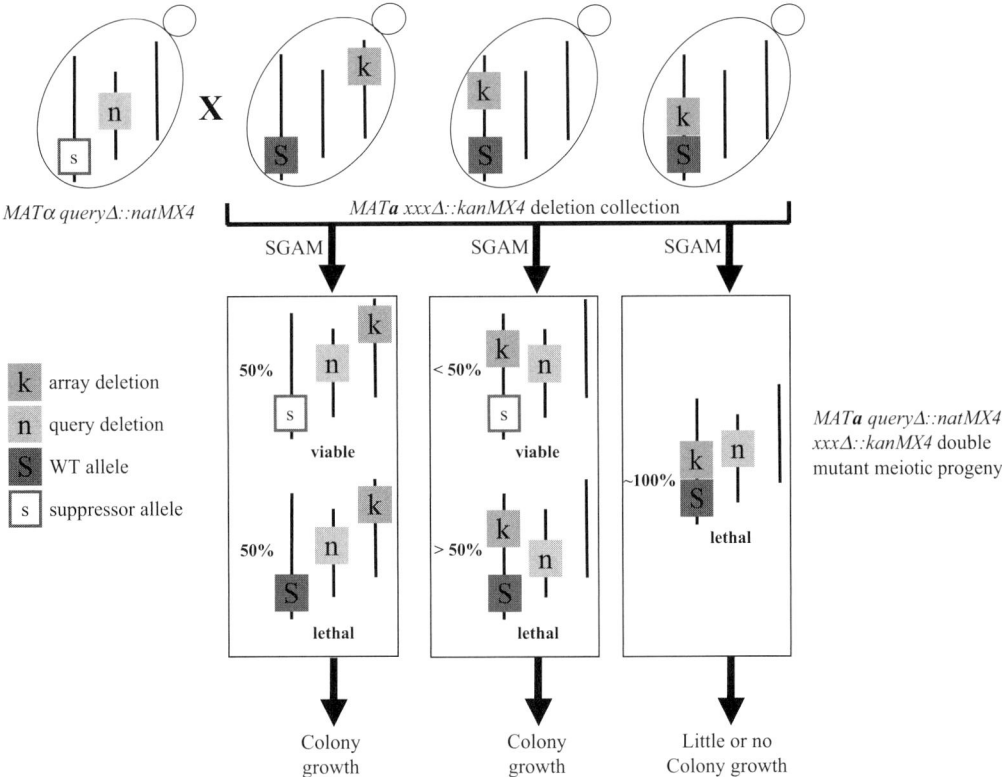

Fig. 2. Genetic mapping using SGAM. In this example, the lethal phenotype associated with deletion of a query gene (n, *green box*) is suppressed by mutation of an unidentified gene (s, *open box*). SGA analysis is used to cross the query mutant harboring a suppressor mutation to the collection of ~5,000 viable deletion mutants (k, *purple box*), all of which contain a wild-type copy of the unidentified gene (S, *red box*). As described in **Subheading 3.3**, double deletion mutants harboring the *NatMX4*-marked query deletion and a *KanMX4*-marked array deletion are selected following mating, meiotic recombination, and germination of haploid spore progeny. In the first example, the unidentified locus and *KanMX4*-marked deletion are located on different chromosomes and, therefore, segregate independently of one another. As a result, 50% of the resultant double deletion mutant meiotic progeny should contain the suppressor mutation locus and grow normally. In the second example, we expect that less than 50% of the *NatMX4*- and *KanMX4*-marked meiotic progeny will be viable because the array deletion mutant and the unidentified locus are linked and suppression of the growth defect is dependent on the distance separating the two genes and the frequency of meiotic recombination in the given chromosomal region. In the final example, the array-deletion strain and wild-type allele are tightly linked and, consequently, the number of viable double mutant progeny will be limited because of the reduced frequency of recombination between the suppressor allele and the *KanMX4*-marked strain. This results in the absence of colony growth. Adapted from *(9)* (*see Color Plates*).

pressor mutation(s) in an unidentified gene (S) is able to rescue the lethality associated with deletion of the query gene *(9)*. The viable query strain harboring the suppressor (queryΔ::*NatMX4s*) is systematically crossed to each of the ~5,000 haploid deletion array strains (*xxx*Δ::*KanMX4*) via SGA and double deletion meiotic progeny (*query*Δ::*NatMX4 xxx*Δ::*KanMX4*) are selected. If the wild-type allele (S) and a particular array deletion mutant

($xxx\Delta$::$KanMX4$) are not linked, 50% of the double deletion meiotic progeny (queryΔ::$NatMX4$ $xxx\Delta$::$KanMX4$) will contain the suppressor mutation (s) and germinate to form a colony. Conversely, recombination frequency is reduced when the deleted array gene ($xxx\Delta$::$KanMX4$) and the wild-type allele (S) are tightly linked, resulting in limited recovery of viable double-deletion meiotic progeny. Hence, mapping the suppressor allele (s) involves identification of a colinear set of linked double mutants that fail to grow and form colonies. The chromosomal location of this linkage group defines the general location of the suppressor locus, which should occur roughly within the middle of the linkage group (*see* **Note 14**) *(9)*.

3.4.1. Identifying Linkage Groups and Putative Suppressor Mutant Loci

1. Perform SGA screen using the "wild-type" control strain (Y8835) following the steps as described (*see* **Subheading 3.3**).
2. Visually inspect the experimental plates and compare to the wild-type control plates. Take note of the double mutant colonies that fail to grow, or appear smaller in size relative to wild-type controls (*see* **Notes 13** and **14**).
3. Record the potential interactions from the first round of screening.
4. Repeat the screen a second time.
5. Generate a set of putative interactions. While these data are expected to contain a substantial number of false positives, this unconfirmed SGA data should be sufficient for mapping purposes (*see* **Notes 15** and **16**).
6. Sort the set of putative interactions according to chromosome number and position (*see* **step 7**).
7. The putative suppressor mutation locus is determined via identification of a linkage group (i.e., failure to observe the formation of a potential collinear set of double mutant deletion strains; **Fig. 3a**; *see* **Note 17**).
8. The gene(s) that lie within the center of the linkage group should comprise the suppressor mutation. (**Fig. 3b**; *see* **Note 18**).

3.5. Confirming Identity of the Mutant Allele(s)

Once the candidate gene(s) is mapped, confirmatory genetic analysis is performed to prove that the mapped mutation is coincident with the suppressor mutation.

1. Sequence genes within the linkage group using standard techniques. Begin with genes situated closest to the center of the defined region, which should reveal the mutation (**Fig. 3b**).
2. Clone the mutant allele(s) from the suppressor strain using standard molecular biology techniques.
3. Identify the true suppressor allele(s) by complementation assays to prove that the mapped mutation is coincident with

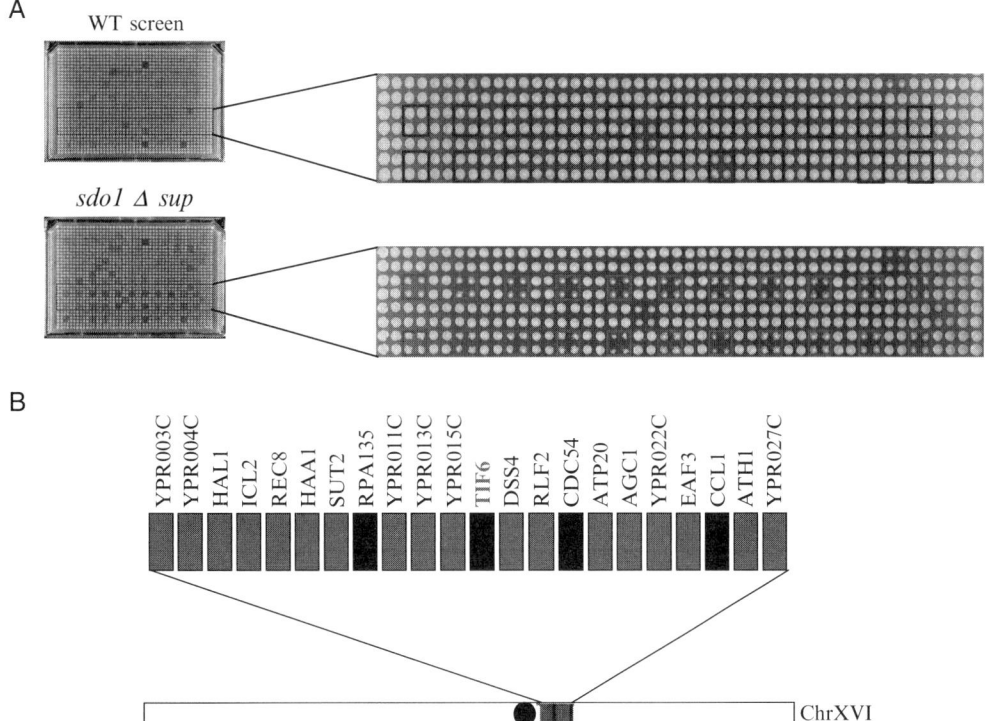

Fig. 3. SGAM identification of a suppressor of sdo1Δ slow growth. The SDO1 human ortholog has been linked to an autosomal recessive disorder known as the Shwachmann–Diamond syndrome. We previously isolated strains harboring a second-site mutation which suppressed the severe growth defect associated with deletion of SDO1 in yeast (11). (a) Example of a SGAM output plate consisting of 384 different sdo1Δ double mutants pinned in replicates of four. The KanMX4-marked array strains are arranged such that mutants corresponding to genes located adjacent to each other are arrayed at alternating positions of the same plate. A linkage group of inviable or slow-growing double mutants was identified following SGAM analysis (red squares) indicating the putative location of the sdo1Δ suppressor. (b) Genes located within the chromosome 15 linkage group are shown. Red bars indicate slow growth or death of the particular deletion when combined with the sdo1Δ mutant. Black bars indicate essential genes not represented in the arrayed collection of KanMX4-marked strains. Sequencing of alleles within this linkage group identified a mutation in the TIF6 gene (located near the center of the linkage group). Indeed, additional experiments confirmed that a dominant TIF6 gain-of-function mutation rescues the slow growth phenotype associated with a sdo1Δ strain. Adapted from (11) (see Color Plates).

the suppressor mutation. Introduce cloned alleles and vector controls into the appropriate deletion mutant strain and identify the cloned allele(s) that restores wild-type growth to that of the original slow-growing deletion mutant.

3.6. SGAM Applications

In principle, SGAM can be applied to uncover modifiers of any allele that leads to a phenotype distinct from the wild-type strain (9). These phenotypes may include conditional alleles of essential genes, extragenic suppressors of null mutants, expression of chromophores (or other reporter genes), as well as cytotoxic drug resistance. In addition to mapping of recessive alleles, SGAM is particularly useful for rapid mapping of dominant mutations that are challenging to clone using standard techniques *(11)*. Moreover, combinations of alleles responsible for complex, multigenic quantitative traits could be mapped in a single SGAM experiment. SGAM has been used successfully to map locations of several extragenic suppressors of yeast deletion alleles *(9–11)*. Notably, SGAM was recently used to identify a dominant suppressor of a gene associated with the Shwachman–Diamond syndrome. This study provided insight into the molecular basis of this disease and emphasized the importance of genetic analysis in model organisms *(11)*.

4. Notes

1. When making up the amino acid supplement mixture, the solid ingredients should be combined and then mixed thoroughly by turning end-over-end for at least 15 min. The resultant mixture can be stored in tinted glass bottles at room temperature.

2. Ammonium sulfate impedes the function of G418 and clonNAT. Hence, synthetic medium containing these antibiotics is made with monosodium glutamic acid as a nitrogen source *(13)*.

3. This medium does not contain any antibiotics such as G418 and clonNAT and therefore ammonium sulfate is used as the nitrogen source.

4. The Singer RoTor DHA benchtop robot uses disposable replicators (RePads, Singer Instruments, UK) and PlusPlates (Singer Instruments, UK) that have a larger surface area but the same external footprint dimensions as OmniTray (Nunc). The BioMatrix robot can be used in conjunction with Omni-Trays for the replica pinning steps involved in SGA analysis. We use 100-mm Petri dishes for the construction of SGA query strains and tetrad analysis, and 60-mm Petri dishes for random spore analysis. We found that approximately 35 and 50 mL of media in OmniTrays and PlusPlates, respectively, yield optimal results. For random spore analysis, approximately 10 mL of media in a 60-mm dish is optimal.

5. The 1.58-mm diameter, flat-tip pins (FP6) can be used as an alternative to the chamfered-tip pins. However, they transfer more cells than the chamfered-tip pins and may not be suitable for producing high-density arrays (1,536 spots/array).

6. Empty tip boxes can be used as a substitute to the reservoirs for bleach, water, and ethanol.

7. We have also constructed ordered arrays comprising temperature-sensitive alleles of essential genes.

8. To ensure that the pins are cleaned properly and to avoid contamination in the wash procedure, the volume of wash liquids in the cleaning reservoirs is calculated to cover the pins sequentially in small increments. For example, only the tips of the pins should be submerged in water in the first step. As the pins are transferred to subsequent cleaning reservoirs and the final ethanol step, the lower halves of the pins should be covered.

9. Deletion mutant arrays are frozen in 96-well microtiter plates. It is feasible to conduct SGA and SGAM at this density (96 mutants/plate). However, screen throughput and accuracy are increased by construction of high-density deletion arrays. Array densities include, 384, 768, or 1,536 colonies/plate. Methods for assembling higher density arrays have been described in detail *(14)*.

10. Pinning the query strain in a 1,536-format on an agar plate is advantageous, as cells are evenly transferred to subsequent mating steps. One query plate should contain a sufficient amount of cells for mating with eight plates of the DMA.

11. The DMA can be reused for three to four rounds of mating.

12. It is important to incubate the sporulation plates at approximately 22–24 °C for efficient sporulation. Following sporulation, these plates can be stored at 4 °C for up to 4 months without significant loss of spore viability, to provide a source of spores for subsequent random spore analysis and tetrad analysis.

13. In addition to visual inspection of the double mutants, we have developed a computer-based scoring system, which generates an estimate of relative growth rates from the area of individual colonies as measured from digital images of the double-mutant plates. Statistical significance can be determined for each strain by comparing the measurements between the mutants and wild-type controls.

14. SGAM provides several advantages as a high-resolution genetic mapping approach. First, although the mutation is mapped by linkage to nonessential genes, it may lie in either nonessential or essential genes. Second, both dominant and recessive mutations can be mapped. Third, as the size of the linked group will be inversely proportional to the frequency of meiosis in the diploid colonies, the assay could be fine-tuned by modulating sporulation efficiency. For instance, hampering sporulation by manipulating temperature and/or incubation times should extend the linked region.

In addition, multigenic traits in which several alleles are required to confer the mutant phenotype can also be mapped *(9)*.

15. Although SGAM is applied for mapping mutations, it is possible to identify synthetic lethal or sick interactions between the query gene deletion mutant and the array of 5,000 viable deletion mutants from the same SGAM screen. However, tetrad analysis is required in order to ensure that observed synthetic lethal/sick interactions are specific to combinations of array and query mutations and not to the unknown suppressor mutation(s).

16. Linkage groups can usually be identified from raw SGA data, and therefore random spore and tetrad analysis is not required to map suppressor loci. However, random spore and tetrad analysis can be used to determine the frequency of recombination between the putative suppressor allele and *KanMX4*-marked array gene deletion strains. Random spore and tetrad analyses have been previously described *(14)*.

17. The number of linkage groups identified by SGAM is dependent on the nature of the genetic trait. If the suppression phenotype is attributable to a single locus, then we expect to find two linkage groups: one corresponding to the *NatMX4*-marked query mutation and a second corresponding to the suppressor allele. We expect to identify more than two linkage groups if the suppression phenotype is a multigenic trait.

18. We have developed an algorithm for detecting significant linkage groups in SGAM *(9)*.

Acknowledgments

We thank Anastasia Baryshnikova, Renee Brost, Julie Guzzo, and Corey Nislow for help with figures and comments on the manuscript. This work was supported by grants to CB from the Canadian Institutes for Health Research and Genome Canada through the Ontario Genomics Institute.

References

1. Boone, C., Bussey, H., et al. (2007). Exploring genetic interactions and networks with yeast. *Nature Reviews Genetics* **8**, 437–449.
2. Giaever, G., Chu, A. M., et al. (2002). Functional profiling of the *Saccharomyces cerevisiae* genome. *Nature* **418**, 387–391.
3. Winzeler, E., Shoemaker, D. D., et al. (1999). Functional Characterization of the *S. cerevisiae* genome by gene deletion and parallel analysis. *Science* **285**, 901–906.
4. Kaelin, W. G. (2005). The concept of synthetic lethality in the context of anticancer therapy. *Nature Reviews Cancer* **5**, 689–698.
5. Hartman, J. L., Garvik, B., et al. (2001). Principles for the buffering of genetic variation. *Science* **291**, 1001–1004.
6. Hartwell, L. (2004). Genetics. Robust interactions. *Science* **303**, 774–775.
7. Tong, A., Lesage, G., et al. (2004). Global mapping of the yeast genetic interaction network. *Science* **303**, 808–813.
8. Tong, A. H. Y., Evangelista, M., et al. (2001). Systematic genetic analysis with ordered arrays of yeast deletion mutants. *Science* **294**, 2364–2368.
9. Jorgensen, P., Nelson, B., et al. (2002). High-resolution genetic mapping with ordered arrays of *Saccharomyces cerevisiae* deletion mutants. *Genetics* **162**, 1091–1099.
10. Chang, M., Bellaoui, M., et al. (2005). RMI1/NCE4, a suppressor of genome instability, encodes a member of the RecQ helicase TopoIII complex. *EMBO J.* **24**, 2024–2033.
11. Menne, T. F., Goyenechea, B., et al. (2007). The Shwachman-Bodian-Diamond syndrome protein mediates translation activation of ribosomes in yeast. *Nature Genetics* **39**, 486–495
12. Goldstein, A. L., McCusker, J. H. (1999). Three new dominant drug resistance cassettes for gene disruption in Saccharomyces cerevisiae. *Yeast* **15**, 1541–1553.
13. Cheng, T. H., Chang, C. R., et al. (2000). Controlling gene expression in yeast by inducible site-specific recombination. *Nucleic Acids Research* **28**, E108.
14. Tong, A., Boone, C. (2006). Synthetic genetic array analysis in Saccharomyces cerevisiae. In Xiao W (ed.) *Yeast Protocols*, 2nd edn. Humana, Totowa, NJ, pp. 171–191.

Chapter 4

Reporter-Based Synthetic Genetic Array Analysis: A Functional Genomics Approach for Investigating the Cell Cycle in *Saccharomyces cerevisiae*

Holly E. Sassi, Nazareth Bastajian, Pinay Kainth, and Brenda J. Andrews

Summary

Temporal control of gene expression is a widespread feature of cell cycles, with clear transcriptional programs in bacteria, yeast, and metazoans. In budding yeast, approximately 1,000 genes are transcribed during a specific interval of the cell cycle. Although a number of factors that contribute to this periodic pattern of gene expression have been studied in *Saccharomyces cerevisiae*, pathways of cell cycle-regulated transcription remain largely undefined. To identify regulators of genes exhibiting cell cycle periodicity, we have developed a functional genomics approach termed reporter-based synthetic genetic array (R-SGA) analysis. Based on synthetic genetic array (SGA) analysis, R-SGA allows rapid and easily automated incorporation of a cell cycle reporter gene into the array of viable haploid yeast gene-deletion mutants. Scoring of reporter activity in mutant strains compared to wild type identifies candidate regulators of the cell cycle gene of interest. In contrast to microarrays, which generally provide information about the expression of all genes under a particular condition (for example, a single gene deletion), R-SGA analysis facilitates the study of the expression of a single gene in all deletion mutants. Our system can be adapted to examine the expression of any gene not only in the context of haploid deletion mutants but also using other array-based strain collections available to the yeast community.

Key words: Yeast, Synthetic genetic array, Reporter, High-throughput plasmid transfer, Cell cycle, Transcription, *lacZ*, *HIS3*, 3-Aminotriazole

1. Introduction

Cells must follow a specific genetic program to ensure high-fidelity chromosome duplication and segregation. The cell division cycle of all eukaryotes consists of four main phases: DNA replication

(S phase) and mitosis (M phase), separated by gap phases (G1 and G2). In *Saccharomyces cerevisiae*, the cell cycle is characterized by dramatic periodic alternations of gene expression, with approximately 15% of all genes fluctuating in successive transcriptional waves *(1–3)*. Key cell cycle regulators, including the mitotic cyclins, transcription factors, and DNA replication complex components, exhibit periodic patterns of expression along with many genes that carry out cell cycle-specific functions. The set of genes causing fitness defects when overexpressed is enriched for cell cycle-regulated genes *(4)*, highlighting the importance of their restricted expression.

Transcriptional profiling *(1–3)*, genome-wide chromatin immunoprecipitation (ChIP-chip) *(5–8)*, as well as bioinformatic *(9, 10)* and mechanistic studies have collectively identified hundreds of potential targets and regulators of cell cycle-regulated transcription factors. However, the global picture of cell cycle circuitry is far from complete. For example, at the G1/S transition alone, only 23% of the promoters adjacent to G1-phase genes and 12% of the promoters adjacent to S-phase genes appear to be bound by SBF or MBF *(5)*, the known transcriptional regulators of this transition.

We have developed reporter-based synthetic genetic array (R-SGA) analysis, a novel method that facilitates the systematic identification of genes that regulate expression from cell cycle-regulated promoters (CCprs) *in vivo*. In R-SGA, a query yeast strain bearing a CCpr fused to a reporter gene is mated to the haploid deletion mutant array *(11–13)*. The query strain contains selectable markers that allow the use of the SGA method *(12, 14, 15)* to obtain haploid cells that contain both the reporter gene and deletion mutation through a series of robotic pinning steps. This plasmid transfer protocol effectively replaces thousands of manual yeast transformations. The resulting strains are assayed for reporter activity; enhanced reporter activity in a particular deletion mutant indicates that the deleted gene is a putative repressor of the cell cycle-regulated gene, while a reduction in reporter activity indicates a candidate activator. A control promoter (Controlpr), which is not cell cycle regulated, is used to identify mutants with nonspecific transcriptional defects. We describe the system using plasmid-based *URA3*-marked *Escherichia coli lacZ* and *S. cerevisiae HIS3* reporters crossed to the array of viable haploid deletion mutants; however, with minor modifications our approach could be adapted to incorporate any reporter in either a plasmid or integrated context, any compatible selectable marker, and any collection of arrayed yeast strains.

2. Materials

2.1. Reporters

Starting vectors for reporter gene construction (*see* **Subheading 3.1**):
1. pΔSS *(16)*.
2. pRS423, pRS426 *(17)*.

2.2. R-SGA Procedure

Yeast gene-deletion mutant collection maintenance and all R-SGA selection steps are carried out in OmniTray plates (Nunc, Catalog #242811) using manual or robotic pinning tools as described in **refs. 12**, **14**, and **15**. Standard yeast manipulations are performed in 92 × 16 mm petri dishes (Sarstedt).

1. The yeast gene-deletion mutant collection is derived from strain BY4741 (*MATa ura3Δ0 leu2Δ0 his3Δ1 met15Δ0*) *(18)*. The collection consists of all *S. cerevisiae* viable haploid deletion mutants in which a given open reading frame (ORF), *XXX*, has been replaced by a kanamycin resistance marker (*kan^r*); *xxxΔ::kan^r* *(11, 13)*. The collection is available from Invitrogen (http://www.invitrogen.com), American Type Culture Collection (http://www.atcc.org), EUROSCARF (http://www.uni-frankfurt.de/fb15/micro/euroscarf/index.html), and Open Biosystems (http://www.openbiosystems.com/GeneExpression/Yeast).

2. Y7039 (*MATα can1Δ::STE2pr-LEU2 lyp1Δ ura3Δ0 leu2Δ0 his3Δ1 met15Δ0*) *(15)* derived from BY4741 *(18)*.

3. Standard yeast transformation reagents.

4. Glucose (Dextrose, Sigma-Aldrich Co.): Prepare a 50% (w/v) solution in distilled water, autoclave, and store at room temperature.

5. 10× amino acid supplements for synthetic media (complete): Add 300 mg isoleucine, 1.5 g valine, 400 mg adenine hemisulfate salt, 200 mg arginine hydrochloride, 200 mg histidine, 1 g leucine, 200 mg uracil, 300 mg lysine monohydrochloride, 200 mg methionine, 500 mg phenylalanine, 2 g threonine, 400 mg tryptophan, and 300 mg tyrosine (Sigma-Aldrich Co.) to a beaker. Bring the volume up to 1 L with distilled water, mix, autoclave, and store at 4°C. 10× amino acid drop-out (DO) solution is a combination of the above ingredients minus the appropriate supplement. For the R-SGA protocol described, 10× –Ura and 10× –Ura/Leu/Arg/Lys are required.

6. Amino acid supplement for enriched sporulation media: Dissolve 0.714 g uracil, 0.714 g histidine, and 3.571 g

leucine in 500 mL of distilled water and filter-sterilize. Store the solution at 4°C.

7. G418 (G-418 sulfate/geneticin, Gibco, Invitrogen Corp.): Dissolve in distilled water at 200 mg/mL, filter-sterilize, and store stock solutions in aliquots at 4°C. Store the powder at room temperature.

8. Canavanine (L-canavanine sulfate salt, Sigma-Aldrich Co.): Dissolve in distilled water at 50 mg/mL, filter-sterilize, and store stock solutions in aliquots at 4°C. Store the powder at 4°C.

9. Thialysine (S-(2-aminoethyl)-L-cysteine hydrochloride/L-4-thialysine hydrochloride, Sigma-Aldrich Co.): Dissolve in distilled water at 50 mg/mL, filter-sterilize, and store stock solutions in aliquots at 4°C. Store the powder at 4°C.

10. SD −Ura (synthetic minimal glucose medium supplemented with 10× −Ura DO solution): Add 6.7 g of yeast nitrogen base without amino acids (BD Difco Laboratories) and 860 mL of distilled water to a 2 L flask, mix, and autoclave. Add 40 mL of 50% glucose and 100 mL of 10× −Ura amino acid DO solution and mix. For plates, add 20 g agar (BD Difco Laboratories) prior to autoclaving.

11. YEPD (yeast extract–peptone–dextrose) plates: Add 120 mg adenine, 10 g yeast extract, 20 g peptone (BD Difco Laboratories), and 20 g agar to a 2 L flask, bring the volume up to 960 mL with distilled water, mix, and autoclave. Add 40 mL of 50% glucose and mix thoroughly.

12. YEPD + G418 plates (*see* **Note 1**): Cool the YEPD medium to ~65°C, add 1 mL of G418 stock solution/L (final concentration 200 mg/L), and mix.

13. (SD/MSG)-Ura + G418 plates (*see* **Note 2**): Add 1.7 g yeast nitrogen base without amino acids and ammonium sulfate (BD Difco Laboratories), 1 g monosodium glutamic acid (MSG), Sigma-Aldrich Co., and 20 g agar to a 2 L flask, bring the volume up to 860 mL with distilled water, mix, and autoclave. Add 40 mL of 50% glucose and 100 mL of 10× −Ura amino acid DO solution. Cool to ~65°C, add 1 mL of G418 stock solution (final concentration 200 mg/L), and mix thoroughly.

14. Enriched sporulation + G418 plates (*see* **Note 1**): Add 10 g potassium acetate (BioShop Canada Inc.), 1 g yeast extract, 0.5 g glucose, and 20 g agar to a 2 L flask. Bring the volume up to 990 mL with distilled water, mix, and autoclave. Add 10 mL of amino acid supplement solution and cool to ~65°C. Add 250 μl of G418 stock solution (final concentration 50 mg/L) and mix thoroughly.

15. SD −Ura/Leu/Arg/Lys + canavanine/thialysine plates: Add 6.7 g yeast nitrogen base without amino acids and

20 g agar to a 2 L flask. Add 860 mL of distilled water, mix, and autoclave. Add 40 mL of 50% glucose and 100 mL of 10× −Ura/Leu/Arg/Lys amino acid DO solution. Cool to ~65°C, add 1 mL of canavanine (can) stock solution (final concentration 50 mg/L) and 1 mL of thialysine (thia) stock solution (final concentration 50 mg/L), and mix thoroughly.

16. (SD/MSG)−Ura/Leu/Arg/Lys + can/thia/G418 plates (*see* **Note 2**). Add 1.7 g yeast nitrogen base without amino acids and ammonium sulfate, 1 g MSG, and 20 g agar to a 2 L flask. Bring the volume up to 860 mL with distilled water, mix, and autoclave. Add 40 mL of 50% glucose and 100 mL of 10× −Ura/Leu/Arg/Lys amino acid DO solution. Cool to ~65°C, add 1 mL of can stock solution (final concentration 50 mg/L), 1 mL of thia stock solution (final concentration 50 mg/L), and 1 mL of G418 stock solution (final concentration 200 mg/L) and mix thoroughly.

2.3. Identification of CCpr-lacZ Reporter Regulators

2.3.1. Array-Format β-Galactosidase Activity Assay

1. SD −Ura plates.
2. Agarose (BioShop Canada, Inc.).
3. DMF (*N,N*-dimethylformamide, Sigma-Aldrich Co.). DMF is toxic and harmful by inhalation or on contact with skin. Work in a fume hood and wear protective clothing.
4. SDS (sodium dodecyl sulfate, Bio-Rad Laboratories, Inc.): Add 100 g to 900 mL of distilled water and heat to 68°C to dissolve. Adjust the pH to 7.2 by adding a few drops of concentrated HCl. Adjust the volume to 1 L with distilled water (10% stock solution) and store in aliquots. A mask should be worn when weighing SDS.
5. X-gal (5-bromo-4-chloro-3-indolyl-β-galactopyranoside, Sigma-Aldrich Co.): Make a 20 mg/mL stock solution by dissolving in DMF. Store in the dark at −20°C.
6. KPO_4 (0.5 M potassium phosphate buffer, pH 7.0): Mix appropriate volumes of 0.5 M K_2HPO_4 (~61.5 mL) and 0.5 M KH_2PO_4 (~38.5 mL) such that the final pH of the solution is 7.0.
7. 50 mL conical tubes (BD Biosciences).
8. Hot plate with magnetic stirrer.

2.3.2. Confirmation of Candidate CCpr-lacZ Regulators

1. YEPD plates.
2. YEPD + G418 plates.
3. Standard yeast transformation reagents.
4. SD −Ura plates.
5. X-gal top agarose solution (*see* **Subheadings 2.3.1** and **3.3.1**).

2.4. Identification of CCpr-HIS3 Reporter Regulators

2.4.1. Array-Format HIS3 Activity Assay

1. 10× –His amino acid DO solution.
2. 3-AT (3-amino-1,2,4-triazole/3-aminotriazole, Sigma-Aldrich Co.): Make a 2.5 M stock in distilled water, filter-sterilize, and store at 4°C. 3-AT is possibly carcinogenic; work in a fume hood and wear protective clothing.
3. SD –His + 3-AT plates: Add 6.7 g yeast nitrogen base without amino acids and 20 g agar to a 2 L flask, bring the volume up to 860 mL with distilled water, mix, and autoclave. Add 40 mL of 50% glucose and 100 mL of 10× –His amino acid DO solution. Cool to ~65°C, add the appropriate volume of 3-AT stock (for 10 or 75 mM final concentration), and mix thoroughly.

2.4.2. Confirmation of Candidate CCpr-HIS3 Regulators

1. YEPD + G418 plates.
2. Standard yeast transformation reagents.
3. 96-well round bottom assay plates (Costar #3795).
4. 96-well microplate lids (Costar #3930).
5. SD –Ura liquid medium.
6. SD –Ura plates.
7. SD –His + 10 mM 3-AT plates.
8. SD –His + 75 mM 3-AT plates.

3. Methods

3.1. Reporters

To generate the parent *HIS3* reporter construct, the *S. cerevisiae* *HIS3* gene was amplified from pRS423 *(17)* using primers HIS3 Left (5′ CGAGGATCCGGCAAAGATGACAGAGCAGAA 3′) and HIS3 Right (5′ GCAGAATTCACCGCATAGATCCGTCGAGT 3′), which contain engineered *Bam*HI and *Eco*RI sites, respectively. The PCR fragment was digested with *Bam*HI and *Eco*RI and cloned into similarly digested pRS426 *(17)*. The *lacZ* reporter constructs used in our R-SGA studies are primarily derivatives of pΔSS *(16)*. Both pΔSS and pHIS3 contain a *URA3* marker, the yeast 2-μm origin of replication (*see* **Note 3**), and an ampicillin resistance marker (*amp^r*).

Any desired CCpr or specific promoter element can be amplified from genomic/other DNA using PCR primers containing engineered restriction sites for subcloning into either the *lacZ* or *HIS3* parent reporter plasmid (*see* **Notes 4** and **5**). Control reporter plasmids used in confirmation assays to identify non-cell-cycle-specific transcriptional effects contain either the *S. cerevisiae RPL39* or *ACT1* promoter (*see* **Note 6**).

3.2. R-SGA Procedure

An overview of the R-SGA method is shown in **Fig. 1a**. All R-SGA selection steps are performed in 768-density array format (*see* **Note 7**). For protocols to generate the 768-array format yeast gene-deletion set from 96-well plates, array maintenance, sterilization of robotic and manual pinning tools, and manual pinning protocols, *see* **refs.** *12*, *14*, and *15*. The following procedure facilitates high-throughput transfer of the CCpr-reporter plasmid from strain Y7039 to each *xxxΔ::kan^r* strain in the yeast gene-deletion set (*see* **Note 8**). Adaptations of the method are described in **Subheading 4** (*see* **Note 9**).

1. To generate the R-SGA query strain, introduce the CCpr-reporter plasmid into strain Y7039 by standard yeast transformation, plate on SD −Ura, and incubate for 2–3 days at 30°C.

2. Patch the transformants onto SD −Ura and incubate for 2 days at 30°C.

3. Inoculate the query strain into 10 mL of liquid SD −Ura. Grow to saturation at 30°C overnight.

4. Pellet the culture (2,100 × g, 5 min), resuspend in 5 mL of fresh SD −Ura, and plate onto two YEPD plates (OmniTray format). Grow for 2 days until dense lawns of yeast have formed (*see* **Note 10**).

5. In parallel, propagate the yeast gene-deletion set in 768-array format at 30°C on YEPD + G418 plates.

6. Pin the query strain containing the CCpr-reporter plasmid from the lawns onto YEPD plates. Pin the yeast gene-deletion set directly on top of the query cells to allow the cells to mate. Incubate for 1 day at 30°C.

7. Pin the resulting *MATa/α* diploids onto (SD/MSG) −Ura + G418 to select for the plasmid and the *xxxΔ::kan^r* gene, respectively. Incubate for 2 days at 30°C.

8. Repeat the above diploid selection pinning step and incubate the plates at 30°C for 1 day.

9. Pin the diploid cells onto enriched sporulation medium + G418. Incubate the sporulation plates for 5–9 days in the dark at 22°C.

10. Pin the spores onto SD −Ura/Leu/Arg/Lys + can/thia plates in duplicate (*see* **Note 11**) to select for *MATa* meiotic progeny carrying the CCpr-reporter plasmid. Incubate the haploid selection plates at 30°C for 2 days.

11. Pin onto (SD/MSG) −Ura/Leu/Arg/Lys + can/thia/G418 to select for *MATa xxxΔ::kan^r* cells carrying the CCpr-reporter plasmid. Incubate the final selection plates at 30°C for 2 days.

12. Repeat the final selection pinning on (SD/MSG) −Ura/Leu/Arg/Lys + can/thia/G418. Incubate the plates for 2 days at 30°C.

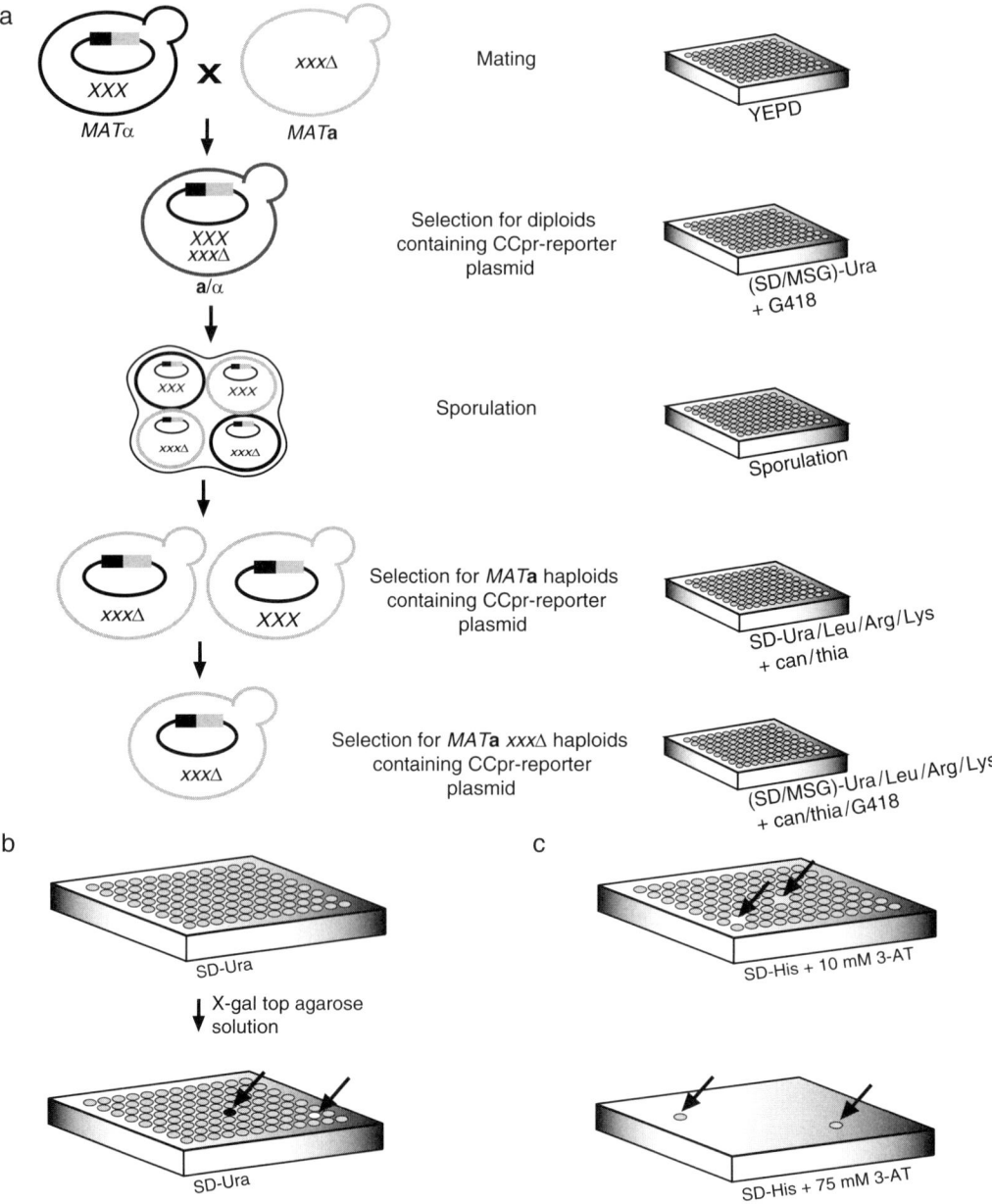

Fig. 1. Overview of the R-SGA methodology. (**a**) R-SGA selection steps. The *MAT*α query strain containing the *URA3*-marked CCpr-reporter plasmid is crossed to the ordered array of *MAT***a** viable haploid gene-deletion mutants on YEPD. Each mutant carries a gene deletion linked to a *kan*r marker that confers resistance to G418. The resulting *MAT***a**/α diploids are pinned onto medium lacking uracil and supplemented with G418 to select the plasmid and gene deletion linked to the *kan*r marker, respectively. Diploids are transferred to enriched sporulation medium to induce sporulation and the formation of haploid meiotic spore progeny. Spores are subsequently transferred to medium lacking uracil, leucine, arginine, and lysine and supplemented with canavanine and thialysine. The omission of uracil maintains plasmid selection. The omission of leucine allows selective growth of *MAT***a** meiotic progeny, since only these cells express the *STE2pr-LEU2* reporter. *can1*Δ and *lyp1*Δ markers and canavanine and thialysine selections are described in Tong and Boone *(15)*. Strains are subsequently pinned onto medium lacking uracil, leucine, arginine, and lysine and supplemented with canavanine, thialysine and G418 to select for *MAT***a** gene-deletion mutant cells containing the CCpr-reporter plasmid. *XXX*: wild-type allele, *xxx*Δ: gene-deletion allele, black oval: *URA3*-marked plasmid, black box: cell cycle-regulated promoter,

See **Subheadings** **2.3** and **3.3** for *CCpr-lacZ* constructs or **Subheadings** **2.4** and **3.4** for *CCpr-HIS3* constructs. **Fig. 1b**, **c** depicts downstream reporter activity assays for *lacZ* and *HIS3* constructs, respectively.

3.3. Identification of CCpr-lacZ Reporter Regulators

3.3.1. Array-Format β-Galactosidase Activity Assay

1. Pin the array of deletion mutants containing the *CCpr-lacZ* plasmids a final time onto SD –Ura. Incubate for 1–2 days at 30°C.

2. Prepare top agarose to cover the genome-wide array (this recipe is enough for 18 OmniTray plates):
   a. Add 1.8 g of agarose (final concentration 0.5%) to a 1 L flask.
   b. Add 334.8 mL of 0.5 M KPO_4.
   c. Mix, boil, and cool to 60°C on a hot plate with stirring.
   d. Add 14.4 mL of DMF and 3.6 mL of 10% SDS and mix thoroughly.

3. For one OmniTray plate, dispense 19.6 mL of the top agarose solution into a 50 mL conical tube and add 400 µl of 20 mg/mL X-gal in DMF. Invert to mix.

4. Pour the X-gal top agarose over the plate on a flat surface, and immediately tip the plate to distribute the solution evenly over the surface of the agar (*see* **Note 12**).

5. Repeat **steps** **3** and **4** for the remaining plates. Incubate the plates at 37°C for 1 h (*see* **Note 13**).

6. Score deletion mutants containing the *CCpr-lacZ* plasmid for β-galactosidase activity. Deletion of putative activators will result in white colonies with no β-galactosidase activity, while deletion of repressors will result in dark blue colonies with enhanced reporter activity (*see* **Note 14**). An example of results produced is shown in **Fig. 2a**. Liquid β-galactosidase assays may be used if quantitative results are desired (**Fig. 2b**).

Fig. 1. (continued) grey box: reporter. Adapted from Tong and Boone *(15)*. (**b**) Detection of *lacZ* reporter activity. Yeast gene-deletion mutants containing a *CCpr-lacZ* plasmid are pinned onto medium lacking uracil. X-gal top agarose solution is poured over the plates and β-galactosidase activity (i.e., the extent of blue color) is scored. Black arrows indicate putative regulators. Black circle: yeast strain exhibiting increased β-galactosidase activity in which a candidate repressor is deleted, white circle: yeast strain exhibiting decreased β-galactosidase activity in which a candidate activator is deleted. (**c**) Detection of *HIS3* reporter activity. Yeast gene-deletion mutants containing a *CCpr-HIS3* reporter are pinned onto medium lacking histidine and supplemented with low or high concentrations of 3-AT. Black arrows on the SD –His +10 mM 3-AT plate indicate strains that are unable to grow, which contain deletions in candidate activators. Black arrows on the SD –His +75 mM 3-AT plate indicate strains that are able to grow in the presence of high concentrations of 3-AT, which contain deletions in candidate repressors.

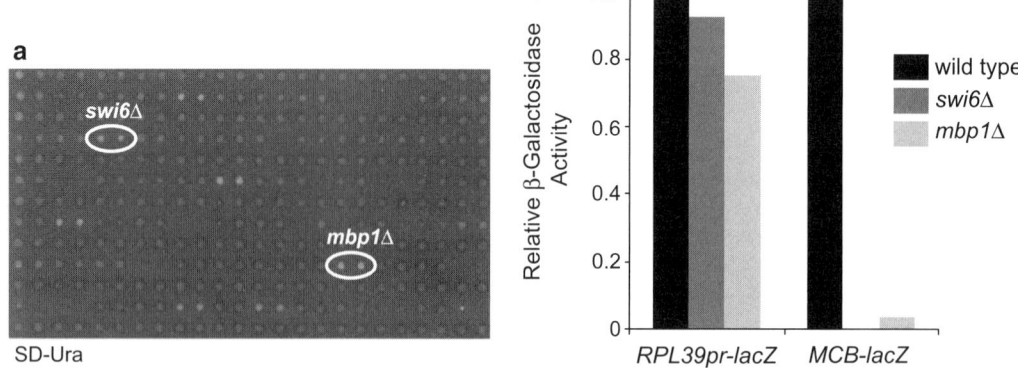

Fig. 2. Identifying regulators of a G1-phase cell cycle promoter element using a *lacZ* reporter. (**a**) Identification of *MCB-lacZ* regulators in array format. A reporter construct containing four copies of the *MluI* cell cycle box (MCB) sequence element fused to *lacZ* **(19)** was crossed to the yeast gene-deletion array, and haploid *MATa xxxΔ::kan^r* strains containing the *MCB-lacZ* plasmid were isolated using the R-SGA selection steps. The strains were pinned onto SD −Ura plates, grown for 2 days, and subsequently covered with top agarose containing X-gal to identify regulators of *MCB-lacZ* expression. As expected, deletion of *SWI6* and *MBP1*, the components of the MBF transcription factor that binds to MCB elements and activates transcription of G1-specific genes, resulted in a failure to produce β-galactosidase. A 384-array format is shown. (**b**) Liquid β-galactosidase assays. *RPL39-lacZ* (*see* **Note 6**) or *MCB-lacZ* reporter plasmids were introduced into *swi6Δ*, *mbp1Δ*, and wild-type strains, and quantitative liquid β-galactosidase assays were performed to confirm screen results. *MCB-lacZ* expression was reduced in *swi6Δ* and *mbp1Δ* strains compared to wild type, and the effect was specific to MCB elements. Values for β-galactosidase activity were calculated relative to wild-type levels for each plasmid (*see Color Plates*).

3.3.2. Confirmation of Candidate CCpr-lacZ Regulators

1. From the original master deletion set plate, streak the *xxxΔ::kan^r* strains corresponding to the mutants identified in the primary screen on YEPD + G418 and streak BY4741 (wild type) on YEPD. Incubate for 2 days at 30°C.

2. Introduce the *CCpr-lacZ* plasmid into the deletion strains and into BY4741 by standard yeast transformation, and plate onto SD −Ura. Incubate for 2–3 days at 30°C.

3. Pick four transformants per gene-deletion strain and patch them onto SD −Ura plates in a 96-density array format. Incubate for 2 days at 30°C.

4. Pin the 96-density array onto a fresh SD −Ura plate. Incubate for 2 days at 30°C. Retain this plate as the master plate of candidate regulator gene-deletion strains identified in the screen.

5. In parallel, make a lawn of strain BY4741 containing the *CCpr-lacZ* plasmid on SD −Ura. Incubate the lawn for 2 days at 30°C.

6. Pin from the lawn onto a fresh SD −Ura plate in 96-array format. Incubate for 2 days at 30°C.

7. Pin the 96-density array of candidate regulator deletion mutants containing the *CCpr-lacZ* plasmid onto a fresh SD −Ura plate in 192-array format.

8. Pin the BY4741 colonies containing the *CCpr-lacZ* plasmid onto the same SD −Ura plate in the offset position in 192-array

format. This will generate a 384-density array in which all candidate regulator gene-deletion strains are adjacent to wild-type colonies (**Fig. 3**). Incubate for 1–2 days at 30°C.

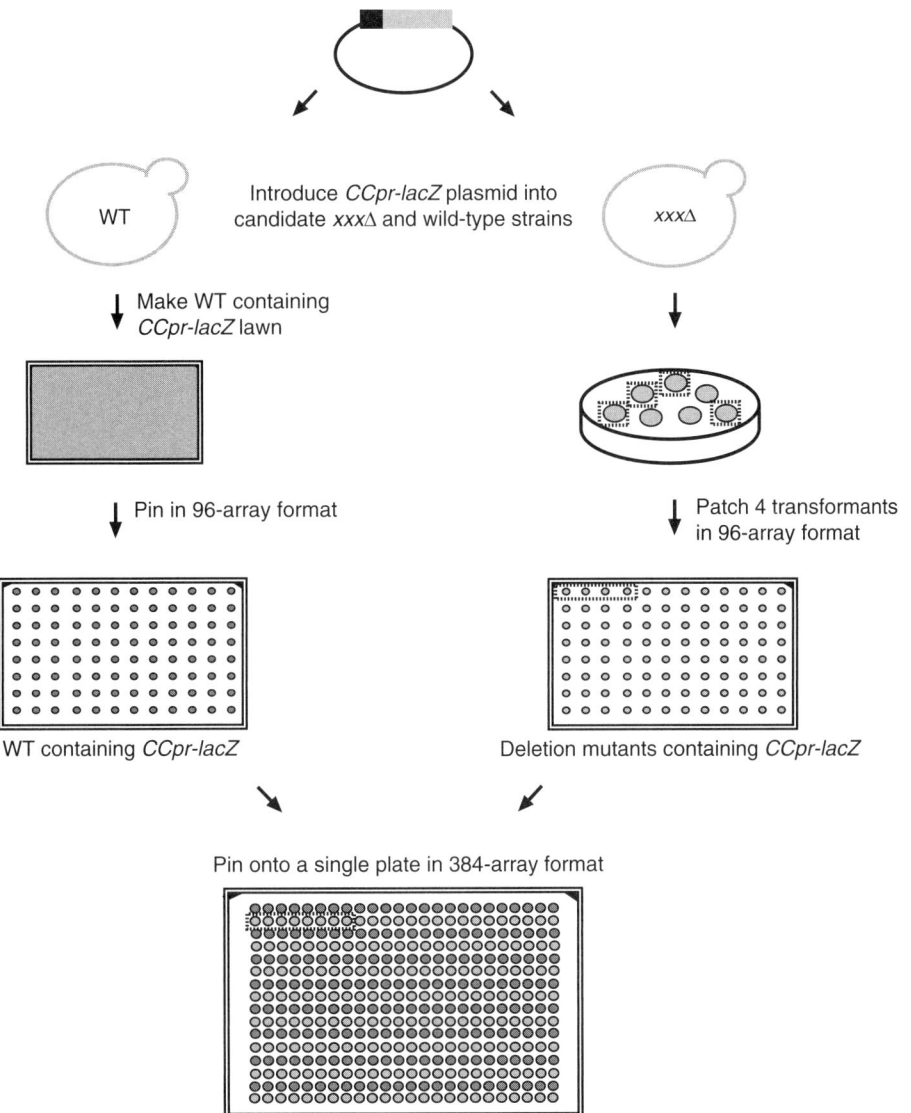

Fig. 3. Building a 384-format array to confirm regulators identified in genome-wide *lacZ* R-SGA screens. The *CCpr-lacZ* plasmid is introduced into the wild-type strain and into all yeast gene-deletion strains identified as candidate regulators in the primary R-SGA screen. A lawn of the wild-type strain containing the *CCpr-lacZ* plasmid is made and a 96-head pinning tool is used to array the strain. In parallel, four independent transformants of each *xxx*Δ*::kan^r* strain containing the *CCpr-lacZ* plasmid are patched in a 96-array format (boxes). These strains are subsequently pinned onto a fresh plate in duplicate offset positions resulting in a 192-format array of 24 distinct plasmid-containing deletion mutants, each represented eight times (box on 384-array plate). A 96-head pinning tool is used to transfer the wild-type strains containing the plasmid onto the same plate in different offset positions, resulting in 192 wild-type colonies. Each deletion strain (blue) is adjacent to a wild-type colony (light grey) on the final 384-format array plate for direct comparison of *CCpr-lacZ* activity. All plates are SD –Ura for plasmid selection. This procedure is carried out in parallel with the *Controlpr-lacZ* construct to determine the specificity of candidate regulators for the *CCpr-lacZ* reporter. Blue rectangle: *lacZ*, WT: wild type (*see Color Plates*).

9. Perform the X-gal overlay assay as described (*see* **Subheadings 2.3.1** and **3.3.1**).

10. Repeat **steps 2–9** using the *Controlpr-lacZ* construct.

11. Score the *xxxΔ::kan^r* and BY4741 strains containing the *CCpr-lacZ* construct by comparing mutants next to wild type to identify high-confidence candidate regulators (*see* **Note 15**).

12. Score the *xxxΔ::kan^r* and BY4741 strains containing the *Controlpr-lacZ* construct in the same way (*see* **Note 16**).

13. Identify the candidate regulators that are specific to the *CCpr-lacZ* construct (*see* **Notes 17** and **18**).

3.4. Identification of CCpr-HIS3 Reporter Regulators

3.4.1. Array-Format HIS3 Activity Assay

1. Pin the array of deletion mutants containing the *CCpr-HIS3* plasmids a final time onto SD –His + 10 mM 3-AT and SD –His + 75 mM 3-AT plates (*see* **Note 19**).

2. Incubate the plates for 2 days at 30°C.

3. Score the SD –His + 10 mM 3-AT plates for colonies that are slow growing or dead compared to those growing on the (SD/MSG) –Ura/Leu/Arg/Lys + can/thia/G418 final selection plates. These strains contain deletions in candidate activators (*see* **Note 20**).

4. Identify colonies on the SD –His + 75 mM 3-AT plates that are able to grow. These strains contain deletions in candidate repressors. Results from a genome-wide R-SGA screen are depicted in **Fig. 4a, b**.

3.4.2. Confirmation of Candidate CCpr-HIS3 Regulators

1. From the original master deletion set plate, streak the *xxxΔ::kan^r* strains corresponding to the mutants identified in the primary screen on YEPD + G418. Incubate for 2 days at 30°C.

Fig. 4. Identifying regulators of a G1-phase cell cycle promoter element using a *HIS3* reporter. (**a**) Modified R-SGA screening approach. In an R-SGA screen designed to identify inhibitors of SBF-dependent transcription, a *cln3Δhis3Δ* query strain bearing an integrated copy of the *HIS3* reporter gene under the control of four consensus Swi4/Swi6-dependent cell cycle box (SCB) elements was crossed to the yeast gene-deletion array (*see* **Note 9**) *(20)*. This query strain is a histidine auxotroph as a result of the absence of *CLN3*, which is required for expression of the *SCB-HIS3* reporter. MAT**a** *xxxΔ::kan^r cln3Δhis3Δ* strains containing the integrated reporter were isolated using a variation of the R-SGA selection steps described. (**b**) Detection of *SCB-HIS3* repressors. MAT**a** *xxxΔ::kan^r cln3Δhis3Δ* strains containing the *SCB-HIS3* reporter were pinned onto SD –His + 30 mM 3-AT (high concentration) to identify strains that express sufficient levels of *HIS3* to confer viability. These are deletion mutants in which the transcriptional defect resulting from *cln3Δ* is suppressed. *WHI5* (boxed) was identified as a repressor of *SCB-HIS3*. Reproduced from **ref. 20** with permission from Elsevier Science. (**c**) Confirmation of *SCB-HIS3* repression by *WHI5*. Serial dilution growth assays confirmed the ability of the *cln3Δhis3Δwhi5Δ* strain containing the *SCB-HIS3* reporter to grow on SD –His + 30 mM 3-AT.

Reporter-Based Synthetic Genetic Array Analysis 67

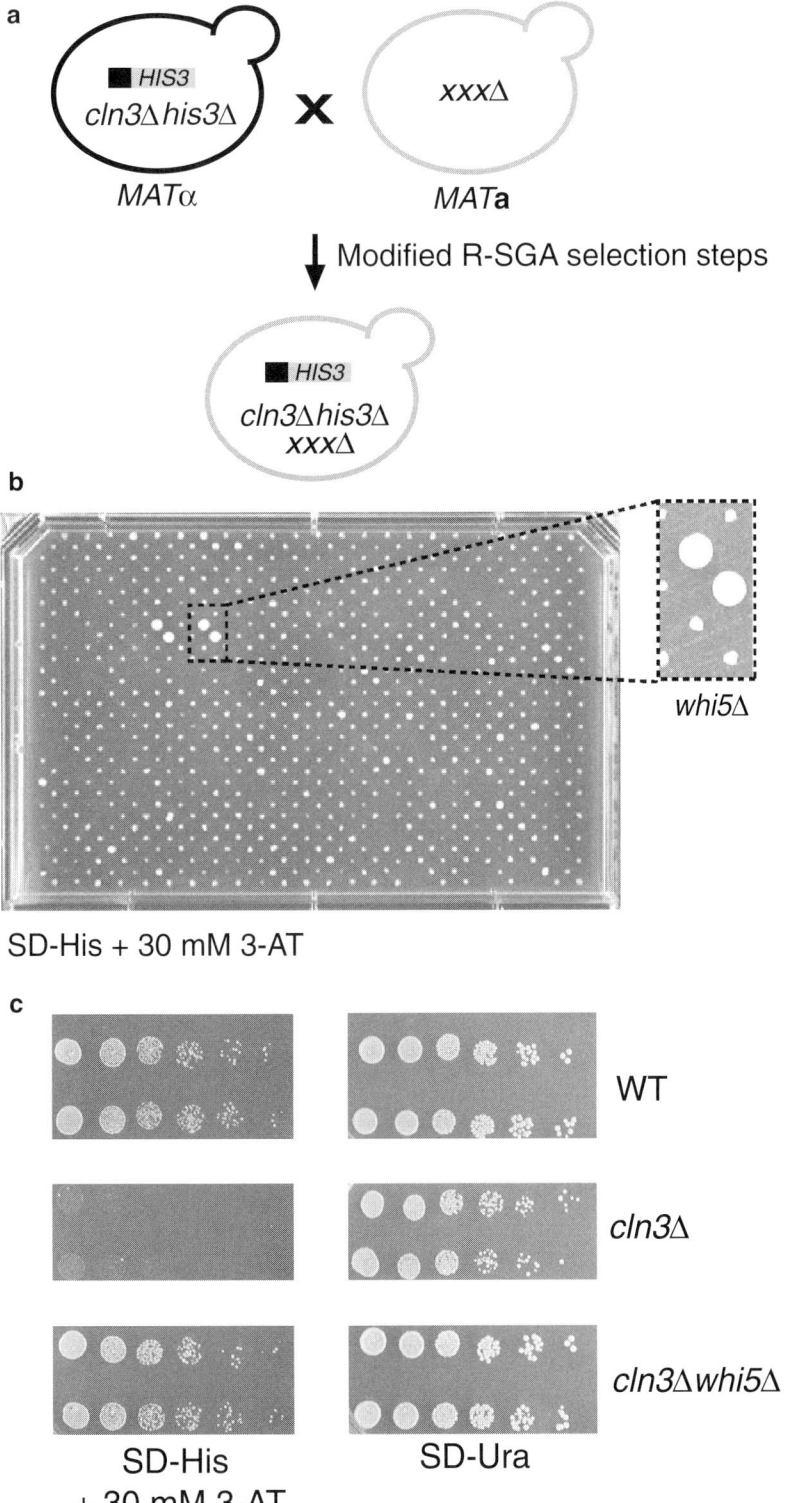

2. Introduce the *CCpr-HIS3* plasmid into the deletion strains by standard yeast transformation, and plate onto SD −Ura. Incubate for 2–3 days at 30°C.

3. Aliquot 200 μl of SD −Ura liquid medium into each well of a 96-well round bottom plate.

4. Inoculate each candidate regulator gene-deletion strain carrying the *CCpr-HIS3* plasmid into the 96-well round bottom plate in duplicate.

5. Grow the cultures for 2 days at 30°C. Mix the cultures by pipetting after the first day.

6. Inoculate 5 μl of each culture into 315 μl of fresh SD −Ura in a 96-well round bottom plate.

7. Perform five 1:5 serial dilutions of each culture in the same 96-well plate (30 μl added to 120 μl of fresh SD −Ura liquid medium each).

8. Spot 4 μl of all dilutions of each strain onto SD −Ura, SD −His + 10 mM 3-AT and SD −His + 75 mM 3-AT plates. Incubate the plates for 2 days at 30°C.

9. Repeat **steps 2–8** using the *Controlpr-HIS3* plasmid.

10. Score strain growth on the SD −His + 3-AT plates compared to the SD −Ura control plate for both the *CCpr-HIS3* and *Controlpr-HIS3* plasmids to identify high-confidence regulators that are CCpr-specific (*see* **Note 21**). **Fig. 4c** shows an example of serial dilution growth assays performed with *xxxΔ::kan^r* strains containing a *HIS3* reporter.

4. Notes

1. The antibiotic G418 is added to YEPD and to enriched sporulation plates (at a lower concentration) to minimize contamination.

2. Ammonium sulfate interferes with the sensitivity of cells to G418 *(21)*. We therefore use MSG as the nitrogen source in our synthetic plates containing G418.

3. In addition to 2-μm reporter constructs, we have used *CEN*-based constructs and integrated CCpr reporters in our R-SGA screens. The reporter context will depend on the promoter under study. Weak promoters may require the use of the 2-μm origin of replication to sufficiently amplify the reporter signal for robust detection, whereas *CEN*-based or integrated CCpr reporters are generally more appropriate for strong promoters.

4. For the *lacZ* construct, promoters can be cloned as *Xho*I–*Xho*I or *Sal*I–*Sal*I fragments. For pHIS3, promoters can be cloned as *Eco*RI–*Bam*HI fragments.

5. In addition to *lacZ* and *HIS3* reporters, we are currently expanding the R-SGA system to include green fluorescent protein (GFP) and derivatives of red fluorescent protein (RFP) reporters.

6. *RPL39* encodes a ribosomal protein, and *ACT1* encodes actin. Expression of these genes is constitutive and independent of cell cycle regulation *(2)*. In our screens, expression of reporters under the control of *RPL39* or *ACT1* promoters is unaffected in the vast majority of mutants defective in known cell cycle transcription factors.

7. We maintain a master copy of the yeast gene-deletion set in 384-array format on YEPD + G418 plates at 4°C. The plates are propagated as needed in 384-format for maintenance and 768-format for screens. Each colony is represented in duplicate in the 768-format, and our current array is 14 plates. Yeast colonies adjacent to empty spots and those on the borders of each plate in a 768-density array have access to more nutrients and generally grow faster and larger than neighboring colonies that are completely surrounded by others. Reporter activity in these larger colonies can be higher than in genetically identical smaller colonies positioned in high-density areas. Steps can be taken to minimize these positional effects: (1) each plate in the 384-density master array can be designed with borders consisting of a neutral strain carrying all the markers required for the R-SGA selection steps, (2) the neutral strain can be positioned at fixed intervals within the array as a reference for wild-type reporter levels, (3) for a given CCpr, known regulators can be positioned at fixed intervals within the array as references for enhanced or reduced reporter activity.

8. Using the R-SGA protocol, deletion strains containing CCpr-reporter plasmids can be recovered 16 days from the generation of the plasmid-bearing starting strain. Approximately 2 h of pinning time is required to replicate one complete yeast gene-deletion set at each selection step. Recently, we have made use of the Singer RoToR HDA platform (Singer Instruments), which cuts pinning time down to approximately 20 min for each selection step per screen.

9. We have screened the yeast overexpression array *(4)* using query strains bearing integrated, nourseothricin (cloNAT) resistance-marked cell cycle reporters, and are currently making adaptations to screen essential genes using the Tet-promoters Hughes collection (yTHC) *(22)*. We used a variation

of the R-SGA method that incorporated a *cln3Δ* allele into the starting strain to identify Whi5 as an inhibitor of G1/S transcription (**Fig. 4**) *(20)*. One could modify the reporter constructs to include complete ORFs, facilitating the identification of regulators of cell cycle protein expression and/or stability. Further, specific promoter elements rather than full promoter sequences could be screened: we have screened both Swi4/6-dependent cell cycle box (SCB) (**Fig. 4**) and *Mlu*I cell cycle box (MCB) (**Fig. 2**) promoter elements using R-SGA.

10. To successfully cross query strains to the array, lawns must be as dense and as even as possible. Two or three lawns are sufficient for mating to the entire yeast gene-deletion array.

11. Spores are pinned in duplicate onto two separate SD –Ura/Leu/Arg/Lys + can/thia plates, effectively generating two copies, or sets. Each set is carried through the remaining haploid selection steps such that four colony replicates of each deletion mutant (two in each set) are ultimately analyzed for reporter activity.

12. Care must be taken when pouring the X-gal top agarose mixture, as excessive force can wash the colonies away. Using 0.5% agarose, solidification of the X-gal solution usually occurs within 30 s.

13. Color development can take anywhere from 15 min to overnight depending on the strength of the promoter and the concentration of X-gal. The array-based β-galactosidase assay conditions described above have been optimized for promoters screened in our laboratory. Prior to assaying *lacZ* reporter activity on a genome-wide scale, it is important to perform a small-scale overlay experiment to determine the optimum assay conditions (concentration of X-gal, incubation time and temperature) and examine basal reporter activity. Pilot experiments should include several transformants and should be done using pinned colonies to mimic the final assay conditions.

14. Under ideal experimental conditions, colonies with wild-type levels of *lacZ* expression will appear light blue, those with reduced expression will appear white, and those with enhanced expression will appear dark blue. In practice, a range of color intensities is usually observed. It is helpful to use known regulators as benchmarks with which to develop scoring criteria to identify new candidates. We have found the identification of activators significantly easier than the identification of repressors using *lacZ* reporters; the ability to identify repressors of a given CCpr will largely depend on the basal activity of the reporter and may not be possible with certain promoters.

15. Plates covered with X-gal top agarose can show uneven distribution of blue color, particularly around the edges. The confirmation plates containing mutant strains next to wild type are therefore especially important.

16. In our screens, non-cell-cycle-specific regulators generally include genes with roles in general transcription, including components of the RNA polymerase II mediator complex (*MED1, NUT* genes, *SRB* genes). We also find genes involved in transcriptional elongation, histone modification, chromatin remodeling, and mRNA stability, likely as a result of their roles in the general regulation of gene expression. Genes that function in cellular respiration (*HAP* genes, *COX* genes), protein translation, trafficking, and degradation are also frequently identified, although the reasoning behind their regulatory effects is less clear.

17. The number of candidate regulators identified in each screen is variable and depends on the promoter under study and the scoring criteria. In a *CLN2pr-lacZ* R-SGA screen, approximately 500 candidate regulators were identified in the primary screen and 235 were confirmed in miniarray format (N. Bastajian, unpublished data). Candidate regulators are not limited to those that bind the promoter of interest; R-SGA has the capacity to identify both direct and indirect regulators. Care should therefore be taken when inferring regulatory relationships.

18. The sensitivity of detecting reporter-based transcriptional effects from whole colonies in array format is dependent on the promoter, the reporter, and the scoring system. In a *CLN2pr-lacZ* screen, we identified known regulators of *CLN2* expression, including Swi4, Swi6, and Bck2 (N. Bastajian, unpublished data).

19. In some cases, CCpr-reporter constructs are somewhat leaky, allowing survival on SD −His even in the absence of activated transcription, probably as a result of basal transcription. To overcome this problem, the competitive inhibitor of the imidazoleglycerol-phosphate dehydratase enzyme encoded by *HIS3*, 3-AT, is added to the plates. The level of background expression varies depending on the reporter construct and genetic background. It is therefore essential that prior to every screen, serial dilution growth assays (*see* **Subheadings 2.4.2** and **3.4.2**) be performed to determine the minimum amount of 3-AT that effectively inhibits growth of the starting strain. Generally, this will lie in the range of 10–100 mM, but in certain cases it may be higher or lower. The advantage of employing this inhibitor is that for any given strain, one can identify genes that both upregulate and downregulate expression of the reporter gene by analyzing

growth at concentrations both above and below the minimum inhibitory concentration.

20. The basis for comparison to the final R-SGA selection plates is that any *xxxΔ::kan^r* strain that is inherently slow growing will produce smaller than average colonies on both the final selection and SD –His + 10 mM 3-AT plates. These strains can then be removed from the list of putative activators. In addition to visual inspection of colony size, we use a computer-based scoring system developed in the Boone Laboratory.

21. In the *HIS3* reporter assays, subtle differences in strain survival observed upon pinning are generally obvious using serial dilution growth assays.

Acknowledgments

We thank Michael Costanzo and Jonathan Millman for sharing unpublished results. We are grateful to Helena Friesen for helpful discussions and critical reading of the manuscript. This work was supported by grants to B.A. from the Canadian Institutes for Health Research, the National Cancer Institute of Canada, and Genome Canada through the Ontario Genomics Institute.

References

1. Cho, R. J., Campbell, M. J., Winzeler, E. A., et al. (1998). A genome-wide transcriptional analysis of the mitotic cell cycle. *Mol. Cell.* **2**, 65–73.
2. Spellman, P. T., Sherlock, G., Zhang, M. Q., et al. (1998). Comprehensive identification of cell cycle-regulated genes of the yeast *Saccharomyces cerevisiae* by microarray hybridization. *Mol. Biol. Cell.* **9**, 3273–3297.
3. Pramila, T., Wu, W., Miles, S., et al. (2006). The Forkhead transcription factor Hcm1 regulates chromosome segregation genes and fills the S-phase gap in the transcriptional circuitry of the cell cycle. *Genes Dev.* **20**, 2266–2278.
4. Sopko, R., Huang, D., Preston, N., et al. (2006). Mapping pathways and phenotypes by systematic gene overexpression. *Mol. Cell.* **21**, 319–330.
5. Iyer, V. R., Horak, C. E., Scafe, C. S., et al. (2001). Genomic binding sites of the yeast cell-cycle transcription factors SBF and MBF. *Nature* **409**, 533–538.
6. Simon, I., Barnett, J., Hannett, N., et al. (2001). Serial regulation of transcriptional regulators in the yeast cell cycle. *Cell* **106**, 697–708.
7. Horak, C. E., Luscombe, N. M., Qian, J., et al. (2002). Complex transcriptional circuitry at the G1/S transition in *Saccharomyces cerevisiae*. *Genes Dev.* **16**, 3017–3033.
8. Lee, T. I., Rinaldi, N. J., Robert, F., et al. (2002). Transcriptional regulatory networks in Saccharomyces cerevisiae. *Science* **298**, 799–804.
9. Ihmels, J., Friedlander, G., Bergmann, S., et al. (2002). Revealing modular organization in the yeast transcriptional network. *Nat. Genet.* **31**, 370–377.
10. Chen, H. C., Lee, H. C., Lin, T. Y., et al. (2004). Quantitative characterization of the transcriptional regulatory network in the yeast cell cycle. *Bioinformatics* **20**, 1914–1927.
11. Winzeler, E. A., Shoemaker, D. D., Astromoff, A., et al. (1999). Functional characterization

of the *S. cerevisiae* genome by gene deletion and parallel analysis. *Science* **285**, 901–906.
12. Tong, A. H., Evangelista, M., Parsons, A. B., et al. (2001). Systematic genetic analysis with ordered arrays of yeast deletion mutants. *Science* **294**, 2364–2368.
13. Giaever, G., Chu, A. M., Ni, L., et al. (2002). Functional profiling of the *Saccharomyces cerevisiae* genome. *Nature* **418**, 387–391.
14. Tong, A. H., Boone, C. (2005). Synthetic genetic array analysis in *Saccharomyces cerevisiae*. In Yeast Protocols, 2nd edn., Methods in molecular biology, vol. 313 (Xiao, W., ed.).. Humana, Totowa, NJ, pp. 171–192.
15. Tong, A. H., Boone, C. (2007). High-throughput strain construction and systematic synthetic lethal screening in *Saccharomyces cerevisiae*. In Yeast Gene Analysis, 2nd edn., Methods in microbiology, vol. 36 (Stansfield, I. and Stark, M., eds.).. Elsevier, Burlington, MA, pp. 369–386; 706–707.
16. Johnson, A. D., Herskowitz, I. (1985). A repressor (*MATα2* product). and its operator control expression of a set of cell type specific genes in yeast. *Cell* **42**, 237–247.
17. Christianson, T. W., Sikorski, R. S., Dante, M., et al. (1992). Multifunctional yeast high-copy-number shuttle vectors. *Gene* 110, 119–122.
18. Brachmann, C. B., Davies, A., Cost, G. J., et al. (1998). Designer deletion strains derived from *Saccharomyces cerevisiae* S288C: a useful set of strains and plasmids for PCR-mediated gene disruption and other applications. *Yeast* **14**, 115–132.
19. Verma, R., Smiley, J., Andrews, B., Campbell, J. L. (1992). Regulation of the yeast DNA replication genes through the *Mlu I* cell cycle box is dependent on SWI6. *Proc. Natl. Acad. Sci. U S A.* **89**, 9479–9483.
20. Costanzo, M., Nishikawa, J. L., Tang, X., et al. (2004). CDK activity antagonizes Whi5, an inhibitor of G1/S transcription in yeast. *Cell* **117**, 899–913.
21. Cheng, T. H., Chang, C. R., Joy, P., et al. (2000). Controlling gene expression in yeast by inducible site-specific recombination. *Nucl. Acids Res.* **28**, E108.
22. Mnaimneh, S., Davierwala, A. P., Haynes, J., et al. (2004). Exploration of essential gene functions via titratable promoter alleles. *Cell* **118**, 31–44.

Chapter 5

The Fidgety Yeast: Focus on High-Resolution Live Yeast Cell Microscopy

Heimo Wolinski, Klaus Natter, and Sepp D. Kohlwein

Summary

Despite its small size of 5–8 μm – only one order of magnitude above the wavelength of visible light – yeast has developed into an attractive system for light microscopic analysis. First, the ease of genetic manipulation and integrative transformation have opened numerous experimental strategies for genome-wide tagging approaches, e.g., with fluorescent proteins (as discussed in several chapters of this issue). Second, the large number of cells that can be simultaneously visualized provides an excellent basis for statistical image analysis, resulting in reliable morphological or localization information. Third, the flexibility of yeast cultivation in terms of biochemical manipulation, rapid cellular growth, mutant isolation or drug susceptibility offers an unprecedented spectrum of possibilities for *in vivo* functional studies, and analysis of cellular dynamics and organelle inheritance. Although yeast in itself is an interesting cellular system, its "prototype character" in understanding cellular metabolism, physiology, and signaling in eukaryotes accounts for its popular use in technology development and biomedical research.

Here we discuss experimental strategies for live yeast cell imaging, geared towards imaging-based large-scale screens. Major emphasis is on the methods for immobilizing cells under "physiological" conditions, with minimum impact on yeast. We also point out potential pitfalls resulting from live cell imaging that once again stresses the necessity for extremely careful experimental design and interpretation of data resulting from imaging experiments. It goes without saying that these problems are not restricted to yeast and are also highly relevant to "large" cells. If an image tells more than a thousand (perhaps misleading?) words, the ease of obtaining "images" thus rather suggests analyzing many thousands of images, to come up with one relevant and biologically significant conclusion.

Key words: Yeast, *Saccharomyces cerevisiae*, Fluorescence microscopy, Green fluorescent protein, High-content imaging, Large-scale microscopy, Live cell microscopy

1. Introduction

The ease of obtaining "images" in microscopy requires consideration of numerous experimental requirements and constraints in order to obtain reliable and meaningful results. This involves: (a) construction of cells expressing fluorescent protein (FP) fusions, (b) cell cultivation prior to microscopy, (c) cell labeling with vital dyes, (d) preparation and maintenance of cells during microscopy, (e) microscope setup and imaging, (f) downstream processing of images.

The emphasis of the methods described below was put on generating (a) a rather simple and reliable setup for standard live yeast cell imaging and imaging-based screens that are (b) also applicable without dedicated imaging systems on "standard" upright and inverted confocal microscopes or other types of high-resolution imaging devices.

The advantage of looking at many – 100s, 1,000s – of yeast cells at once provides a significant advantage over larger cells, to obtain reliable localization patterns. However, It becomes immediately evident that yeast cell populations are never homogeneous as a result of replicative or chronological aging, different stages of the cell cycle etc. This may not necessarily impact interpretation of global localization patterns in most cases, but obviously may affect the interpretation of localization patterns that are changing during the cell cycle or the aging process. Observation of such fluctuations of protein localization patterns obviously yields very important functional information. However, there is also the need to clearly discriminate "physiological" patterns from induced deficiencies during microscopy, which requires very careful control of the microscopy setup. In some examples we will highlight – unexpected – observations for rapidly induced changes in organelle morphology or protein localization due to changes in the microenvironment during microscopy. Labeling of cells with vital dyes is particularly error-prone as cellular age and physiology may significantly affect uptake of dyes and labeling efficiency, e.g., through the activity of pleiotropic drug resistance pumps that may take care, more or less efficiently, of xenobiotic vital dyes. For instance, elimination of the Pdr1 transcriptional regulator yields drastically stimulated uptake and staining intensity with Nile Red, a popular lipophilic dye used to detect lipidic structures in yeast *(1)*. Additionally, the illumination of the specimen itself may have a detrimental effect on the physiology and cellular morphology. Thus, reducing the intensity and exposure time to (laser) light is of utmost importance in live cell imaging, representing challenges for efficient labeling or detectable levels of expression, and optimized sensitivity of the microscopy equipment (*see* **Note 1**).

Preparation problems multiply for cell handling in "high-throughput" setups, i.e., for imaging-based screens. In such experiments, mutants may be identified with altered organelle morphology or subcellular distribution of a particular FP-tagged "query" construct that defines a specific subcellular structure. Thus, factors that are important for organelle morphology, protein trafficking, protein expression and localization may be identified that may not be accessible with any other type of screen. Vital dyes appear as an easy solution to analyze mutant collections, but are obviously restricted to organelles that can be specifically labeled; further considerations involve the impact of dye uptake and labeling efficiency in a particular mutant background that may obscure the imaging pattern. Alternatively, FP-tagged query constructs expressed from plasmids can be rather easily introduced into the mutant collections, by mass transformation or cytoduction (R. Rothstein, personal communication). Main disadvantages of plasmid-borne constructs, however, are the requirement for selective conditions for plasmid maintenance, limiting its practical application, and, more severely, a rather inhomogeneous expression of plasmid-borne FP fusions in the cell population; this inhomogeneity in fluorescence intensity may obscure subtle changes in localization patterns, thus leading to false-positive and false-negative results. Also, such labeling patterns are not amenable to computer-based statistical image analysis (*see* **Note 2**).

The large number of cell types – in the range of 6,000 – that are to be imaged for mutant screens requires some compromising, in particular at the level of cell preparation. The time requirements for imaging challenges the logistics for cell handling in order to analyze cell populations that are cultivated for a comparable time window (*see* **Note 3**).

A critical issue involves the maintenance of "physiological" conditions during the microscopy. A rather simple agar-based method for cell immobilization and maintenance is described in detail below. This method allows for preparation of slides with typically 96 (up to 2× 96) colonies on standard microscope slides, both for upright and inverse microscopes. The spacing of colonies is compatible with the 1,536 colony format on standard rectangular plates that can be ideally and precisely handled by the Singer Rotor replicating robot. A few examples with "critical" proteins or subcellular structures that obviously react highly sensitively to the microenvironment on the slide may illustrate the reliability of this immobilization method. It must be stressed, however, that conditions under the microscope may never completely represent conditions in liquid culture or on the surface of solid media plates.

The selection of the correct microscope settings obviously depends on the signal intensity of the fluorescent sample (*see* **Note 4**).

The optical path in general and the objectives in particular are major determinants of the resolution that can be achieved. For yeast imaging typically oil immersion objectives with high numerical apertures (NA = 1.4) are used. For optimum resolution, thus, the mounting medium of the specimen should match the refractive index of the immersion fluid (*see* **Note 5**).

The downstream processing of images obviously depends on the specific experimental problem; typically, selection of images occurs "by eye," but more standardized procedures for microscopic image acquisition need to be considered, i.e., for the evaluation of the dynamic range, saturation, and contrast of the digital image (*see* **Note 6**).

Only recently, methods have been implemented for quantitative feature extraction of yeast fluorescence images *(2–4)*.

Repositories for image data are available through the Saccharomyces Genome Database (http://www.yeastgenome.org/) at:

YeastGFP database (http://yeastgfp.ucsf.edu) *(5)*;

Yeast Protein Localization database (http://YPL.uni-graz.at) *(6, 7)*;

Organelle.db (http://organelleview.lsi.umich.edu) *(8)*;

TRIPLES (http://ygac.med.yale.edu/triples/) *(9)*.

Morphological data of yeast mutants, rather than localization data, are available at the Saccharomyces Cerevisiae Morphological Database (http://yeast.gi.k.u-tokyo.ac.jp/) *(10)*.

2. Materials

2.1. Yeast Strains

Wild type strains BY4741 (*MAT*a *his3Δ1 leu2Δ0 met15Δ0 ura3Δ0*) and BY4742 (*MAT*α *his3Δ leu2Δ0 lys2Δ0 ura3Δ0*) and strains Acc1-GFP, CoxIV-GFP, and Elo3-GFP harboring chromosomally integrated GFP fusion constructs are obtained from Euroscarf (Institute of Microbiology, Johann Wolfgang Goethe-University Frankfurt, Germany) as well as from Invitrogen, Inc. *(5)*. Strains Y7092 (*MAT*α *can1Δ::STE2^{pr}-Sp_his5 lyp1Δ his3Δ1 leu2Δ0 ura3Δ0 met15Δ0*) and Y8205 (*MAT*α *can1Δ::STE2^{pr}-Sp_his5 lyp1Δ::STE3pr-LEU2 his3Δ1 leu2Δ0 ura3Δ0*) used for construction of query strains were provided by Charlie Boone (Donnelly Centre for Cellular and Biochemical Research, University of Toronto). Template plasmids for fluorescence protein fusions can be obtained through NCRR Yeast Resource Center, Seattle (http://depts.washington.edu/yeastrc/pages/micro.html) or from the authors.

2.2. Cell Culture

1. Bacto™ yeast extract.
2. D(+)-glucose monohydrate.
3. Bacto™ peptone.
4. Bacto™ agar.

For preparation of yeast media, yeast extract (1% w/v) and peptone (2% w/v) are weighed in an appropriate flask and filled up with distilled water to 9/10 of the final volume. Autoclave at 121°C for 20 min; add 1/10 volume of 10× glucose solution (2% w/v final concentration) by sterile filtration to the media, mix well and dispense into sterile flasks or glass tubes. For solid media plates 2% Bacto™ agar is added to the solution prior to autoclaving. For preparation of agar plates and agar-coated slides without nutrients 2% (w/v) Bacto™ agar in distilled water is autoclaved at 121°C for 15 min.

For small scale cultivation and vital staining, cells are cultivated in 6- or 12-well plates in 5–2 ml of the medium under shaking and temperature control (30°C) on a thermomixer.

Plates for selection of positive transformants contain 200 mg/l G418 sulfate (Sigma-Aldrich, Inc.) or 200 mg/L nurseothricin (Werner BioAgents, Inc.). Media for the array-based introduction of chromosomally integrated GFP fusions into the deletion mutant collection are prepared according to Tong and Boone *(11)*.

2.3. Fluorescence Labeling

1. MitoTracker® Red CM-H_2XRos (Invitrogen, Inc., Cat.-No. M7513).
   Prepare stock solution of 1 mg/ml MitoTracker® Red CM-H_2XRos in dimethylsulfoxide (DMSO). Store 20 μl aliquots of the stock solution in 500 μl reaction tubes at –20°C. One microliter of MitoTracker® Red CM-H_2XRos was added to 1 ml of a cell suspension in a 1.5-ml reaction tube (final concentration: 1 μg/ml). Labeling was performed for 10 min at RT without subsequent washing of cells.
2. DASPMI (4-(4-(dimethylamino)styryl)-*N*-methylpyridinium iodide (Invitrogen, Inc., Cat.-No D288): For preparation of stock solution and labeling see MitoTracker Red.
3. FM4–64 *(12)* (Invitrogen, Inc., Cat.-No. T13320). For preparation of stock solution and labeling see MitoTracker Red. Labeling was performed for 60 min at RT.
4. Nile Red (Invitrogen, Inc., Cat.-No. N1142). For preparation of stock solution and labeling see MitoTracker® Red. Centrifuge Nile Red stock solution briefly in a mini centrifuge prior to cell staining, to remove precipitates that may have formed.
5. Formaldehyde solution min. 37% (Merck, Inc.).

2.4. Cell Preparation

1. Singer RoToR High Density Array robot (Singer Instrument Co.Ltd).
2. Cultivation plates (Plus plates; Singer Instrument Co. Ltd).
3. Plastic string or metal wire (0.2 mm diameter).
4. 4× magnifying glass with stand and illumination feature.
5. Scalpel No. 10 (Martor Inc., Germany).
6. Low-melting agarose.

2.5. Microscopy Hardware and Imaging Software

1. Microscope slides 76 × 26 mm.
2. Coverslips 50 × 24 mm #1.
3. Leica TCS SP2 confocal microscope with differential interference contrast (DIC) optics, acousto optical tunable filter (AOTF), acousto optical beam splitter (AOBS); spectral detection. Märzhäuser MCX-2 stage controller (Märzhäuser, Inc.). Leica ScanWare™ and Leica TCS SP2 microscope control and imaging software v2.60.1537.
4. Leica TCS4d confocal microscope with DIC optics, AOTF.
5. Laser sources: 488 nm (argon-laser; GFP, DASPMI excitation), 543 nm (green helium-neon laser; MitoTracker® Red CM-H_2XRos, Nile Red excitation); ArKr laser (488 nm; 543 nm).
6. Leica objectives: HCX PL APO 100×/1.4–0.7 OIL CS; HCX PL APO 63×/1.32–0.6 OIL CS; HCX PL APO 63×/1.3 GLYC CORR, Leica HCX PL APO 63×/1.2 W CORR.
7. Bioptechs Objective Heater System (Bioptechs, Inc.).
8. Adobe Photoshop™ CS (Adobe, Inc., USA). Image J, public domain image processing software (author: Wayne Rasband, National Institute of Mental Health, Bethesda, Maryland, USA). amira™ 4.0 (Mercury Computer Systems, Inc.).

3. Methods

3.1. Introduction of Chromosomally Tagged GFP Fusion into the Deletion Mutant Collection

GFP fused genes are introduced into a mutant collection according to the "synthetic genetic array (SGA)" protocol, as described by Tong et al. for the systematic construction of double mutant strains *(11, 13)*. GFP fusions in appropriate query strains suitable for the SGA protocol are constructed as follows (**Fig. 1**):

3.1.1. Construction of the GFP Query Strain

The GFP cassette is integrated into the query strain background required for the SGA protocol, harboring the following markers: *MATα can1Δ::STE2^{pr}-Sp_his5 lyp1Δ his3Δ1 leu2Δ0 ura3Δ0*

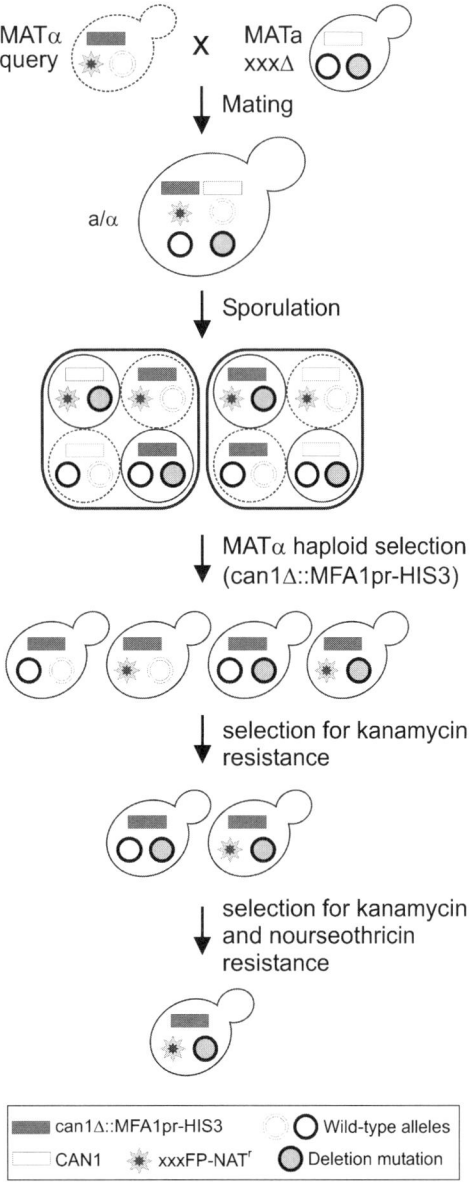

Fig. 1. Schematic representation of introducing a chromosomally integrated GFP-labeled query construct into the deletion mutant collection. (Adapted by permission from Macmillan Publishers Ltd (**ref**. *14)*, copyright 2007.) This method is applicable for both N- or C-terminally tagged variants, and relies on the basic protocol by Tong and Boone (*11,13*).

met15Δ0. If the integrated GFP-cassette linked with the Clon-NAT marker is already available in a different background, the query strain is generated by mating with a switching strain harboring the required markers, and subsequent sporulation and selection of haploid progeny with the correct genotype. A switching strain obtained from Charlie Boone (Donnelly Centre for Cellular and Biochemical Research, University of Toronto), Y8205 *MATα can1Δ::STE2^{pr}-Sp_his5 lyp1Δ::STE3pr-LEU2 his3Δ1 leu2Δ0 ura3Δ0* allows for selection of suitable query strains on selective media:

1. Mate GFP-expressing strain (*MAT*a) with strain Y8205.
2. Pick zygotes.
3. Sporulate on acetate plates.
4. Select α-haploid progeny on leucine-free plates containing thialysine, canavanine.
5. Select for a query strain on leucine-free plates that contain thialysine, canavanine, nurseothricin.
6. Test for GFP expression (microscopy) and correct integration of markers by PCR.

If the integrated GFP-cassette is marked with the *kanMX* resistance marker instead of nurseothricin resistance (clonNAT), an appropriate query strain can be obtained by marker switching *(11)* (*see* **Note 7**). The PCR product is transformed into the recipient strains by standard lithium acetate transformation.

Construction of C-Terminal GFP Fusions (see Note 8)

The primers for amplification of the GFP-ClonNAT cassette contain 50 bases homologous to the 3′-region of the gene of interest. The template plasmid is pKN082 (Natter, unpublished). Sequences for binding on the vector template are 5′-GTGAG-CAAGGGCGAGGAG-3′ and 5′-CAGTATAGCGACCAGCAT-TCAC-3′.

1. Amplify GFP-ClonNAT cassette using appropriate primers for insertion 3′ to the reading frame.
2. Transform PCR product into the SGA-compatible query strain (*MATα can1Δ::STE2^{pr}-Sp_his5 lyp1Δ his3Δ1 leu2Δ0 ura3Δ0 met15Δ0*).
3. Select transformants on plates containing nurseothricin.
4. Test for GFP expression (microscopy) and correct integration of markers by PCR.

Construction of N-Terminal Fusions

Construction of N-terminally GFP-tagged proteins that are expressed under endogenous promoter control requires a two-step cloning strategy. We prefer to use the endogenous promoter for reasons discussed in the introduction. We constructed a vector, pCGCNm, bearing the ClonNAT cassette followed by a multiple cloning site and monomeric GFP (Natter, unpublished). The promoter region of the gene of interest is cloned into the MCS of pCGCNm. The entire cassette, NAT^R-Promoter-mGFP, is amplified with primers containing 50 bases, homologous regions to the 5′-region of the target gene and the annealing sequences

Color Plates

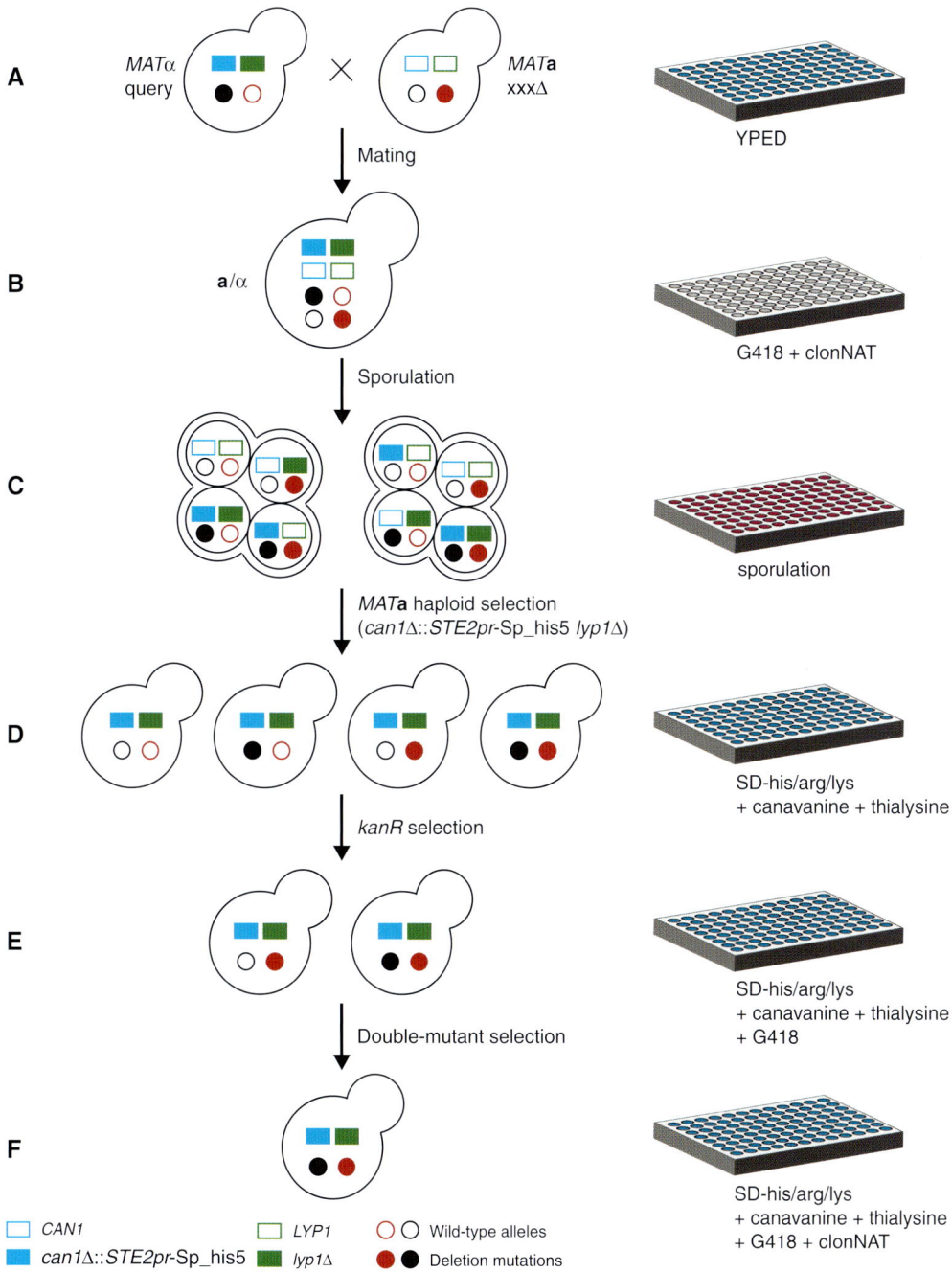

Fig. 1. Synthetic genetic array (SGA) methodology. (*For complete caption go to* Page 45)

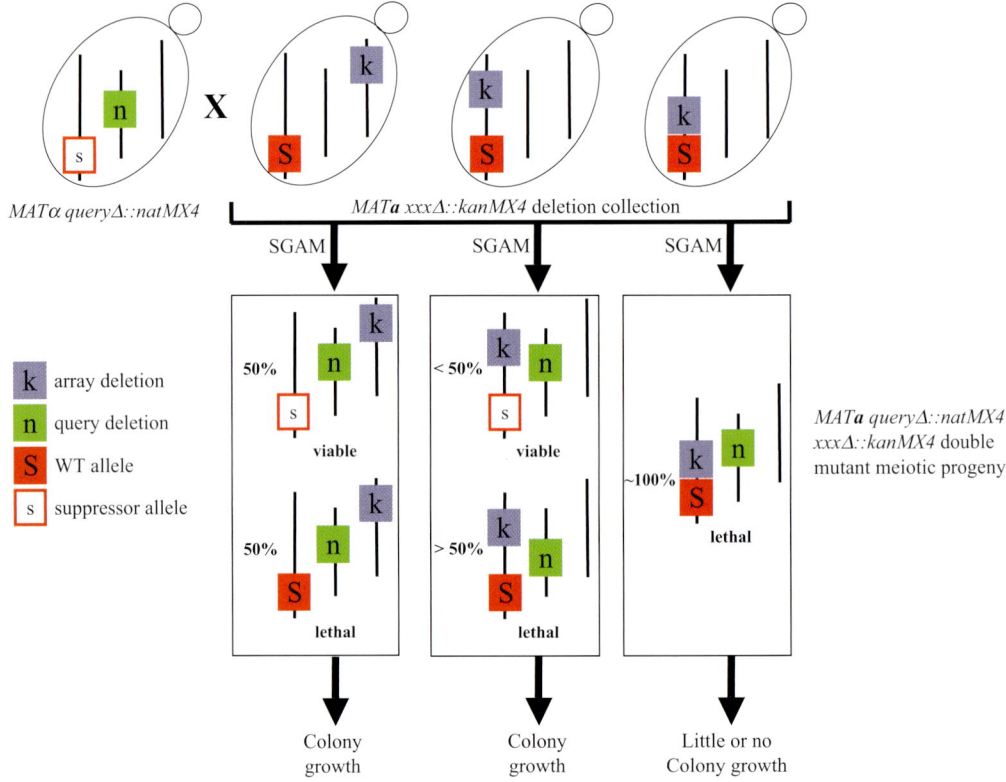

Fig. 2. Genetic mapping using SGAM. (*For complete caption go to* Page 47)

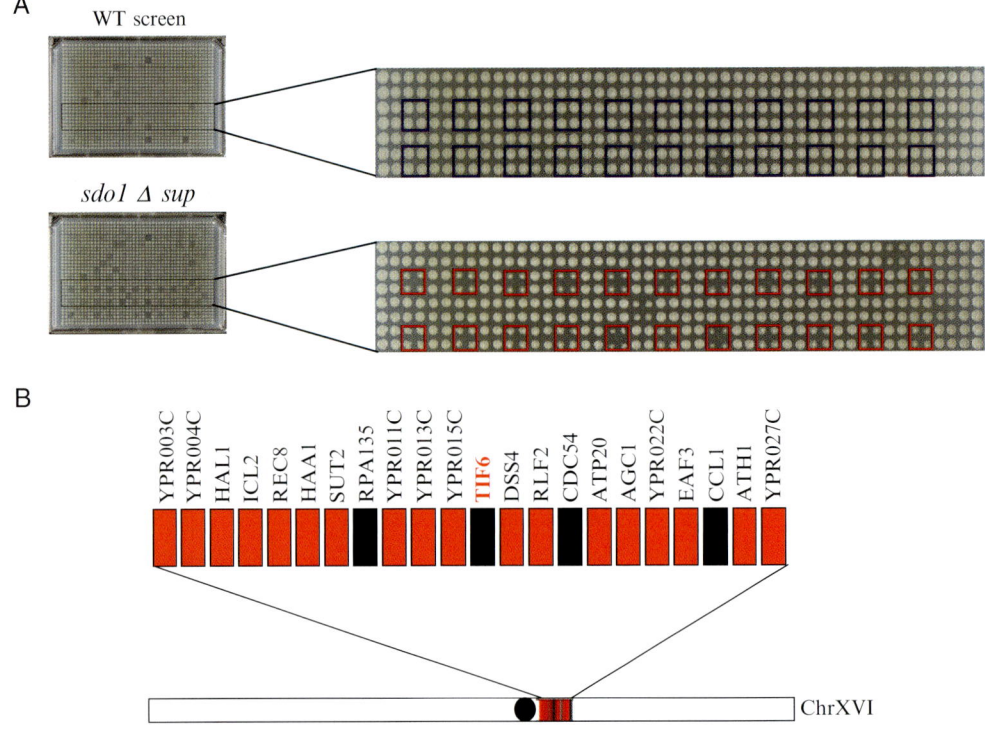

Fig. 3. SGAM identification of a suppressor of *sdo1Δ* slow growth. (*For complete caption go to* Page 49)

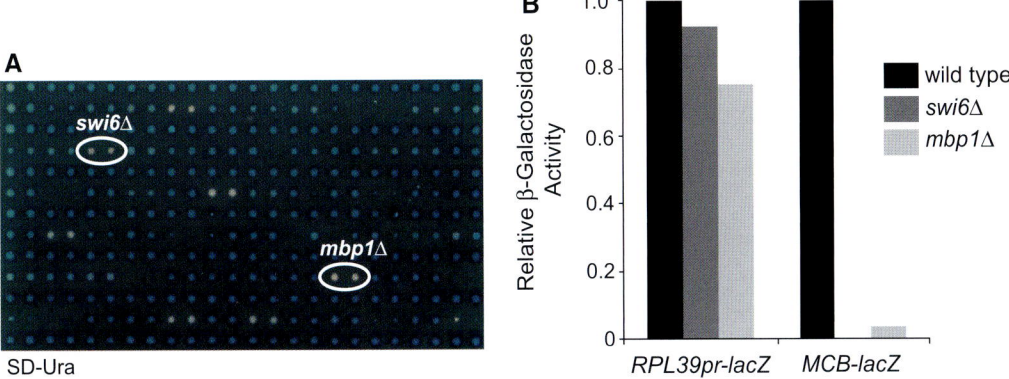

Fig. 2. Identifying regulators of a G1-phase cell cycle promoter element using a *lacZ* reporter. (*For complete caption go to* Page 64)

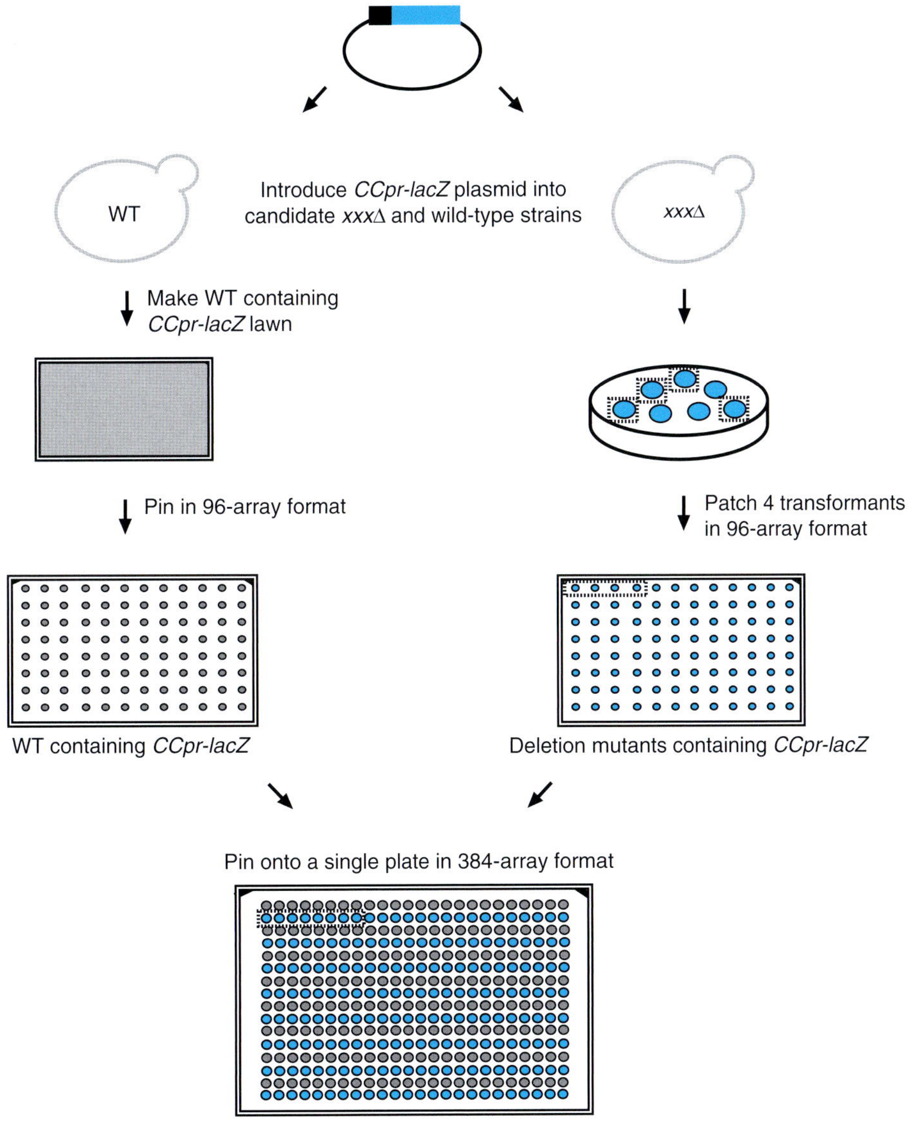

Fig. 3. Building a 384-format array to confirm regulators identified in genome-wide *lacZ* R-SGA screens. (*For complete caption go to* Page 65)

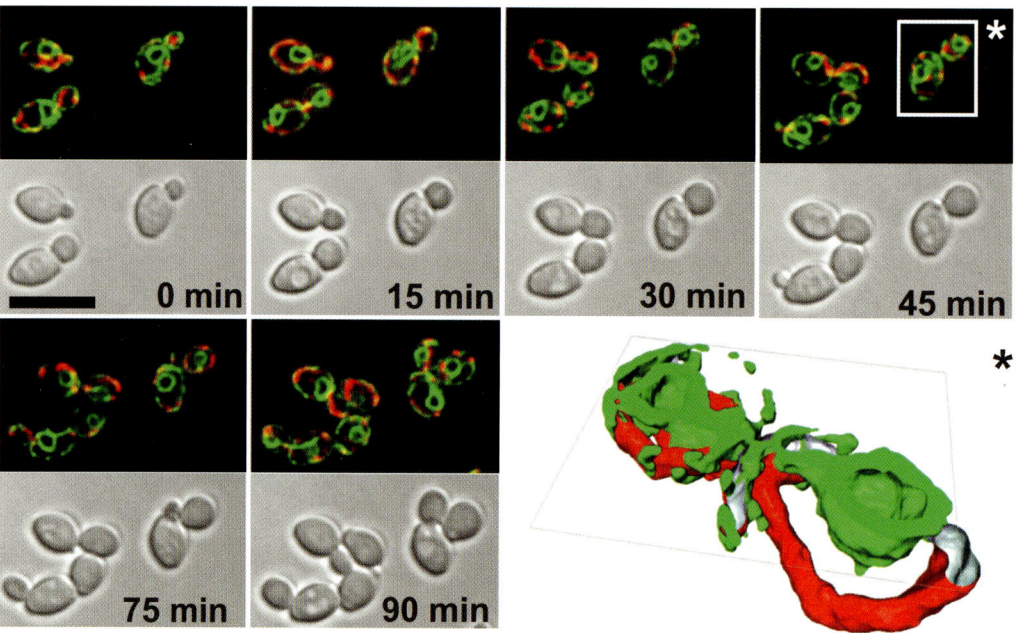

Fig. 3. Four-dimensional imaging of multi-labeled yeast cells during cellular growth. Endoplasmic reticulum was labeled with Elo3-GFP (*green*). Mitochondrial structures were labeled with MitoTracker® Red CM-H_2XRos (*red*). Cells were cultivated on agarose/complete media covered slides. Selected single optical sections (fluorescence and transmission images) of the acquired 4D data set (25 optical sections each) are shown. The image at the bottom right represents a polygon-based 3D reconstruction (longitudinal cut through the surface model) of the masked cell shown in the *top panel* at *right*. Bar, 10 μm. (*For complete caption go to* Page 86)

5′-ACATGGAGGCCCAGAATACCC-3′ (fwd) and 5′-CTTG-TACAGCTCGTCCATGCC-3′ (rev), respectively.

1. Amplify the promoter sequence of the gene of interest, using flanking primers with compatible restriction sites for insertion into the MCS of plasmid pCGCNm.
2. Amplify ClonNAT-gene promoter-GFP-cassette using appropriate primers for insertion 5′ to the reading frame.
3. Transform PCR product into the SGA-compatible query strain (*MATα can1Δ::STE2^{pr}-Sp_his5 lyp1Δ his3Δ1 leu2Δ0 ura3Δ0 met15Δ0*).
4. Select transformants on plates containing nurseothricin.
5. Test for GFP expression (microscopy) and correct integration of markers by PCR.

3.2. Cell Cultivation (see Notes 9 and 10)

1. Replicate mutant cells on suitable agar media plates (Plus plates, Singer Instrument Co. Ltd.) to a density of 1,536 colonies using a Singer Rotor HDA robot.
2. Grow cells for 24 h at 30°C.
3. Replicate cells to fresh plates.
4. Repeat replication step to generate very small colonies ("dilution"), on desired media plates (*see* **Note 10**).

3.3. Immobilization of Cells for Live Cell Imaging

3.3.1. Setup for Vital Dye Labeling

1. Boil 2.5% low melting temperature agarose in aqua dest. or in appropriate growth media in a microwave oven (600 W) until the agarose solution is completely clear. Alternatively, use 2% Bacto™ agar instead of agarose (*see* **Note 11**).
2. Transfer 5 ml of melted agarose solution into a 15-ml plastic tube and equilibrate at ~50°C in a heated water bath.
3. Add 5 µl of the fluorescence dye from stock solution (final concentration typically 1 µg/ml) and vortex for 30 s.
   (a) Optionally add 135 µl formaldehyde solution (37% in water; final concentration 1% v/v) to the agar solution (Note: Addition of a low concentration of formaldehyde kills the cells and enables more homogeneous labeling with vital dyes due to inactivation of drug pumps. Cell morphology as imaged by DIC is not compromised in these preparations).
4. Place a standard microscope slide (76 × 26 mm) on a flat surface and pipette 3 ml of the agarose solution on the microscope slide until it is completely covered.
5. Allow gelation of the agarose on the slide for 15 min at RT.
6. Add 1 µl of (labeled) yeast cell suspension to the center of the agarose layer. Mount the cell preparation with a large (50 × 24 mm) coverslip (*see* **Note 12**).

3.3.2. Setup of Yeast Cell Arrays for Large-scale High-content Imaging (Fig. 2)

1. Put a sheet of plastic (2 mm) as solid support for the agar-layer into a Plus plate.
2. Pour agar solution (cooled to ~50°C) into the plate.
3. Allow gelation of the agar for 15 min at RT.
4. Replicate cell colonies in the 1,536 format with a Singer Rotor replicator.
5. Remove agar block and dissect into 16 pieces with 96-colonies each, preferably by cutting with a thin plastic string or metal wire.

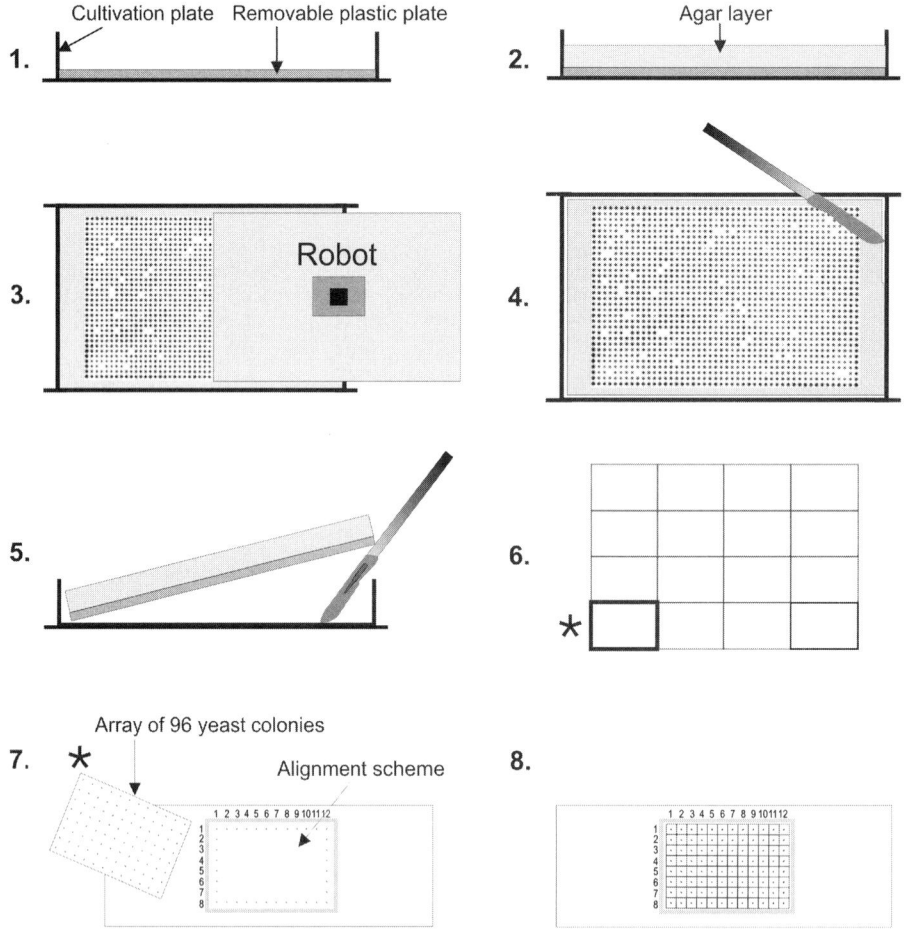

Fig. 2. Preparation of yeast cell-arrays for large-scale, high-content microscopy. *1*, Put a sheet of plastic (2 mm) as solid support for the agar-layer into a Plus plate. *2*, Pour 50 ml agar solution (cooled to ~50°C) into the plate (~5 mm thickness). *3*, Replicate cell colonies in the 1,536 format with a Singer Rotor replicator. *4/5*, Remove agar block together with solid support. *6*, Dissect agar block into 16 pieces with 96 colonies each. *7*, Mount agar block with 96 colonies onto standard glass slide with an alignment scheme (printed with a laser printer on transparency film), for reproducible positioning on the microscope. *8*, Cut horizontal and vertical slits between the yeast colonies prior to covering colonies with a cover slip. This prevents air bubbles and mixing of colonies.

6. Cut horizontal and vertical slits separating the colonies.

7. Mount agar block with 96 colonies on standard glass slide with the aid of an alignment scheme (printed with a laser printer on transparency film), for easier and reproducible positioning on the microscope stage.

8. Cover colonies with a large (50 × 24 mm) cover slip.

9. Press the coverslip carefully onto the agar surface to remove air bubbles.

10. Perform imaging, using manual or automatic positioning, software-based autofocus and multiple section image acquisition.

11. Carry out image storage, data processing and subsequent off-line image evaluation (*see* **Note 13**).

3.4. Microscopy Setup

3.4.1. Transmission Microscopy: Differential Interference Contrast

1. Pipet ~1 µl of cell suspension on standard microscope slides or on agarose-coated slides.

2. Cover cells with a large coverslip and mount slides under the microscope; set temperature controller of the objective heater to the desired temperature.

3. Visualize cells using DIC optics. Note that the condenser position needs to be adjusted for cells mounted on agarose-covered slides, due to the extended distance, for optimized illumination of the specimen.

3.4.2. Fluorescence Microscopy (*see* **Note 14**)

1. Pipet ~1 µl of cell suspension on standard microscope slides or on agarose-coated slides.

2. Cover cells with a large coverslip and mount slides under the microscope; set temperature controller of the objective heater to the desired temperature.

3. Optimize laser intensity, fluorescence filter (fluorescence emission range) and photomultiplier settings, using the lowest possible laser intensity.

4. Allow the cells to adjust to the conditions under the microscope (*see* **Note 15**).

5. Re-adjust focus and image cells in areas not previously exposed to the laser light.

This experimental setup allows reliable 3D imaging of double-labeled cells (ELO3-GFP, MitoTracker) over several hours (**Fig. 3**).

3.5. Specific Applications: FRAP and FLIP (Fig. 4)

Laser scanning microscopy allows for application of specific techniques, such as fluorescence recovery after photobleaching (FRAP) or fluorescence loss in photobleaching (FLIP), to investigate organelle and protein dynamics. Modern microscopes provide features to restrict laser illumination to regions of interest in the

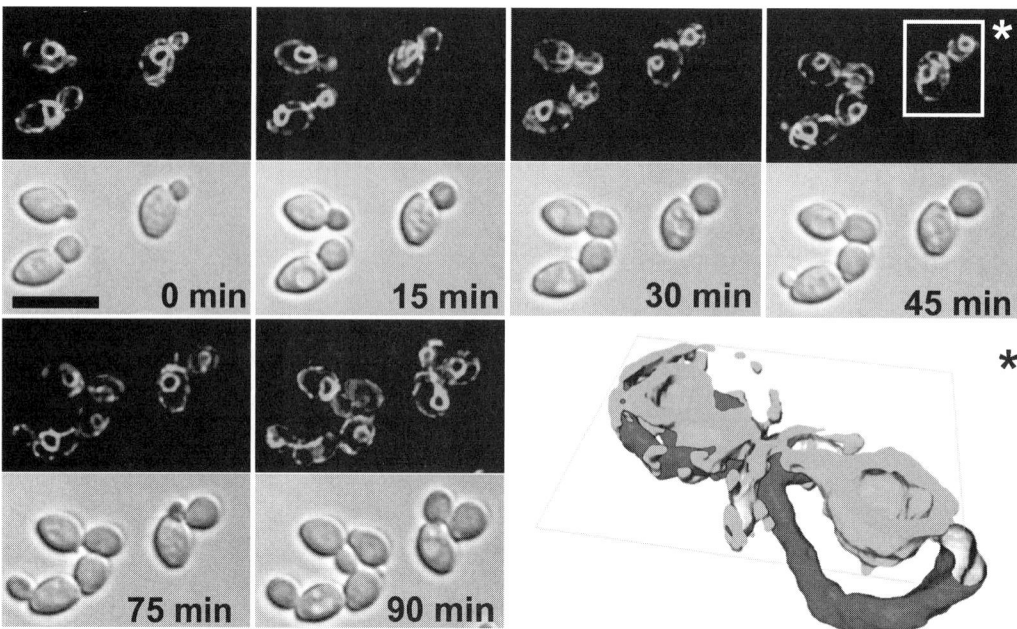

Fig. 3. Four-dimensional imaging of multi-labeled yeast cells during cellular growth. Endoplasmic reticulum was labeled with Elo3-GFP (*green*). Mitochondrial structures were labeled with MitoTracker® Red CM-H$_2$XRos (*red*). Cells were cultivated on agarose/complete media covered slides. Selected single optical sections (fluorescence and transmission images) of the acquired 4D data set (25 optical sections each) are shown. The image at the bottom right represents a polygon-based 3D reconstruction (longitudinal cut through the surface model) of the masked cell shown in the *top panel* at *right*. Bar, 10 μm (*see Color Plates*).

nanometer scale, which are required for bleaching subcellular yeast structures at the limit of optical resolution. (*see* **Note 16**).

1. Precultivate GFP-fusion expressing cells in appropriate media under vigorous shaking at 30°C.
2. Concentrate cells by centrifugation and mount ~1 μl of cell suspension on standard glass slides; cover with a large coverslip.
3. Define scanning area with low intensity laser light. Include control cells within the field of view that are not going to be bleached.
4. Define region of interest (ROI) for bleaching. Optionally define a ROI for detection of fluorescence intensity outside the bleaching ROI (FLIP).
5. Acquire a series of images before bleaching using fast scanning mode.
6. Apply short bleach pulse (FRAP, FLIP) or longer bleach pulses (FLIP).
7. Acquire a series of post bleach images using the same settings as applied for prebleach scanning.

8. Normalize experimental data (background subtraction, correction for photobleaching during measurement).
9. Estimate quantitative parameters (mobile and immobile fraction, recovery half-time, diffusion coefficient).)

3.6. Potential Pitfalls of Live Cell Fluorescence Microscopy (see Note 17)

The popular use of some fluorescence dyes has prompted us to discuss potential pitfalls resulting from inappropriate preparation or imaging techniques. Examples illustrate the impact of imaging conditions on organelle morphology or protein localization. Thus, various settings need to be tested to obtain reliable imaging information; the following scenarios may aid in designing appropriate control setups.

3.6.1. Fluorescence Activation of DASPMI (Fig. 5; See Note 18)

1. Label cells grown in complete media to stationary growth phase for 10 min in suspension with 1 µg/ml DASPMI and mount on agarose-coated slides.
2. Set (increase) zoom factor to obtain a field of view (image) of ~40 × 40 µm (100× magnification objective).
3. Image cells using 488-nm excitation and 580–650-nm emission with ~25% total laser output for 5 s.
4. Scan field with reduced zoom factor. Pre-scanned area appears at least fivefold brighter.

3.6.2. Preparation-dependent Labeling with MitoTracker® Red CM-H_2Xros (Fig. 6; See Note 18)

1. Cultivate cells in complete media to stationary growth phase (48 h).
2. Label cells for 10 min in suspension with 1 µg/ml MitoTracker® Red CM-H_2XRos and mount on agar-coated slides, or
2b. Mount 1 µl on standard glass slides.
3. Image cells 5 min after preparation, using 543-nm excitation and 550–650-nm detection.
4. Depending on preparation conditions, fluorescence appears in mitochondria (agarose-coated slide) or spread throughout the cytosol (standard slide).

3.6.3. Laser-induced Morphological Alterations of FM4–64-Labeled Vacuoles (Fig. 7; See Note 20)

1. Cultivate cells to log phase in complete media (8 h).
2. Mount cells on slides coated with agar containing 1 µg/ml FM4–64.
3. Incubate for 60 min, to allow for incorporation of dye into vacuolar membranes.
4. Image cells at 488-nm excitation and 500-nm long-pass filter detection, with 25% max laser power; high-speed scanning and 8× averaging (1 s/frame).
5. Sequential recording of images displays extensive morphological alterations (fusion) of vacuoles, and appearance of rounded vacuoles with multiple inclusions.

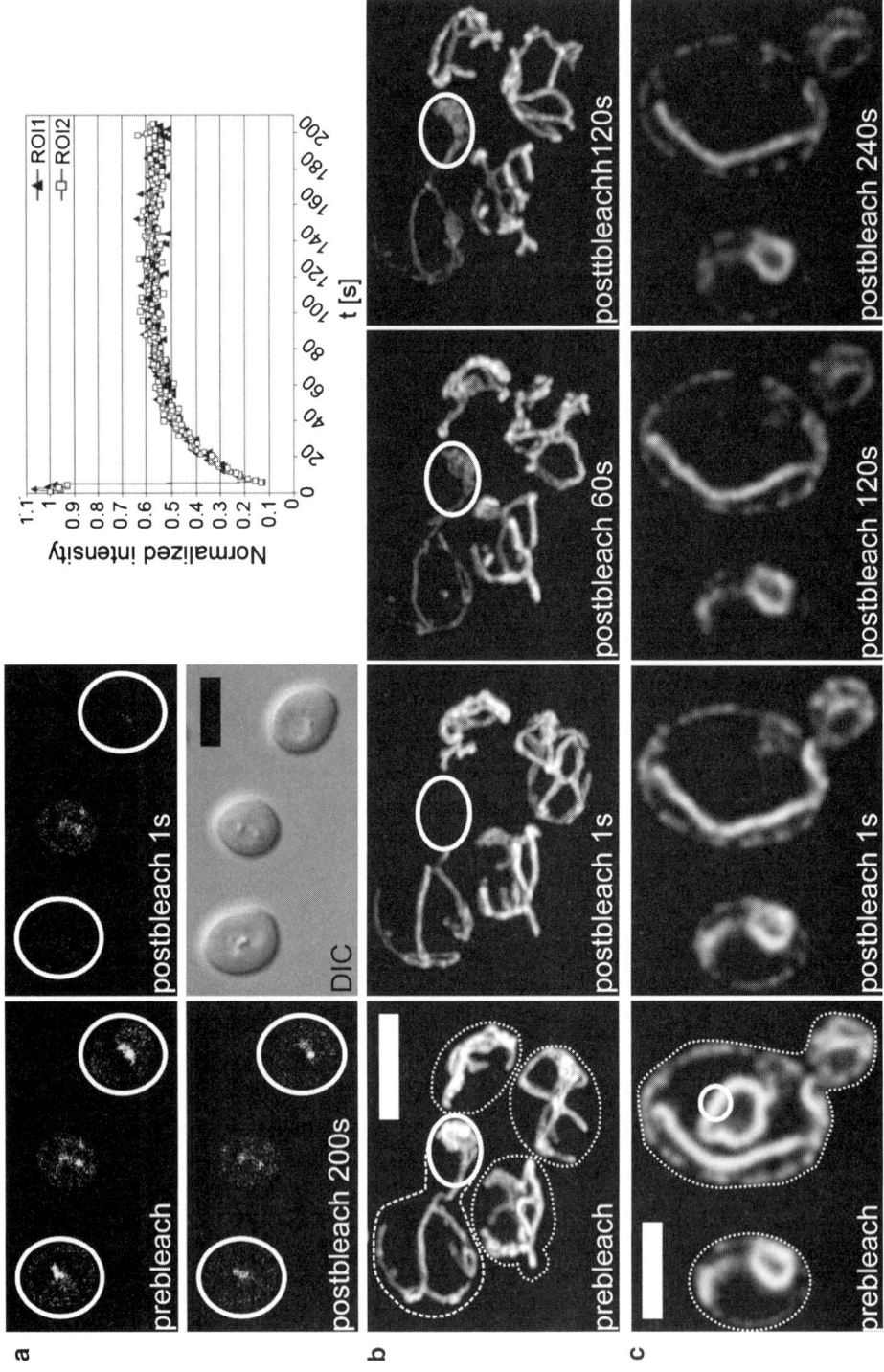

Fig. 4. Visualizing protein and dye dynamics using bleaching technology. (**a**) Quantitative FRAP analysis of the influx of Nile Red molecules into yeast cells. Lipid droplets in stationary phase yeast cells were labeled with the lipophilic dye, Nile Red. Cells were mounted on standard microscope slides without washing, leaving some (nonfluorescent) dye in the surrounding media. Entire yeast cells were simultaneously illuminated with high laser intensity for 2 s (region of interests (ROIs) are labeled by *white circles*). The time-lapse record revealed a significant and uniform recovery of Nile Red fluorescence over time, indicating a continuous uptake of the dye into yeast cells. The cell in the middle of the image represents an unbleached control cell. (**b**) Qualitative FRAP analysis of Cox4-GFP labeled mitochondrial structures. Fluorescently labeled mitochondria in a bud were bleached with high laser intensity for 3 s (ROI indicated by *white circle*). 60 s after bleaching of mitochondrial Cox4-GFP in the bud, a significant recovery of fluorescence was detected in the bleached area, with a concomitant decrease of Cox4-GFP fluorescence in the mother cell. Images represent maximum-intensity projections of multiple optical sections acquired with high scan speed. Note the significant morphological alterations of mitochondrial structures of the unbleached control cells over time. (**c**) Qualitative FLIP analysis of Elo3-GFP labeled endoplasmic reticulum. A small part of the fluorescently labeled nuclear ER (ROI indicated by *white circle*) was continuously bleached for 10 s. The first postbleach image (1 s) shows a loss of fluorescence of the entire nuclear ER. In contrast, cortical ER structures remain unaffected by the bleaching process indicating the existence of a barrier for Elo3-GFP diffusion between nuclear and cortical ER. The gradual loss of fluorescence intensity of the non-bleached control cell is caused by photobleaching during measurement and has to be considered for a quantitative analysis of fluorescence recovery. Bar 5 μm.

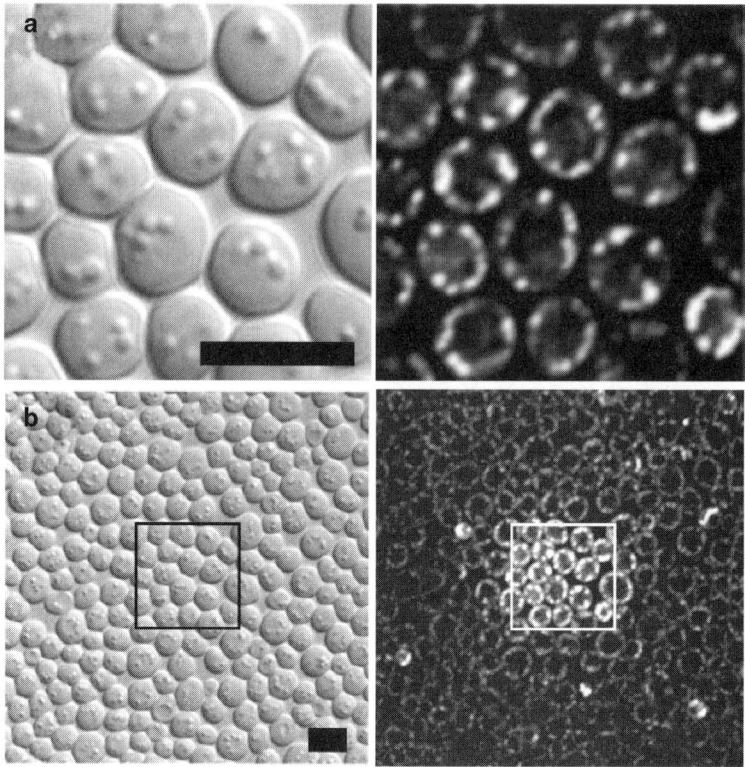

Fig. 5. Laser-induced increase of DASPMI fluorescence. (**a**) Yeast cells were cultivated in complete media to stationary growth phase and labeled with 1 μ/ml DASPMI. A field of view was imaged with increased zoom factor at 488 nm excitation; the laser source was set to ~20% of total output power. (**b**) Subsequent scan of the same field of view with decreased zoom factor. The scan revealed a significant increase of DASPMI fluorescence in cells of the previously scanned area. The rectangular mask indicates the scanned region shown in (**a**). Single optical sections; fluorescence and DIC transmission images. Bar 10 μm.

3.6.4. Preparation-Dependent Localization of Acc1-GFP (Fig. 8, See Note 21)

1. Cultivate cells expressing chromosomally integrated Acc1-GFP to late log or to stationary phase.
2. (a) Mount log phase or stationary phase cells on standard microscope slides under cover slips or
   (b) Mount log phase cells on agar-coated slides.
3. Image GFP-distribution over time, using standard settings (488-nm excitation, 500–550-nm band-pass detection). Immediately after preparation, Acc1-GFP is present exclusively in the cytosol of log-phase cells imaged both on agar-coated slides and on standard microscope slides.

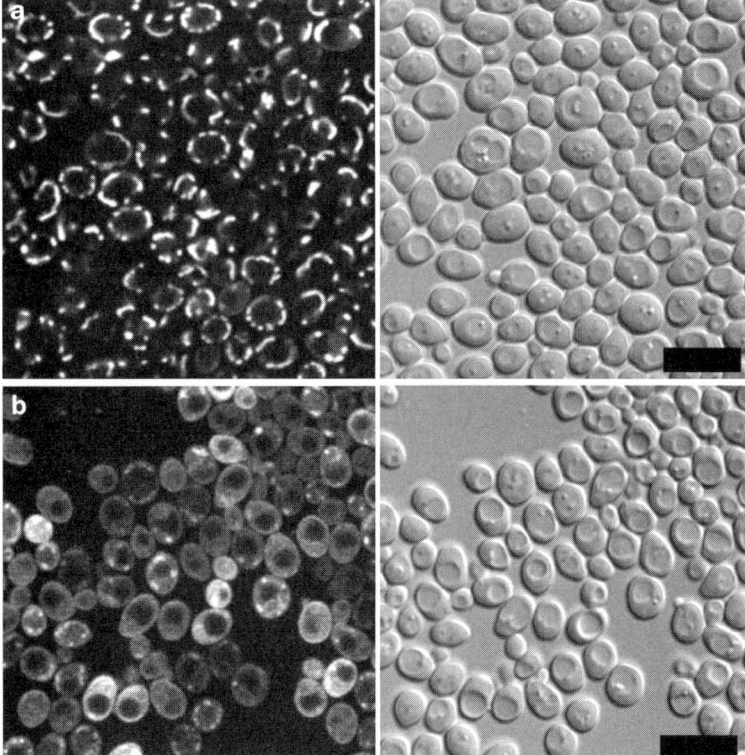

Fig. 6. Labeling "artifacts" of a vital dye due to different preparative conditions. Yeast cells were cultivated to early stationary growth phase (12–24-h cultures) and labeled with MitoTracker® Red CM-H$_2$XRos for 10 min. (**a**) After labeling, cells were mounted on agar-coated slides and imaged after 15 min. The single optical section shows fluorescence distribution typical for active mitochondria, at the cellular periphery. (**b**) After labeling, cells were mounted on a standard microscope slide without agar support and imaged after 5 min. In these cells, the vital dye brightly fluoresces in the cytosol and mitochondria are barely detectable. Bar 10 μm.

4. After 5 min, on standard slides, Acc1-GFP forms foci, which are present in more than 50% of the cells after 10 min; cells prepared on agarose-coated slides retain cytosolic Acc1-GFP localization over extended periods of time.
    (a) Inclusion of an air bubble in a cell preparation on standard slides shows that foci are absent about 50 μm distal to the air bubble, demonstrating the sensitivity and dependence of Acc1-GFP localization on the microenvironment, during microscopy.

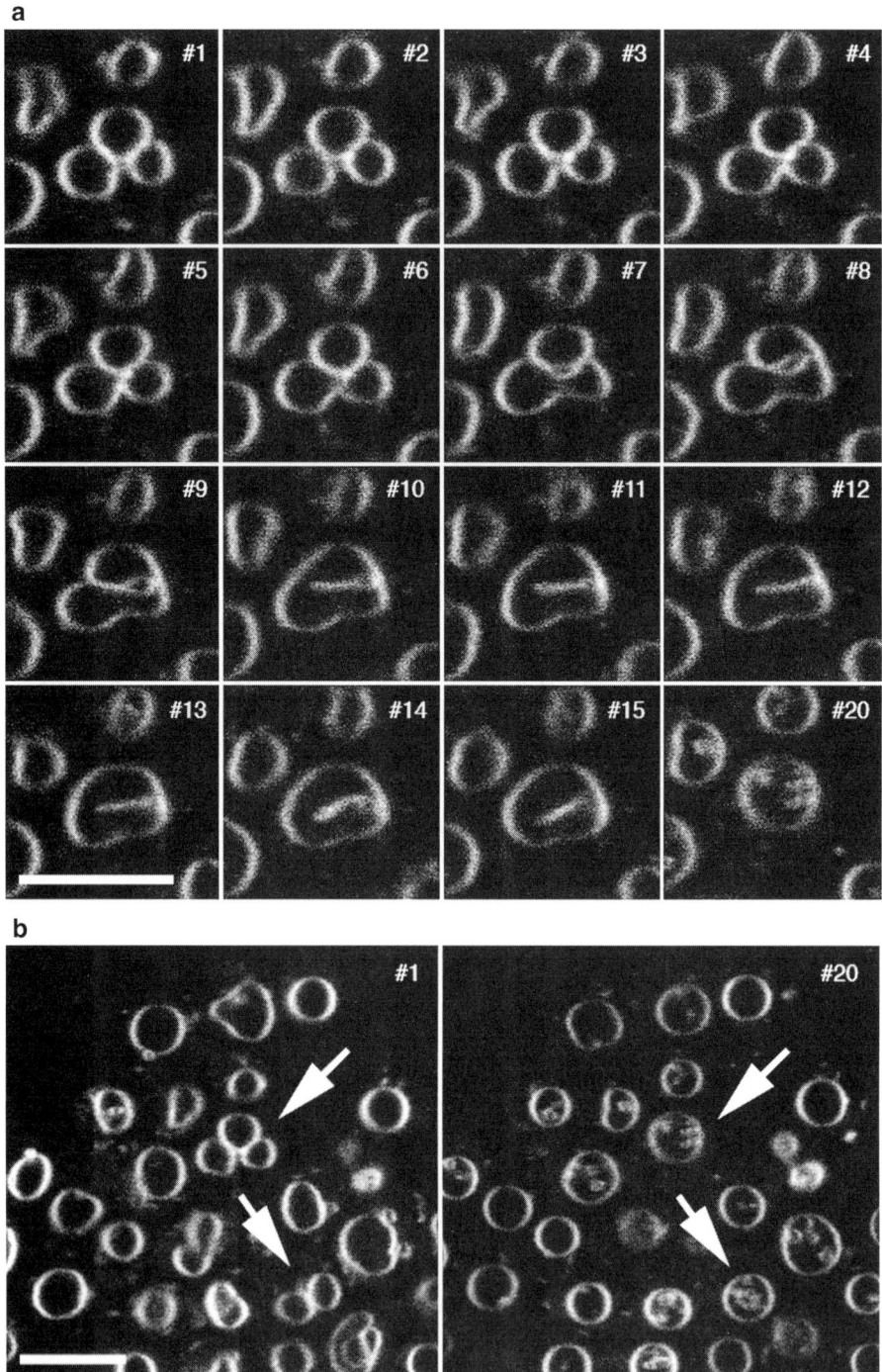

Fig. 7. Illumination of labeled vacuoles induces rapid morphological changes. Log-phase cells were mounted on slides coated with agar containing 1 μg/ml FM4–64. After incubation for 30 min in the dark, cells were sequentially imaged using fast-scan mode and 8× averaging (about 2 s/frame ~180 ms dwell time in the displayed scanning areas; *#1–#20* indicate the frame sequence). Morphological alterations of some vacuoles (marked by *white arrows* in the lower magnification panel) become evident after seven scans (total laser dwell time in that area 1.2 s). After 20 scans, vacuoles appear completely rounded in all cells of the scanned area. Bar 5 μm.

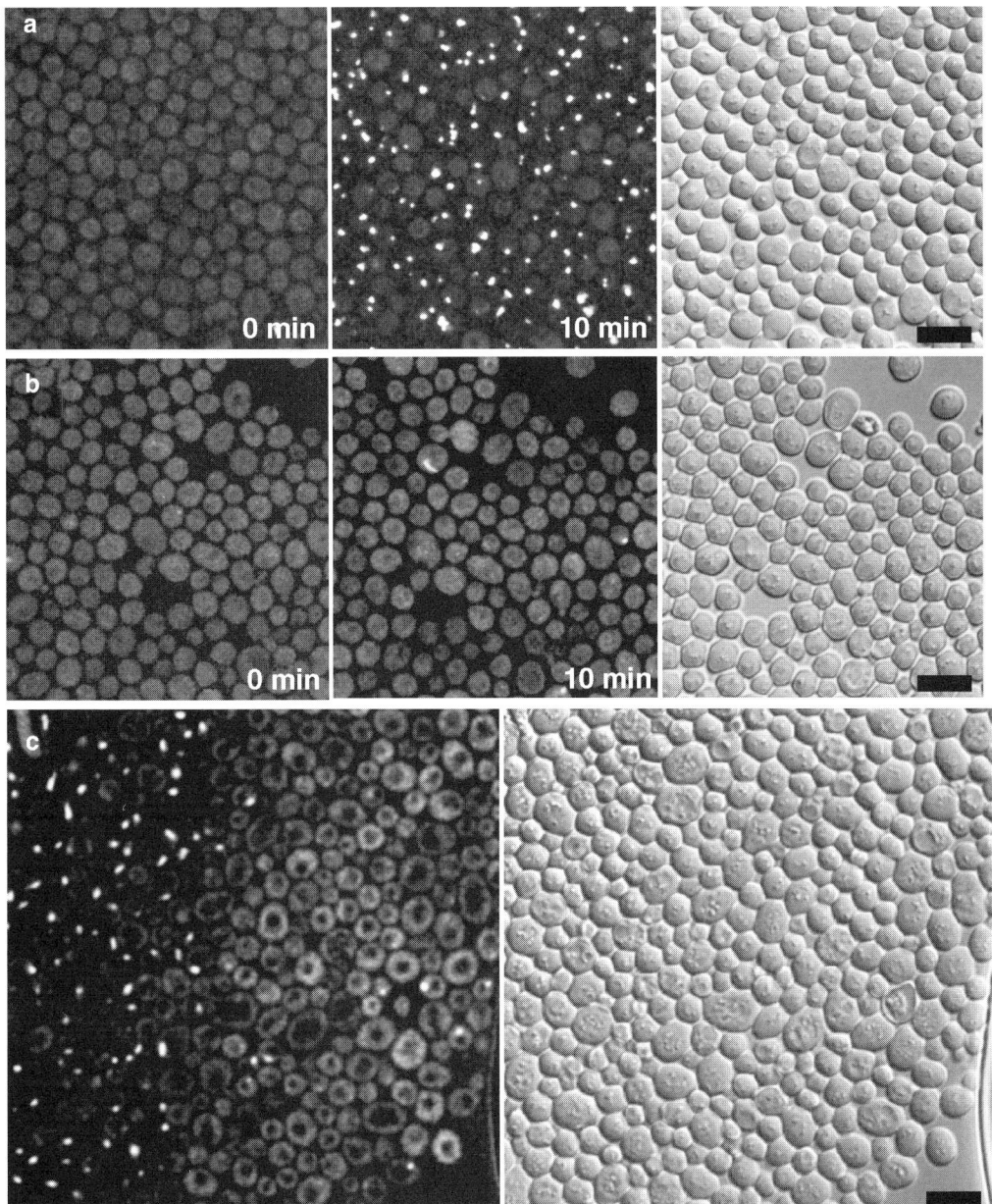

Fig. 8. Changes of Acc1-GFP localization due to the microenvironment on the slide. (**a**) Yeast cells cultivated in complete media to late logarithmic growth phase were mounted on a standard microscope slide and imaged over time. Immediately after preparation, Acc1p-GFP shows a cytosolic distribution. Within minutes, Acc1-GFP localization changes and appears in bright punctae (foci) in most cells, when incubated on a standard slide. (**b**) Same cell suspension as in (**a**), but mounted on agar-coated slides (YPD). Aggregation of Acc1p-GFP is prevented under these microscopic conditions. (**c**) Relocalization of Acc1p-GFP due to an oxygen gradient within the preparation. Stationary phase cells are mounted on standard slides next to an air bubble. Cells in close vicinity to the air bubble (within ~50–60 μm) disassemble Acc1-GFP foci and the protein becomes dispersed in the cytosol, whereas cells in greater distance to the air bubble retain Acc1-GFP foci, after 10 min of incubation. Maximum-intensity projections of multiple (9–13) optical sections. DIC transmission images on the right and fluorescence images *middle* and *left*. Bar 10 μm.

4. Notes

1. If only a few samples of cells are to be imaged, we recommend a protocol for cell preparation based on a method first described by Allan *(15)* and adapted for microscopy *(1)*. This protocol involves isolation of virgin daughter cells from stationary phase cultures by density gradient centrifugation. This procedure results in a highly homogeneous population of cells that may enter vegetative growth in a synchronous way.

2. There is a strong bias in the visual inspection towards "bright" cells but serious consideration must be given as to the significance of this observation, in particular to whether the overall pattern of protein distribution in a cell depends on the level of fluorescence intensity. We find it re-assuring that localization patterns appear in most cases independent of the intensity (expression) level, demonstrating typically robust targeting. Also, plasmid loss under nonselective conditions is often over-emphasized; during the period of microscopic experiments, we find it very useful to cultivate plasmid-harboring cells in complete media (2–6 h), in order to stimulate growth and FP formation. Plasmid-loss is negligible during that period but the FP signal typically improves significantly.

   A strategy for introducing a chromosomally integrated FP-tagged "query" sequence into the deletion strain collection based on "synthetic genetic array" technology *(11, 13)* is clearly the method of choice to screen mutant collections for alterations of specific localization patterns.

3. If the screen is designed to identify cells in the mutant collection with altered localization patterns or morphologies, one may consider fixing the cells prior to microscopy, in order to freeze cell morphologies at a similar stage of development, and thus allowing for comparison of cells in a similar state of growth. This trick is also of advantage for screening mutant populations with vital dyes since mild fixation typically eliminates differences in label uptake. However, imaging based screens are also suitable to identify alterations in organelle dynamics and inheritance, which obviously requires more time-consuming repetitive analyses of the same population of living cells.

4. In order to reduce the impact of light on the cells, illumination should be as short as possible and with lowest intensity possible; factors, which are mainly determined by the intensity of the fluorescence signal and the sensitivity of the microscope system. Confocal microscopy setups – which we prefer

for high-resolution yeast imaging – now provide increased scanning speed and improved (cooled) photodetectors which enable high-speed, high-resolution imaging with excellent signal-to-noise ratios and low electronic background. In our hands, confocal systems are advantageous due to their short dwell-times in a particular area of the specimen, thus reducing bleaching or physiological interference. The rather slow imaging speeds of conventional confocal systems (typically ~1 s/frame) compared to video/deconvolution systems (~10 ms/frame) are now challenged by the introduction of resonant scanner systems that provide video-rate imaging capabilities with typical confocal resolution.

5. Since live yeast cells are typically maintained in aqueous media during microscopy there is a significant loss in brightness (due to diffraction), and resolution. Water or glycerol immersion objectives with an adjustable collar to correct for variations in coverslip thickness (which is critical especially for water and glycerol immersion objectives) are preferable for imaging yeast cells in aqueous solutions. Even with a similar nominal numerical aperture for various objectives, resolution is significantly determined by the type of immersion fluid and demonstrates significant advantages of water or glycerol immersion lenses in comparison to typical (and cheaper) oil immersion lenses.

6. For publication, only minimal enhancements of images are considered appropriate, such as adjustments of brightness and contrast. Other types of image manipulation, such as background correction, nonlinear image adjustments etc. are considered inappropriate and not acceptable for publication, unless explicitly explained (see e.g., editorial policies, e.g., J. Cell Biology: http://www.jcb.org/misc/ifora.shtml#image_aquisition) *(16)*.

7. The gene for nurseothricin resistance (ClonNAT) contains a GC rich region. When standard PCR conditions with proof reading enzymes fail, we use a PCR kit for GC rich regions (e.g., GC-rich PCR System, Roche Diagnostics, Germany) and a two-step PCR protocol with the following cycling conditions: 8×(95°C/30 s – 55°C/20 s – 72°C/3 min), 22×(95°C/30 s – 65°C/20 s – 72°C/3 min) in a GeneAmp 9600 thermocycler (Applied Biosystems).

8. To reduce potential problems resulting from (weak) GFP dimerization we are routinely using monomeric GFP *(17)* which we have introduced into vector pFa6a-GFP(S65T)-*kanMX4* *(18)*. In this plasmid, the *kanMX* cassette was replaced by the ClonNAT cassette, resulting in the vector pKN082. Chromosomal integration of the GFP-ClonNAT PCR fragment is performed as described.

9. Growth conditions largely determine staining efficiency of vital dyes and GFP expression levels. Thus, great care has to be taken to maintain reproducible growth conditions. We have previously described a method for cell preparation for small scale imaging experiments with vital dyes, which is based on a centrifugation-based method for isolation of unbudded daughter cells from stationary phase cultures that are particularly suited for microscopic analysis *(1)*. Precultivation of cell colonies for large-scale investigations, i.e., imaging-based mutant screenings using vital dyes or GFP-labeled cells (*see* **Subheading 3.1**), must by nature be less laborious and is, therefore, restricted to (robotics-based) replicating to appropriate media plates, with the disadvantage of representing more heterogeneous cell populations.

10. Make sure that colonies resulting from SGA protocol are not contaminated with unlabeled cells remaining from an incomplete elimination by the various selection steps. The repetitive replication process may help to obtain more homogeneous cell populations and eliminate dead cells.

11. Complete YPD is rich in fluorescent components, which may obscure weak GFP fluorescence signals. Thus, minimal media are preferable for fluorescence microscopy.

12. Long-term observation of cells requires adequate immobilization, which is readily achieved using the agar technique. The cells are typically arranged as a monolayer in a liquid film right underneath the cover slip. Bleaching experiments (FRAP) may require mounting of cells without agarose (agar) in order to avoid drifting of the focal plane (*see* also **Note 15** below). Agarose-immobilization is required for long-term live cell imaging and for maintaining "physiological conditions").

13. Images are typically stored using the Leica microscope software or a relational database. Image processing and evaluation follows in-house protocols and software tools, i.e., at least two independent viewers and automated evaluation of fluorescence intensity distribution. Very recently, first applications of automated image analysis and feature extraction have been published *(2)*.

14. Typically, objectives with 63× or 100× magnification and high numerical apertures (1.2–1.4) are to be used for yeast imaging. Despite comparable (or perhaps even somewhat lower) numerical apertures, glycerol and water immersion optics with adjustable collar ring may yield better resolution than oil immersion lenses for live yeast cell imaging. This is due to the aqueous environment and the more glycerol-compatible diffraction index of the cells. The impact of non-refractive index-matched optics becomes noticeable already

a few μm below the cover slip, about half way through the yeast cells. Spectral fluorescence detection allows for optimized adjustment of the detection wavelength range and efficient suppression of background fluorescence.

15. Cell preparation typically poses some stress on the cells that may result in altered labeling. Also, the agar layer may shrink somewhat due to water evaporation, which results in cell displacement.

16. Although photobleaching is a widely used approach to address protein and organelle dynamics several reports have been published pointing out limitations, pitfalls, and potential artifacts generated by this technology that need to be carefully considered *(19–21)*.

17. Microscopists are aware of multiple experimental problems during live cell imaging. Most prominently, fluorescence *may bleach* due to excess light absorption by the dye and subsequent chemical modification, resulting in a loss of the ability to emit photons. Other phenomena like *fluorescence saturation*, a process that limits fluorescence emission, is frequently observed as well. As a consequence, fluorescence emission may be much higher at lower excitation intensity, stressing that "more light" does not necessarily result in more fluorescence. Obviously, numerous types of molecules in an unstained cell may absorb photons as well, which may significantly interfere with a cell's physiology and ultimately result in slower growth and potentially in *cell death*. In this respect, UV light is much more potent and cytotoxic, and dyes absorbing in the red or far-red spectrum are, therefore, becoming more and more popular. The application of two-photon microscopy, which potentially circumvents UV toxicity by using high-intensity pulsed IR lasers that may also excite UV-dyes, however, is obviously limited (in yeast) due to excess heat generation. As a general rule, use of fluorescence dyes typically renders cells much more sensitive to light, thus dictates to keep excitation light intensity, illumination time and concentration of the label as low as possible.

18. We have frequently observed that over time (i.e., by time-lapse imaging) fluorescence intensity of a specimen may increase, rather than decrease (bleaching). Typically, such phenomena are not readily observed in epifluorescence, which relies on simultaneous illumination of an entire field of view over extended periods of time (seconds). Fluorescence activation may be due to light-induced stimulation of dye uptake (i.e., by interfering with the cell's physiology), or alterations of dye concentration and interaction with other molecules within a particular structure.

19. The microenvironment during imaging is of critical importance for obtaining reliable imaging information. Especially, mitochondria or other oxygen-dependent cellular entities may react very sensitively to induced anaerobiosis in standard preparations.

20. Vacuoles are a very dynamic membrane-bound organelle, undergoing rapid fusion and fission processes. FM4–64 has been used extensively to monitor endocytosis *(12)* and vacuolar dynamics.

21. Cells prepared for microscopy are subject to microenvironmental conditions that may affect organelle morphology and protein localization. Thus, cell preparation and maintenance under the microscope need careful consideration in all live cell imaging experiments.

Acknowledgements

Work in our laboratory is supported by grants from the Austrian Federal Ministry for Science and Research (GEN-AU program, projects GOLD – Genomics of Lipid-associated Disorders, and FLUPPY – Fluorescence-based analysis of protein–protein interactions in yeast), and the Austrian Science Fund, FWF (project SFB Lipotox F3005).

References

1. Wolinski H., Kohlwein S.D. (2008). Membrane Trafficking, in Methods in Molecular Biology Vol 457, A. Vancura, ed. Humana Press, pp. 151–163.
2. Chen S.C., Zhao T., Gordon G.J., Murphy R.F. (2007). Automated image analysis of protein localization in budding yeast. *Bioinformatics (Oxford, England)* **23(13)**, 66–71.
3. Carpenter A.E., Jones T.R., Lamprecht M.R., et al. (2006). CellProfiler: image analysis software for identifying and quantifying cell phenotypes. *Genome Biology* **7(10)**, R100.
4. Lamprecht M.R., Sabatini D.M., Carpenter A.E. (2007). CellProfiler: free, versatile software for automated biological image analysis. *BioTechniques* **42(1)**, 71–5.
5. Huh W.K., Falvo J.V., Gerke L.C., et al. (2003). Global analysis of protein localization in budding yeast. *Nature* **425(6959)**, 686–91.
6. Kals M., Natter K., Thallinger G.G., Trajanoski Z., Kohlwein S.D. (2005). YPL.db2: the yeast protein localization database, version 2.0. *Yeast* **22(3)**, 213–8.
7. Natter K., Leitner P., Faschinger A., Wolinski H., McCraith S., Fields S., Kohlwein S.D., et al. (2005). The spatial organization of lipid synthesis in the yeast *Saccharomyces cerevisiae* derived from large-scale green fluorescent protein tagging and high resolution microscopy. *Molecular and Cellular Proteomics* **4(5)**, 662–72.
8. Wiwatwattana N., Landau C.M., Cope G.J., Harp G.A., Kumar A. (2007). Organelle DB: an updated resource of eukaryotic protein localization and function. *Nucleic Acids Research* **35**(Database issue), D810–4.
9. Kumar A., Cheung K.H., Tosches N., et al. (2002). The TRIPLES database: a community resource for yeast molecular biology. *Nucleic Acids Research* **30(1)**, 73–5.
10. Saito T.L., Ohtani M., Sawai H., et al. (2004). SCMD: *Saccharomyces cerevisiae* morphological database. *Nucleic Acids Research* **32**(Database issue), D319–22.
11. Tong A.H., Boone C. (2006). Synthetic genetic array analysis in *Saccharomyces cerevisiae*. *Methods in Molecular Biology* **313**, 171–92.

12. Vida T.A., Emr S.D. (1995) A new vital stain for visualizing vacuolar membrane dynamics and endocytosis in yeast. *Journal of Cell Biology* **128(5)**, 779–92.
13. Tong A.H., Evangelista M., Parsons A.B., et al. (2001). Systematic genetic analysis with ordered arrays of yeast deletion mutants. *Science* **294(5550)**, 2364–8.
14. Boone C., Bussey H., Andrews B.J. (2007). Exploring genetic interactions and networks with yeast. *Nature Reviews* **8(6)**, 437–49.
15. Allen C., Buttner S., Aragon A.D., et al. (2006). Isolation of quiescent and nonquiescent cells from yeast stationary-phase cultures. *Journal of Cell Biology* **174(1)**, 89–100.
16. Rossner M., Yamada K.M. (2004). What's in a picture? The temptation of image manipulation. *Journal of Cell Biology* **166(1)**, 11–5.
17. Zacharias D.A., Violin J.D., Newton A.C., Tsien R.Y. (2002) Partitioning of lipid-modified monomeric GFPs into membrane microdomains of live cells. *Science* **296(5569)**, 913–6.
18. Wach A., Brachat A., Alberti-Segui C., Rebischung C., Philippsen P. (1997). Heterologous HIS3 marker and GFP reporter modules for PCR-targeting in *Saccharomyces cerevisiae*. *Yeast* **13(11)**, 1065–75.
19. Lippincott-Schwartz J., Altan-Bonnet N., Patterson G.H. (2003). Photobleaching and photoactivation: following protein dynamics in living cells. Nature Cell Biology, Suppl:S7–14.
20. Pucadyil T.J., Chattopadhyay A. (2006). Confocal fluorescence recovery after photobleaching of green fluorescent protein in solution. *Journal of Fluorescence* **16(1)**, 87–94.
21. Sbalzarini I.F., Mezzacasa A., Helenius A., Koumoutsakos P. (2005). Effects of organelle shape on fluorescence recovery after photobleaching. *Biophysical Journal* **89(3)**, 1482–92.

Chapter 6

A Genomic Approach to Yeast Chronological Aging

Christopher R. Burtner, Christopher J. Murakami, and Matt Kaeberlein

Summary

Yeast is a useful model organism to study the genetic and biochemical mechanisms of aging. Genomic studies of aging in yeast have been limited, however, by traditional methodologies that require a large investment of labor and resources. In this chapter, we describe a newly-developed method for quantitatively measuring the chronological life span of each strain contained in the yeast ORF deletion collection. Our approach involves determining population survival by monitoring outgrowth kinetics using a Bioscreen C MBR shaker/incubator/plate reader. This method has accuracy comparable to traditional assays, while allowing for higher throughput and decreased variability in measurement.

Key words: Longevity, Aging, Chronological life span, Yeast, Bioscreen, Stationary phase

1. Introduction

Along with the nematode *Caenorhabditis elegans* and the fruit fly *Drosophila melanogaster*, the budding yeast has emerged as one of the primary invertebrate models used by scientists studying the molecular biology of aging *(1)*. Two aging paradigms have been developed in yeast: replicative and chronological (**Fig. 1**). Replicative aging refers to the mitotic capacity of a yeast mother cell and is defined as the number of daughter cells produced by a mother cell prior to senescence *(2)*. Chronological aging describes the survival of yeast cells in a nondividing, quiescent-like state and is generally defined by the length of time a cell can survive in stationary phase *(3)*. The ability to specifically monitor the effect of genetic and environmental perturbations

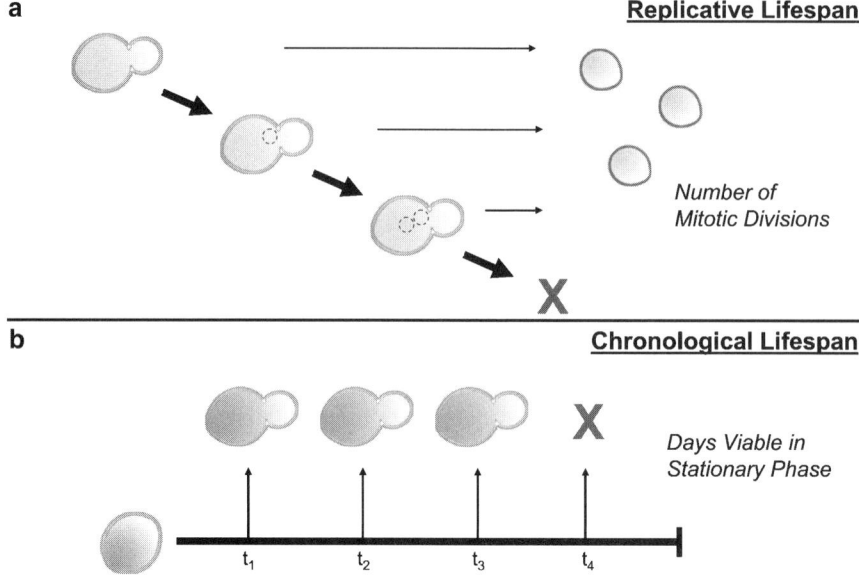

Fig. 1. Schematic for Yeast Replicative and Chronological Aging. (**a**) Replicative life span in yeast is measured by the number of mitotic divisions that can arise from a single mother cell. Replicative viability is calculated as the mean number of daughters produced from mothers of a particular strain background before senescence. (**b**) Chronological life span as measured by the length of time cells in a stationary culture can remain viable. Viability is calculated by the fraction of the culture able to reenter the cell cycle after an extended state of quiescence. Figure modified from *(16)*.

on the longevity of both dividing and nondividing cell types is a major advantage of yeast as a model for studies of aging.

Several dozens of genetic and environmental factors have been identified which influence longevity in yeast, at least some of which play a similar role in multicellular eukaryotes *(4)*. For example, dietary restriction, which has long been known to increase life span in multicellular eukaryotes, such as worms, flies, and rodents, also increases both replicative life span *(5–7)* and chronological life span *(8, 9)* in yeast. Similarly, decreased TOR signaling (a response to dietary restriction) is also known to increase both chronological *(10)* and replicative life span *(11)* in yeast, as well as life span in worms and flies. Additional genes that play an apparently conserved role in modulating longevity in yeast and other organisms include *SIR2(12)*, *SCH9(13–15)*, and protein kinase A *(7, 15)*. The fact that multiple interventions promoting longevity in yeast also promote longevity in multicellular organisms strongly suggests that yeast can be a useful model for at least some aspects of aging in higher eukaryotes *(16)*.

Although studies of aging in yeast have been extremely informative regarding genetic factors that influence longevity, these studies have been limited by the relatively tedious and time consuming nature of the traditional assays for measuring replicative and chronological life span. Higher-throughput approaches have recently been developed for life span determination, however, and are being applied toward genomic screens for increased longevity. An ongoing screen of the yeast ORF deletion collection utilizes an iterative method for identifying single-gene deletion mutants that have increased replicative life span *(11)*. More recently, a qualitative high-throughput method was described for assaying chronological life span in 96-well plates *(10)*. Both of these methods have been described in detail elsewhere *(1, 4, 10, 11, 17)*, and we refer the interested reader to these references.

This chapter will describe a newly developed approach for measuring chronological life span that greatly improves upon the quantitative power of prior methods, while still providing relatively high-throughput capacity. This method utilizes the kinetics of cell growth exhibited by a subset of the aging culture to monitor viability of the population over time (e.g., chronological aging in stationary phase). The validity of such an approach relies on the assumption that, given fixed environmental growth conditions, doubling time, and initial growth state, the relative OD of two (or more) different yeast cultures at any point during subsequent outgrowth will be determined by the relative number of viable cells present at the initial time point. The fraction of viable cells in a particular sample relative to the control sample can be determined based on the corresponding growth curves by the formula:

$$v_n = \frac{1}{2^{(\Delta t_n / \delta)}},$$

where v is the viability in sample n relative to control, Δt_n equals the time shift between the outgrowth curves of sample n and the control sample (**Fig. 2**), and δ equals the doubling time of the

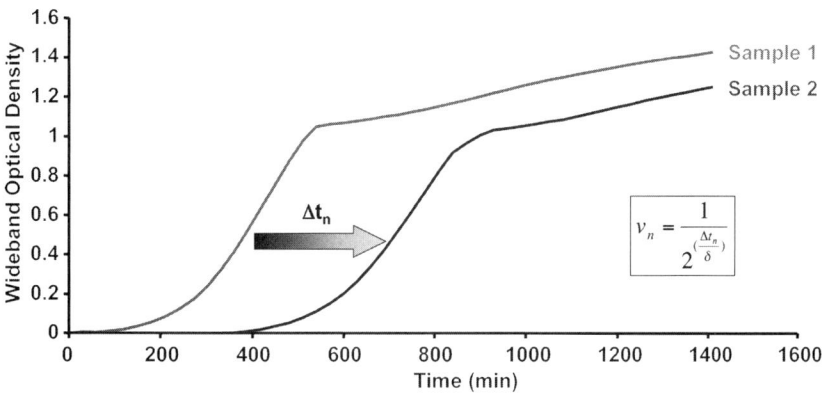

Fig. 2. Determining relative viability from outgrowth curves. The relative number of viable cells in two samples can be determined from the kinetics of outgrowth based on the given formula, where v_n is the relative viability of the two samples, δ is the doubling time of the strain (assumes samples A and B have identical doubling times), and Δt_n is the time shift between the two outgrowth curves.

strain (determined by the maximal slope of the semi-log plot of OD as a function of time).

The primary advance associated with this new protocol is the use of a Bioscreen C MBR (Growth Curves, USA) machine to monitor outgrowth based on optical density (OD). The Bioscreen C MBR machine is an automated shaker/incubator/plate reader useful for determining growth kinetics of yeast strains under a variety of conditions *(18, 19)*. This method has been demonstrated to be accurate by comparison with the traditional colony forming unit assay (CFU), with an increase in precision as determined by lower variance in measurement *(20)*. We are currently using the method described in the following sections to quantitatively determine the chronological aging properties for each strain in the yeast ORF deletion collection *(21)*. This protocol could be easily adapted for other studies of yeast aging or studies examining the sensitivity of yeast cells to a variety of chemical or environmental stressors.

2. Materials

1. Preparation of YPD agar plates (500 mL): In a 1-L Erlenmeyer flask combine 5 g yeast extract (BD, Bacto™ Yeast Extract) and 10 g peptone (BD, Bacto™ Peptone) in 240 mL diH$_2$O and add a stir bar. In another 1-L Erlenmeyer flask, combine 10 g agar with 235 mL diH$_2$O. Cover both flasks with aluminum foil, and autoclave for 45 min. Allow the yeast extract/peptone (YEP) solution to cool on a stir plate with gentle agitation, and slowly add the agar solution to the YEP when the temperature of the agar is warm to the touch (~55°C). Add 40 mL of filter sterilized 50% glucose to the medium (2% final concentration), and dispense approximately 25 mL into individual 100-mm Petri dishes.

2. Preparation of liquid YPD medium (500 mL): In a 2-L flask, combine 450 mL of diH2O, 5 g yeast extract (BD, Bacto™ Yeast Extract) and 10 g peptone (BD, Bacto™ Peptone), and autoclave for 45 min with a loose cap. Add 20 mL of 50% filter sterilized glucose (2% final concentration), and allow to completely cool to room temperature before use.

3. Preparation of liquid Synthetic Defined (SD) medium: Basic medium (B) is composed of 1.7 g/L Yeast Nitrogen Base -AA/-AS (BD, Difco™ Yeast Nitrogen Base) and 5.0 g/L

Table 1
Preparation of amino acid powder mix and 10× stock for SD medium

Component	Powder mix (g)	10× Stock (g/L)
Adenine	2.5	0.4
L-arginine	1.2	0.2
L-aspartate	6	1
L-glutamate	6	1
L-histidine	6	0.2
L-leucine	18	0.6
L-lysine	1.8	0.3
L-methionine	1.2	0.2
L-phenylalanine	3	0.5
L-serine	22.5	3.75
L-threonine	12	2
L-tryptophan	2.4	0.4
L-tyrosine	1.8	0.3
L-valine	9	1.5
Uracil	6	0.2

Add the gram quantities shown for each component and mix well. Make 10× stock by dissolving 16.55 g of powder mix per liter of Basic medium

$(NH_4)_2SO_4$. The medium is then autoclaved and can be stored at room temperature. The amino acid powder mix is made by combining the individual amino acids (Sigma, St. Louis, MO) in gram amounts listed in the first column of **Table 1**. A 10× amino acid stock is made by adding 4.14 g of the powder mix to a final volume of 250 mL B, and filter sterilized. The 10× amino acid stock should be stored at 4°C and kept from exposure to light. To make 500 mL of liquid SD medium, add 50 mL of 10× amino acid stock and 20 mL of filter sterilized 50% glucose to 430 mL of B. The final concentration of each component in the SD medium is provided in **Table 2**.

Table 2
Final composition of synthetic defined medium used for yeast chronological life span analysis of the deletion collection

Carbon source		Base pre-mix	
Glucose	20 g/L	Potassium phosphate	1 g/L
		Magnesium sulfate	500 mg/L
Amino acids		Sodium chloride	100 mg/L
L-arginine	20 mg/L	Calcium chloride	100 mg/L
L-aspartate	100 mg/L	Biotin	0.002 mg/L
L-glutamate	100 mg/L	Pantothenate	0.4 mg/L
L-histidine	20 mg/L	Folate	0.002 mg/L
L-leucine	60 mg/L	Inositol	2 mg/L
L-lysine	30 mg/L	Niacin	0.4 mg/L
L-methionine	20 mg/L	PABA	0.2 mg/L
L-phenylalanine	50 mg/L	Pyridoxine, HCl	0.4 mg/L
L-serine	375 mg/L	Riboflavin	0.2 mg/L
L-threonine	200 mg/L	Thiamine, HCl	0.4 mg/L
L-tryptophan	40 mg/L	Riboflavin	0.2 mg/L
L-tyrosine	30 mg/L	Thiamine, HCl	0.4 mg/L
L-valine	150 mg/L	Boric acid	0.5 mg/L
		Copper sulfate	0.04 mg/L
Other components		Potassium iodide	0.1 mg/L
Adenine	40 mg/L	Ferric chloride	0.2 mg/L
Uracil	20 mg/L	Manganese sulfate	0.4 mg/L
Ammonium sulfate	5 g/L	Sodium molybdate	0.2 mg/L
		Zinc sulfate	0.4 mg/L

3. Methods

3.1. Preparation of Strains for Life Span Analysis

The Bioscreen C MBR machine can accommodate two 100-well Honeycomb plates simultaneously (**Fig. 3**). In our experience, we have found that chronological life span determination of triplicate

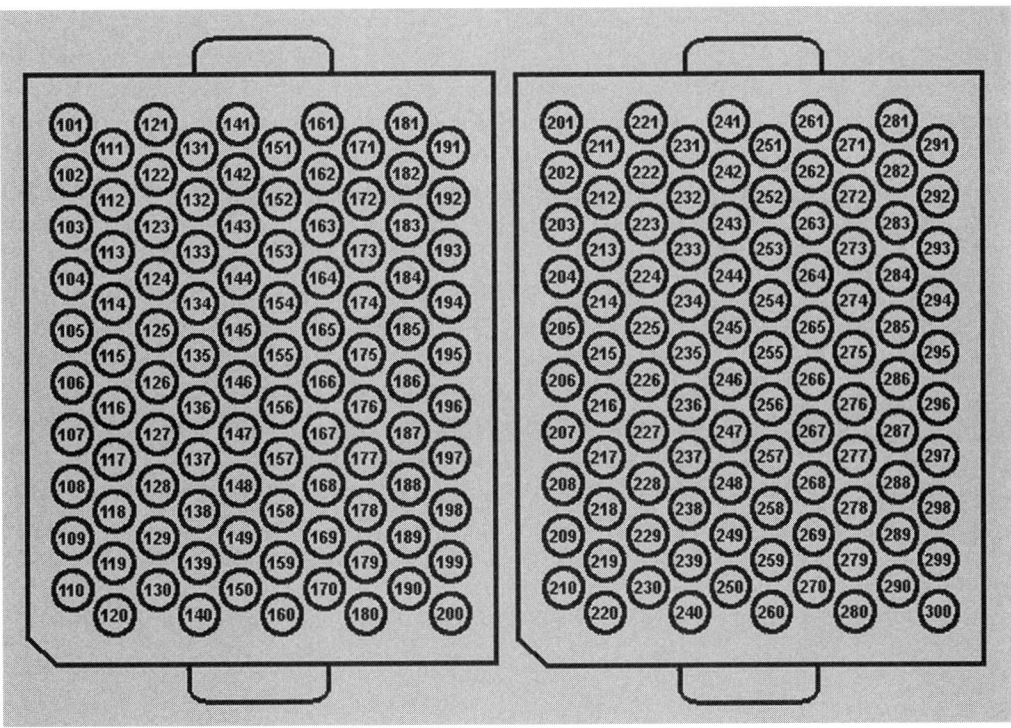

Fig. 3. Honeycomb plates. The Bioscreen C MBR machine uses 100-well Honeycomb plates designed to provide uniform heating in each well. Two plates can be accommodated per run, allowing for a maximum throughput of 200 assays per day per machine.

biological replicates is sufficient to discriminate between relatively small changes in survival. Therefore, we determine the chronological life span of 64 deletion strains per experiment, allowing for triplicate wild type controls to be included in each Honeycomb plate. Based on differential survival in 96-well plates versus culture tubes, we have chosen to age the strains in culture tubes on a rotating drum. Alternative culture conditions could easily be adapted.

1. Cells from 64 deletion strains are obtained from the Yeast ORF Deletion Collection and patched onto a YPD plate. The patch is streaked for single colonies, and allowed to grow at 30°C for 48 h.
2. A single colony from each strain is inoculated into a culture tube containing 5 mL liquid YPD. The liquid cultures are grown at 30°C overnight in a roller drum.
3. To generate the aging cultures for the Bioscreen experiments, 50 µL of the YPD overnight culture is inoculated into 5 mL of SD medium in triplicate (1:100 dilution). Aging cultures are kept in the rotating roller drum at 30°C for the duration of the experiment.

3.2. Obtaining Bioscreen Outgrowth Curves at Each Age Point

To measure the loss in viability over time, an aliquot of the aging culture is periodically challenged for its ability to re-enter the cell cycle in the presence of a rich medium. This is accomplished by inoculating a small volume from the aging culture into fresh medium and monitoring outgrowth using a Bioscreen C MBR machine. The software interface included with the Bioscreen C MBR, called *Easy Bioscreen Experiment*, allows the user to define the temperature of the incubation, the interval at which OD measurements are taken, and the degree and duration of agitation (shaking) of the Bioscreen plate, to keep the culture aerated and the cells in suspension. The design of the Honeycomb plate is optimized to keep heat equally distributed across the plate, so all wells are incubated similarly. The OD readings are given as an output in a tab-delimited format, which can be plotted to visualize outgrowth curves for every aging yeast strain (with a maximum of 200 strains per experiment).

1. In the Easy Bioscreen Experiment window, define the settings for the incubation:
    a) Indicate the number of samples to be measured (200).
    b) In the Filter pulldown menu, select Wideband (420–580 nm).
    c) Set the temperature to 30°C.
    d) Select the Settings menu, and choose Shaking. Set the shaking amplitude to high in the pulldown menu, and check the box below for continuous shaking.
    e) Set measurement interval to 30 min.
    f) Set experiment length to 24 h.
2. Prepare the Bioscreen Honeycomb plates by filling each well with 145 µL liquid YPD (*see* **Note 3**). A standard multichannel pipettor will not align perfectly with the spacing of the wells in the Honeycomb plate, but we find that we can use five tips at a time to load the plate. Alternatively, an expandable multichannel pipettor can be used.
3. Remove aging cultures from the roller drums within the 30°C incubators.
4. Just prior to inoculation, vortex each culture briefly to ensure all cells are evenly suspended. Inoculate 5 µL of each culture into one well of the Honeycomb plate using appropriate sterile technique. Note: For consecutive age-points within the same experiment, maintain a consistent order of inoculation from aging cultures into the Honeycomb plate. In other words, the same aging culture should always be loaded in the same well # at each age-point. This greatly simplifies subsequent data analysis, as described below.
5. After the plate is loaded, place the Honeycomb plate into the Bioscreen, with well #1 in the upper-left corner of the plate

holder. Begin the run by clicking Start in the *Easy Bioscreen Experiment* window. The Bioscreen machine will progress through a series of calibration steps before taking the first absorbance reading (*see* **Notes 1** and **2**).

6. Repeat these steps at day 2, 4, 6, 9, 11, and 13 of the experiment.

3.3. Quantifying Doubling Time and Normalizing Data for Each Age-Point

The Bioscreen C MBR machine gives the output of the OD data as a tab delimited text file that is compatible with Microsoft Excel and a variety of other software. The first column of the output data corresponds to the time at which each OD reading was obtained. Subsequent columns correspond to the OD measurement obtained from each well of the Honeycomb plates at each time point. We use a MATLAB script to analyze the raw data as described below and plot graphs in Microsoft Excel.

1. Obtain the data from the Bioscreen C MBR machine as a tab-delimited text file.
2. Subtract the OD corresponding to YPD without inoculated cells from each measured OD value. The OD of YPD can be obtained from a well containing YPD without inoculated cells (generally ~0.14) and may vary slightly from experiment to experiment.
3. Calculate doubling time (δ) for each column from the maximal slope of the equation defined by the natural logarithm of OD as a function of time. In practice, for each column, we calculate a linear equation of ln(OD) versus time for every two consecutive time points and use the median of the five lowest doubling time values as an estimate of growth rate. Since the length of time between each measurement is constant, the doubling time corresponding to any two consecutive OD measurements can be calculated as,

$$\frac{\ln(2)}{\left(\frac{\ln(OD_2) - \ln(OD_1)}{t_2 - t_1}\right)},$$

where OD_1 and OD_2 are any two consecutive OD readings, and t_2 and t_1 are two consecutive time points ($t_2 - t_1 = 30$ min, in this case). For most strains, the value obtained will be a doubling time of 85–95 min.

4. After doubling time has been calculated for each column, normalize the data for each column by removing the first entry in each column and subtracting the value of the second entry from every entry in the column. This controls for potential differences in initial cell density or optical density variation among the wells.

3.4. Determining Chronological Life Span from Bioscreen Outgrowth Data

After the raw data for each age-point in the experiment has been treated as described in **Subheading 3.4**, a survival plot can be generated for each well. This requires merging the data files from individual age-points in a manner such that each well (corresponding to one biological replicate of one yeast strain) can be analyzed independently over multiple age-points. For simplicity, the following steps assume that the data for a single sample (which will correspond to a single well # if the Honeycomb plate was loaded properly at each age-point) has been organized in a spreadsheet such that the first column contains the time values, the second column contains the initial age-point data (e.g., day 2), the third column contains the second age-point (e.g., day 4), and so forth.

1. Plot the data. This will allow visualization of the shift in the outgrowth curve as a function of age (**Fig. 4a, b**) (*see* **Note 5**).

2. For each age-point, calculate the time shift (Δt) in the outgrowth curve relative to the initial age point. In our experience, there is relatively little variation in Δt for any two outgrowth curves between OD values of 0.1 and 0.5. Calculation of Δt can be accomplished by choosing one or more fixed OD values, estimating the time at which each growth curve reached each OD, and determining the difference in those time values. The time that a particular growth curve achieves a fixed OD value can be calculated from the linear regression equation corresponding to ln(OD) as a function of time between the two time-points bracketing the fixed OD value.

3. From these data a mortality curve can be generated. Define the first age-point as 100% survival. For each subsequent age-point, calculate the percent survival based on the time shift Δt_n, by the formula

$$s_n = \frac{1}{2^{(\Delta t_n / \delta_n)}},$$

where s_n is the fractional survival at age-point n, Δt_n equals the time shift between the growth curve at age-point n and the growth curve at the initial age-point (e.g., day 2), and δ_n equals the doubling time of the strain (**Fig. 4c**) (*see* **Note 6**).

3.5. Statistical Analysis of Changes in Chronological Survival

1. When performing a screen for chronological survival it is desirable to quantify the difference in life span between two strains. There are two general approaches for accomplishing this. One could measure the median and maximum life spans, typically defined as the age at which 50% of the population is no longer viable and the median of the upper decile of survival (~95% senescence), respectively. However, these two measurements can vary independently of each other, and, in our experience, maximum life span measurement can be difficult to determine

A Genomic Approach to Yeast Chronological Aging   111

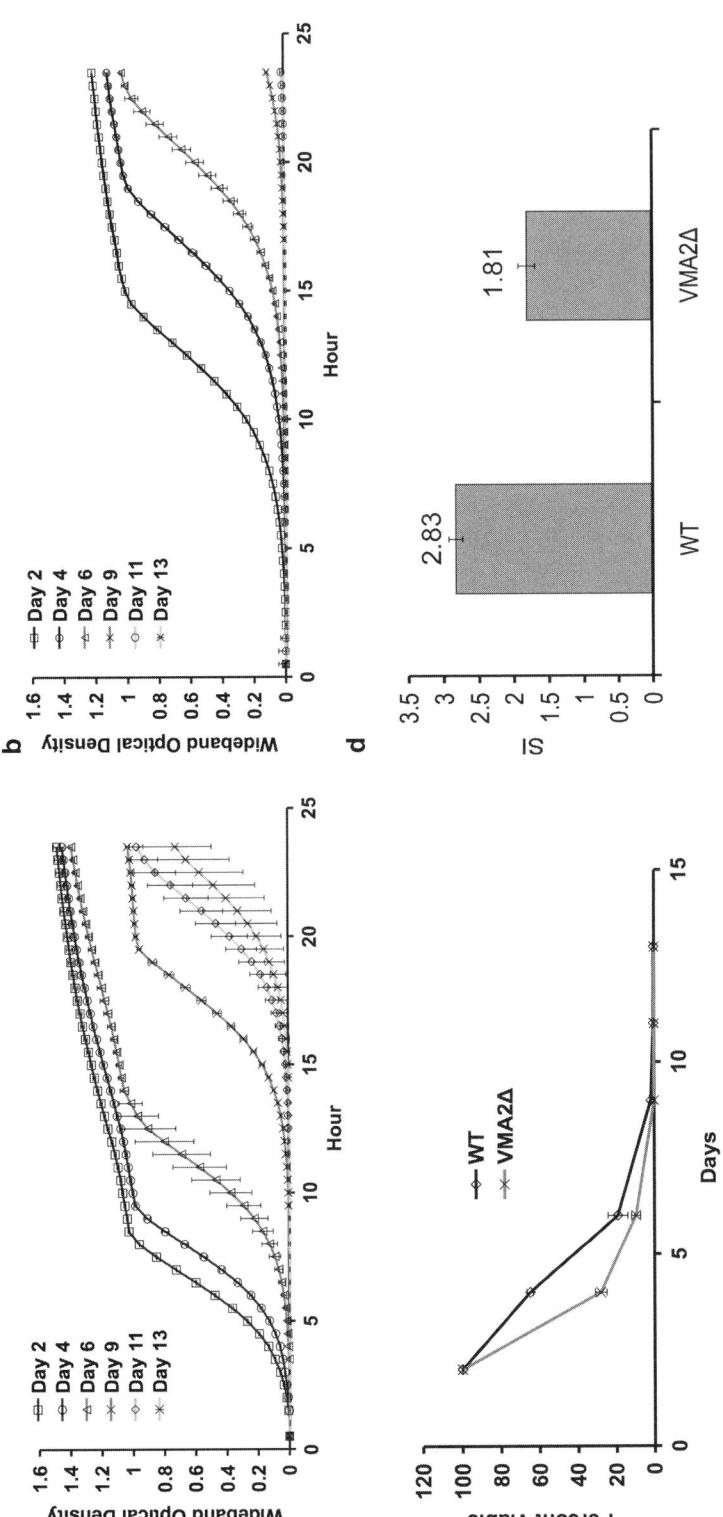

Fig. 4. Representative chronological life span outgrowth and survival curves. Outgrowth curves generated by the Bioscreen C MBR and normalized according to **Subheading 3.4** for the (**a**) WT BY4743 diploid and (**b**) an isogenic homozygous deletion strain from the ORF Deletion Collection (*vma2Δ*). (**c**) Mortality curves for each of the strains were created according to **Subheading 3.5**. Deletion of Vma2, a subunit of vacuolar ATPase, results in a shorter life span compared to the wild type. (**d**) Survival integral for BY4743 and *vma2Δ*. Error bars are standard deviation of three biological replicates.

in some chronological aging experiments. Thus, we utilize an alternative approach to estimate survival in which the integral of the survival curves (survival integral, SI) are compared *(10)*.

2. Once survival data has been generated for each well over the entire course of the experiment, the area under the survival curve (SI) can be estimated by the formula:

$$SI = \sum_{2}^{n} \left(\frac{s_{n-1} + s_n}{2} \right)(age_n - age_{n-1}),$$

where age_n is the age-point (e.g., *refs. 2, 4, 6, 9, 11,* and *13*) and s_n is the survival value at that age-point (*see* **Note 4**). For instance, if viability is 100% on day 2 and 75% on day 4, then the area under the survival curve between those two age-points is 1.75. The sum of this calculation for each pair of consecutive age-points in the experiment is the SI (**Fig. 4d**).

3. Once the SI has been determined for each sample, it is straightforward to calculate statistical values such as mean, median, and variance for each set of triplicate biological replicates. A t-test or similar analysis can be used for pairwise comparison of individual deletion strains against wild type (or other deletion strains), in order to determine whether there is a significant difference in SI.

4. Notes

1. We use YPD medium for outgrowth in the Bioscreen C MBR machine. This is done because growth rate is faster in YPD, allowing for a greater sensitivity to detect shifts in the growth curves. In addition, we have observed that outgrowth curves generated from cells grown in SD are more variable.

2. Occasionally the Bioscreen C MBR machine will display an F1.b2 error message indicating insufficient light from the bulb. This error indicates that the bulb may need to be replaced. It is a good idea to keep several spare bulbs on hand.

3. Although well-to-well variation is very small, over time, a fine plastic dust is produced by the shaking of the Bioscreen Honeycomb well plates. We have found that this particulate matter can obscure the optics, and produce noisy or inaccurate OD readings during the incubation. We therefore recommend cleaning the tray with ethanol and kimwipes regularly to prevent this problem.

4. Periodically during a chronological aging experiment, a subset of the cells will re-enter the cell cycle in the aging culture,

a phenomenon referred to as "gasping" *(22)*. This will be observed in the Bioscreen method as a leftward shift in the outgrowth curve at one or more later age-points, corresponding to an increase in the number of viable cells in the culture over time. Cultures where gasping has occurred should be removed from the analysis, unless viability is already low enough that subsequent age-points will not significantly influence SI.

5. Validation of chronological life span data obtained from the Bioscreen method can be periodically performed by plating for colony forming units from aging cultures and n.onitoring the change in colony formation. To date, we have not observed any significant differences between quantifying viability by colony forming units versus the Bioscreen method.

6. If relative chronological life span is determined by comparing the integral of the survival curves, it is important to maintain a fixed protocol of age-points. For our initial screen of the deletion collection, we monitor survival at day 2, 4, 6, 9, 11, and 13. The age-points used are somewhat arbitrary, but consistency across experiments is essential for fair comparison. These age-points work well for analysis of the ORF deletion collection. In some cases, (e.g., growth in 0.05% glucose), survival is still close to 100% after 14 days, so additional or alternative age-points may be desirable.

Acknowledgments

The development of the method described here was supported by a pilot project grant to MK from the University of Washington Nathan Shock Center for Excellence in the Basic Biology of Aging Grant 5P30 AG013280. CRB is supported by National Institutes of Health Training Grant 5P30 AG013280.

References

1. Kaeberlein, M. (2006). In *Handbook of models for human aging*. (Conn, P. M., Ed.), Elsevier Press, Boston, pp. 109–120.
2. Mortimer, R. K., and Johnston, J. R. (1959). Life span of individual yeast cells. *Nature*. **183**, 1751–1752.
3. Fabrizio, P., and Longo, V. D. (2003). The chronological life span of *Saccharomyces cerevisiae*. *Aging Cell*. **2**, 73–81.
4. Kaeberlein, M. (2006). Genome-wide approaches to understanding human ageing. *Hum Genomics*. **2**, 422–428.
5. Jiang, J. C., Jaruga, E., Repnevskaya, M. V., and Jazwinski, S. M. (2000). An intervention resembling caloric restriction prolongs life span and retards aging in yeast. *Faseb J*. 14, 2135–2137.
6. Kaeberlein, M., Kirkland, K. T., Fields, S., and Kennedy, B. K. (2004). Sir2-independent life span extension by calorie restriction in yeast. *PLoS Biol*. 2, E296.
7. Lin, S. J., Defossez, P. A., and Guarente, L. (2000). Requirement of NAD and SIR2 for life-span extension by calorie restriction in Saccharomyces cerevisiae. *Science*. 289, 2126–2128.

8. Fabrizio, P., Gattazzo, C., Battistella, L., Wei, M., Cheng, C., McGrew, K., and Longo, V. D. (2005). Sir2 blocks extreme life-span extension. *Cell.* **123**, 655–667.

9. Smith Jr, D. L., McClure, J. M., Matecic, M., and Smith, J. S. (2007). Calorie restriction extends the chronological lifespan of *Saccharomyces cerevisiae* independently of the Sirtuins. *Aging Cell.* **6**, 649–62.

10. Powers, R. W., 3rd, Kaeberlein, M., Caldwell, S. D., Kennedy, B. K., and Fields, S. (2006). Extension of chronological life span in yeast by decreased TOR pathway signaling. *Genes Dev.* **20**, 174–184.

11. Kaeberlein, M., Powers, R. W., 3rd, Steffen, K. K., Westman, E. A., Hu, D., Dang, N., Kerr, E. O., Kirkland, K. T., Fields, S., and Kennedy, B. K. (2005). Regulation of yeast replicative life span by TOR and Sch9 in response to nutrients. *Science.* **310**, 1193–1196.

12. Kaeberlein, M., McVey, M., and Guarente, L. (1999). The SIR2/3/4 complex and SIR2 alone promote longevity in *Saccharomyces cerevisiae* by two different mechanisms. *Genes Dev.* **13**, 2570–2580.

13. Fabrizio, P., Liou, L. L., Moy, V. N., Diaspro, A., SelverstoneValentine, J., Gralla, E. B., and Longo, V. D. (2003). SOD2 functions downstream of Sch9 to extend longevity in yeast. *Genetics.* **163**, 35–46.

14. Fabrizio, P., Pletcher, S. D., Minois, N., Vaupel, J. W., and Longo, V. D. (2004). Chronological aging-independent replicative life span regulation by Msn2/Msn4 and Sod2 in *Saccharomyces cerevisiae*. *FEBS Lett.* **557**, 136–142.

15. Fabrizio, P., Pozza, F., Pletcher, S. D., Gendron, C. M., and Longo, V. D. (2001). Regulation of longevity and stress resistance by Sch9 in yeast. *Science.* **292**, 288–290.

16. Kaeberlein, M., Burtner, C. R., and Kennedy, B. K. (2007). Recent developments in yeast aging. *PLoS Genet.* **3**, e84.

17. Kaeberlein, M., and Kennedy, B. K. (2005). Large-scale identification in yeast of conserved ageing genes. *Mech Ageing Dev.* **126**, 17–21.

18. Malathi, K., Higaki, K., Tinkelenberg, A. H., Balderes, D. A., Almanzar-Paramio, D., Wilcox, L. J., Erdeniz, N., Redican, F., Padamsee, M., Liu, Y., Khan, S., Alcantara, F., Carstea, E. D., Morris, J. A., and Sturley, S. L. (2004). Mutagenesis of the putative sterol-sensing domain of yeast Niemann Pick C-related protein reveals a primordial role in subcellular sphingolipid distribution. *J Cell Biol.* **164**, 547–556.

19. Tsuchiya, M., Dang, N., Kerr, E. O., Hu, D., Steffen, K. K., Oakes, J. A., Kennedy, B. K., and Kaeberlein, M. (2006). Sirtuin-independent effects of nicotinamide on lifespan extension from calorie restriction in yeast. *Aging Cell.* **5**, 505–514.

20. Murakami, C. J., Burtner, C. R., Kennedy, B. K., and Kaeberlein, M. (2008). A method for high-throughput quantitative analysis of yeast chronological life span. *J Gerontol A Biol Sci Med Sci.* **63**, 113–21.

21. Winzeler, E. A., Shoemaker, D. D., Astromoff, A., Liang, H., Anderson, K., Andre, B., Bangham, R., Benito, R., Boeke, J. D., Bussey, H., Chu, A. M., Connelly, C., Davis, K., Dietrich, F., Dow, S. W., El Bakkoury, M., Foury, F., Friend, S. H., Gentalen, E., Giaever, G., Hegemann, J. H., Jones, T., Laub, M., Liao, H., Davis, R. W., and et al. (1999). Functional characterization of the *S. cerevisiae* genome by gene deletion and parallel analysis. *Science.* **285**, 901–906.

22. Fabrizio, P., Battistella, L., Vardavas, R., Gattazzo, C., Liou, L. L., Diaspro, A., Dossen, J. W., Gralla, E. B., and Longo, V. D. (2004). Superoxide is a mediator of an altruistic aging program in Saccharomyces cerevisiae. *J Cell Biol.* **166**, 1055–1067.

Chapter 7

Chemogenomic Approaches to Elucidation of Gene Function and Genetic Pathways

Sarah E. Pierce, Ronald W. Davis, Corey Nislow, and Guri Giaever

Summary

The ~6,000 strains in the yeast deletion collection can be studied in a single culture by using a microarray to detect the 20 bp DNA "barcodes" or "tags" contained in each strain. Barcode intensities measured by microarray are compared across time-points or across conditions to analyze the relative fitness of each strain. The development of this pooled fitness assay has greatly facilitated the functional annotation of the yeast genome by making genome-wide gene-deletion studies faster and easier, and has led to the development of high throughput methods for studying drug action in yeast. Pooled screens can be used for identifying gene functions, measuring the functional relatedness of gene pairs to group genes into pathways, identifying drug targets, and determining a drug's mechanism of action. This process involves five main steps: preparing aliquots of pooled cells, pooled growth, isolation of genomic DNA and PCR amplification of the barcodes, array hybridization, and data analysis. In addition to yeast fitness applications, the general method of studying pooled samples with barcode arrays can also be adapted for use with other types of samples, such as mutant collections in other organisms, siRNA vectors, and molecular inversion probes.

Key words: *S. cerevisiae*, Drug-target identification, Functional assays, Chemogenomics, Gene networks, Genomics, DNA barcodes

1. Introduction

The yeast deletion collection is a set of single-gene deletion strains in *S. cerevisiae* that covers almost every open reading frame (ORF) in the genome (96%) *(1)*. The deletion collection was created to provide a pre-constructed, saturated set of gene-deletion mutants that can be used in the place of random-mutant libraries and individually constructed strains. In addition to fulfilling this goal, the collection has also enabled the development of methods for studying all ~6,000 deletion strains in a single culture *(1–3)*. These methods are made possible by the inclusion of unique 20 bp DNA "barcodes" or "tags" in each strain (**Fig. 1**). that enables relative strain concentrations to be compared by hybridizing pooled samples to a microarray that contains the tag compliments (**Fig. 2**). The development of this pooled fitness assay has allowed genome-wide screens to be performed in thousands of conditions, providing a wealth of functional information about the yeast genome *(1–15)*.

Pooled fitness assays have a variety of uses. Gene function can be studied by screening the collection in a condition whose impact on the cell is already known (**Fig. 3a**). For example, if a strain deleted for gene X is sensitive to a DNA damaging agent, it suggests that gene X is involved in DNA damage response *(9,15, 18)*. This is a stronger test of gene function than examining changes in conditional expression because the phenotypic effect of each gene deletion is measured directly. This claim is supported by the observation that genes required for growth in a particular condition are often not overexpressed in that condition, and that a change in expression does not always indicate functional involvement *(1, 4, 15, 16, 19)*.

Pooled assays can also be used to study drug function. The function of a drug can often be determined by identifying the gene deletions that confer increased drug sensitivity or resistance

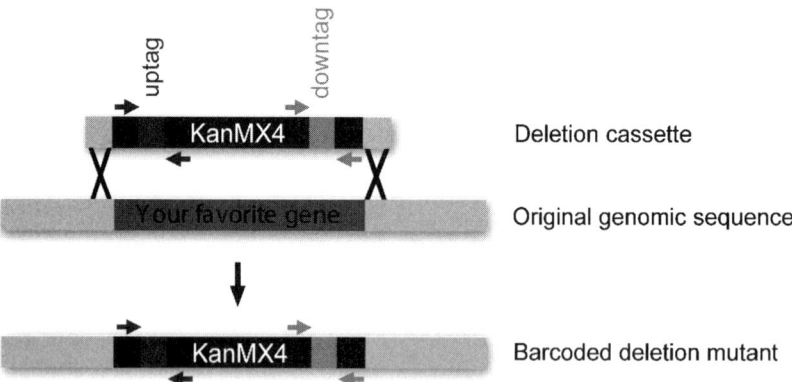

Fig. 1. Deletion cassette used for constructing strains in the yeast deletion collection. Each cassette carries the G418-resistance marker KanMX4, which is required for the selection of transformants. The marker is flanked by two unique "barcode" or "tag" sequences, which are called the "uptag" and the "downtag." These barcodes were designed to be maximally distinct, have uniform hybridization properties, and to carry no homology to the yeast genome (*see* **Note 12**). The two barcode sequences are flanked by four universal primer sites (*arrows*) that are common to all strains, and allow the barcodes to be amplified from a pooled culture. Uptags and downtags are amplified separately to avoid primer crosstalk. Integration into the genome is directed by two 45-bp homology regions which mediate replacement of the targeted gene by mitotic recombination (shown as *black cross marks*). Modified from ref. 38.

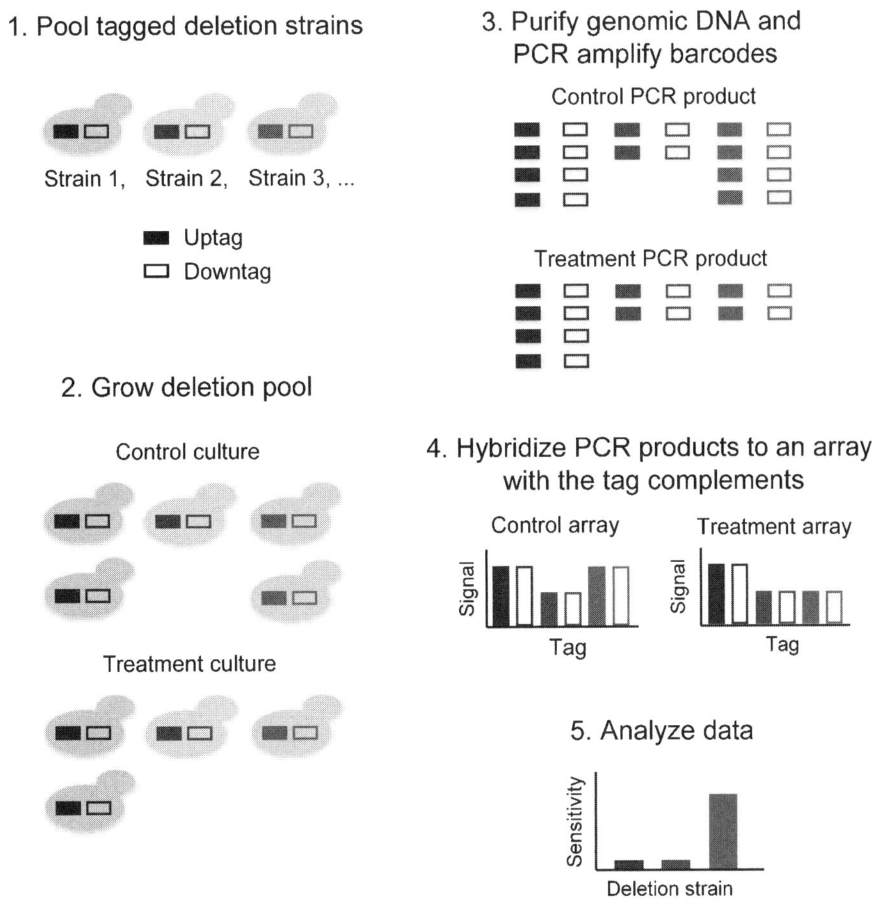

Fig. 2. Overview of the pooled fitness assay. Fitness profiling of pooled deletion strains involves five main steps: (1) Strains are first pooled at approximately equal abundance. (2) The pool is grown competitively in the condition of choice and a control condition. If a gene is sensitive to the treatment condition, the strain carrying this deletion will grow more slowly and become under-represented relative to the control culture (strain 3). Resistant strains will grow faster and become over-represented (not shown). (3) Genomic DNA is isolated from cells harvested at the end of pooled growth, and barcodes are amplified from the genomic DNA with universal primers. (4) PCR products are then hybridized to an array that detects the tag sequences, giving tag intensities for the two samples. (5) The treatment and control sample are then compared to determine the relative fitness of each strain. Note that only strain 3 is called as sensitive to the condition. While strain 2 grows more slowly than strain 1 in the treatment, this growth difference in not of interest because it matches that seen in the control. Modified from ref. 38.

because many genes in the yeast genome have annotated functions, making it likely that any given drug will perturb deletion strains with existing functional information (**Fig. 3b**). For example, if drug X perturbs deletion strains known to be sensitive to DNA damage, it suggests that drug X is a DNA damaging agent *(13)*. Sensitivity data can also be used to identify subtle differences between compounds with similar modes of action, as well as secondary effects that may remain unknown even if the drug's primary function is well characterized *(9, 13, 16)*.

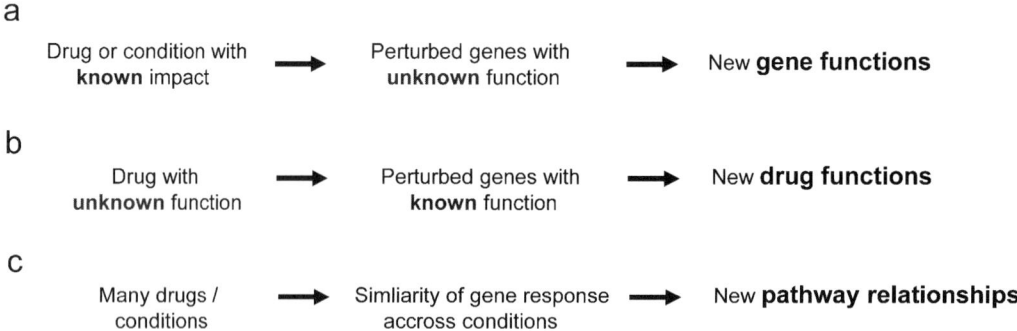

Fig. 3. Applications of homozygous deletion screening. Homozygous deletion screens can be used for studying gene function, drug function, and gene relationships. (**a**) Screening the homozygous deletion collection in a condition with a known impact on the cell provides functional information about the sensitive and resistant genes identified. (**b**) Screening a drug of unknown function will typically perturb some genes with existing functional annotations. These identified genes give information about the drug's mode of action. (**c**) Genes that are functionally related will typically be sensitive in similar conditions. The combined results of many homozygous screens can therefore be used to give information about the functional relatedness of gene pairs. Functional relatedness is estimated by measuring the similarity of a gene-deletion's response across many conditions.

In addition to generating information about single genes and single drugs, large sets of experiments can be used to study the relationships between genes and the relationships between drugs (**Fig. 3c**). For a given gene pair, the correlation of fitness values across conditions can be used to estimate the functional relatedness of the two genes, enabling large functional datasets to identify functional relationships between genes. Similarly, for a given drug pair the correlation of fitness values across genes is a measure of the similarity of their mechanism of action *(15)*.

Pooled analysis of the heterozygous deletion strains can be used to identify novel drug targets (**Fig. 4**) *(6, 7, 10, 17)*. The reasoning behind this technique is as follows: if deletion of the drug target results in reduced growth, then heterozygous deletion of the gene encoding this target will often confer increased sensitivity to the drug. This is because the gene copy number can affect protein abundance *(20)*, and if the heterozygote has a reduced level of the target protein, a lower dose of drug will be required to deplete enough of the protein's activity to mimic the phenotype of the gene's homozygous deletion, which is reduced growth. The heterozygote will therefore require a lower dose of drug than the wild-type strain to show the slow-growth phenotype that result from full inactivation of the target protein. This effect enables all genes whose full deletion confers growth sensitivity to be screened for possible drug targets in a single pooled experiment.

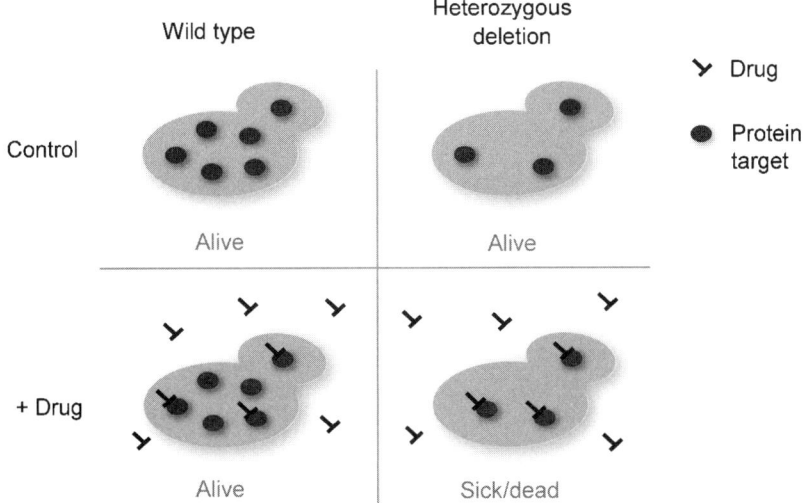

Fig. 4. Heterozygous profiling for drug target identification. Screening the heterozygous deletion collection can identify drug targets for genes whose homozygous deletion causes a growth defect. This method exploits the fact that a gene's copy number is related to its expression level. Here a drug with a specific protein target is shown in *black*, and the corresponding protein target is shown as a *circle*. At an intermediate concentration, the drug will inactivate most target function in the heterozygote, but not in strains with two functional gene copies. This will result in a growth defect for the strain that carries a heterozygous deletion of the gene encoding the drug's target. This phenotype mimics the growth defect caused by the corresponding homozygous deletion. Modified from ref. 38.

In addition to the applications mentioned above, these protocols can also be used to adapt the DNA barcode approach to other types of samples. Barcoded sample-tracking is a robust method that can be applied to a variety sample types. For example, DNA barcodes have already been used to tag mutant collections in other organisms *(21, 22)*, collections of siRNA vectors *(23–30)*, and DNA probes such as molecular inversion probes (MIPs) *(31–34)*.

The pooled fitness assay involves five main steps (**Fig. 2**, for an experimental timeline *see* **Fig. 5**). First, the deletion collection is grown on solid media in an arrayed format and the resulting cells are pooled and frozen to make cell aliquots that will be used for starting growth experiments. Cells are then grown in the desired conditions, typically for 5–20 generations. The barcodes from the resulting cell samples are prepared for hybridization by isolation of the genomic DNA and PCR amplification with the common barcode primers. Amplification is performed in two separate reactions to prevent crosstalk between the uptag and downtag primers. The resulting PCR products are then hybridized to tag microarrays. The barcode intensity data is then analyzed to determine differences in strain representation between pairs of samples.

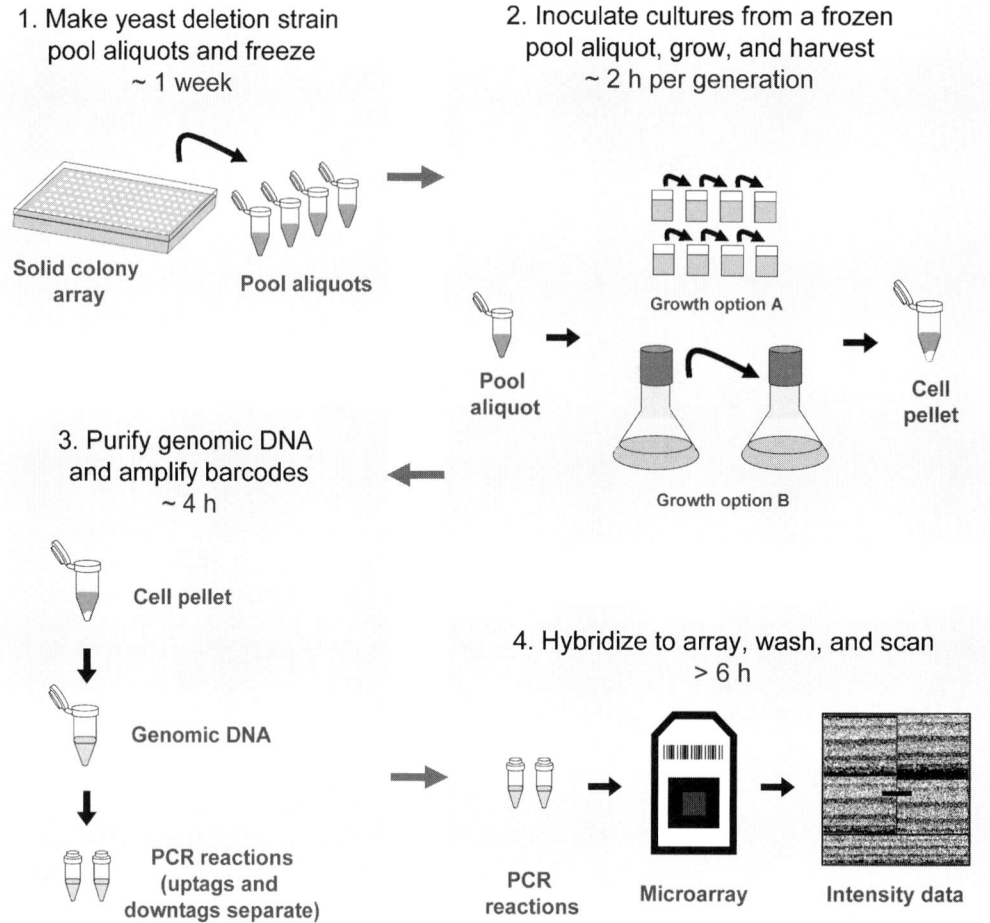

Fig. 5. Timeline for a pooled growth experiment. (1) Pooled cells are generated by transferring a copy of the deletion collection from solid media to liquid and then freezing small aliquots. This step is time consuming, but is performed infrequently because many aliquots can be generated at once. (2) Cultures are inoculated using a frozen cell aliquot, and then grown for the desired number of generations. The specific amount of time needed for growth will vary depending on the number of generations, and on the growth rate of the culture in the treatment condition. The example shown is for 20 generations of growth either as (a) 700-μl cultures in a 48-well plate, shaken and grown in a spectrophotometer, or (b) 50-ml cultures grown in flasks. (3) Genomic DNA is purified from the harvested cells using a standard column-based purification kit, and tags are then PCR amplified from the purified genomic DNA. The uptags and downtags are amplified separately to avoid cross-reactions between the uptag and downtag primer pairs. (4) PCR products from both reactions are hybridized to a single array. After hybridization, the array is washed and scanned. The time needed for this step will vary depending on the hybridization time chosen. Modified from ref. 38.

2. Materials

2.1. Preparing Pooled Cell Aliquots

1. Frozen glycerol stocks of the yeast deletion collection in 96-well microtiter plates (OpenBiosystems, Part Nos. YSC1056 and YSC1055).

2. Nunc Omni Trays (VWR, Catalog No. 62409-600).

3. 96-Well pin tool (V&P Scientific, Catalog No. VP407A).

4. EtOH for flame-sterilizing pin tool (Gold Shield, Catalog No. DSP-CA-151).

5. Incubator set to 30°C for growing yeast plates.

6. Spectrophotometer.

7. 1,000× G418 Stock (200 mg/ml): Dissolve 5 g G418 (Agri-Bio, Catalog No. 3000) in 25 ml dH$_2$O. Filter sterilize using a 0.2-μm filter and a syringe. Shield from light by wrapping bottle in foil. Store at 4°C.

8. YPD + 200 μg/ml G418 rectangular plates: Mix 10 g yeast extract (Becton Dickinson and Co., Catalog No. 212750), 20 g peptone (Becton Dickinson and Co., Catalog No. 211677), 20 g dextrose Becton Dickinson and Co., Catalog No. 215530), 20 g agar (Becton Dickinson and Co., Catalog No. 214010), and 1 l dH$_2$O to a 2-l flask with a stir bar. Autoclave. Allow media to cool to approximately 50°C with gentle stirring. Add 1 ml 1,000× G418 Stock. Stir gently for an additional 1 min to ensure drug is evenly mixed. Pour into Nunc Omni trays, 50 ml per tray. Sufficient for approximately 20 plates. Store at 4°C.

9. YPD liquid + 200 μg/ml G418: Mix 10 g yeast extract (Becton Dickinson and Co., Catalog No. 212750), 20 g peptone (Becton Dickinson and Co., Catalog No. 211677), 20 g dextrose Becton Dickinson and Co., Catalog No. 215530), and 1 l dH$_2$O to a 1-l bottle. Autoclave. Allow media to cool to approximately 50°C. Add 1 ml 1,000× G418 Stock. Store at 4°C.

2.2. Growing Pooled Cultures

1. Temperature controlled shaker for 250-ml flasks or spectrophotometer that allows incubation and shaking of plates, such as the TECAN GENIOS SpectraFLUOR Plus (Tecan, Part No. F129005, Note: many spectrophotometers do not shake plates hard enough for growing yeast).

2. 250-ml culture flasks (if growing cultures in flasks).

3. 48-Well plates (Greiner, Part No. 677102; if growing cultures in plates).

4. Plate roller (Sigma, Catalog No. R1275) for sealing 48-well culture plates (if growing cultures in 48-well plates).

5. Adhesive plate seals (ABgene, Catalog No. AB-0580) for sealing 48-well culture plates (if growing cultures in 48-well plates).

6. Desired media and compounds for growing cultures.

2.3. Purification and Amplification of Barcodes from Cell Samples

1. YeaStar Genomic DNA Kit (Zymo Research, Catalog No. D2002).
2. Taq DNA polymerase (Invitrogen, Catalog No. 10342).
3. dNTPs (Invitrogen, Catalog No. 10297).
4. 10× PCR reaction buffer ($MgCl_2$) (Invitrogen, Part No. 52724).
5. 50 mM $MgCl_2$ (Invitrogen, Part No. 52723).
6. Up primer mix: Dissolve Uptag (5′ – GAT GTC CAC GAG GTC TCT – 3′) and Buptagkanmx4 (5′ biotin – GTC GAC CTG CAG CGT ACG – 3′) each in dH_2O at 100 pmol/μl, then mix in a 1:1 ratio for a final concentration of 50 pmol/μl each. Store at –20°C.
7. Down primer mix: Dissolve Dntag (5′ – CGG TGT CGG TCT CGT AG – 3′) and Bdntagkanmx4 (5′ biotin – GAA AAC GAG CTC GAA TTC ATC G – 3′) each in dH_2O at 100 pmol/μl, then mix in a 1:1 ratio for a final concentration of 50 pmol/μl each. Store at –20°C.
8. Thermocycler with heated lid.

2.4. Array Hybridization and Scanning

1. Genflex Tag 16K Array v2 (Affymetrix, Part No. 511331).
2. Hybridization Oven 640 (Affymetrix, Part No. 800138).
3. GeneChip Fluidic Station 450 (Affymetrix, Part No. 00-0079).
4. GeneArray Scanner 3000 (Affymetrix, Part No. 00-0212).
5. 0.5-ml microfuge tubes suitable for boiling (Eppendorf 0.5-ml Safe-Lock microcentrifuge tubes, Sigma, Catalog No. T8911).
6. Microcentrifuge tube holders (Sigma, Catalog No. Z708372).
7. Ice bucket.
8. Boiling water bath with floating rack for 0.5-ml tubes.
9. Teeny Tough-Spots (Diversified Biotech, Catalog No. TS-TNY; or any other small labels suitable for preventing evaporation from Affymetrix array gaskets).
10. Denhardt's Solution, 50× Concentrate (e.g., Sigma, Catalog No. D-2532).
11. Streptavidin, R-phycoerythrin conjugate (SAPE) – 1 mg/ml (Invitrogen, Catalog No. S-866). Store at 4°C. <u>Do not freeze</u>.
12. B213 oligonucleotide: Dissolve B213 control (5′ biotin – CTGAACGGTAGCATCTTGAC – 3′) in dH_2O at 100 pm/μl to make a concentrated stock, and then use the stock to make 1 fmol/μl working solution. Store at –20°C.

13. Mixed oligonucleotides: Dissolve each of the following eight oligos in dH$_2$O at 100 pmol/μl: Uptag (5′ – GAT GTC CAC GAG GTC TCT – 3′), Dntag (5′ – CGG TGT CGG TCT CGT AG – 3′), Uptagkanmx (5′ – GTC GAC CTG CAG CGT ACG – 3′), Dntagkanmx (5′ – GAA AAC GAG CTC GAA TTC ATC G – 3'), Uptagcomp (5′ – CTA CAG GTG CTC CAG AGA – 3′), Dntagcomp (5′ – GCC ACA GCC AGA GCA TC – 3′), Upkancomp (5′ – CAG CTG GAC GTC GCA TGC – 3′), Dnkancomp (5′ – CTT TTG CTC GAG CTT AAG TAG C – 3′). Mix an equal volume of each of the eight oligos for a final concentration of 12.5 pmol/μl each. Note that the Uptag oligo is also used in Up primer mix and the Dntag oligo is also used in Down primer mix, so take care to leave enough for use in both mixes. Store at –20°C. 5 M NaCl Dissolve 58.4 g NaCl in 200 ml dH$_2$O. Store at room temperature (18–25°C).

14. 0.5 M EDTA: Dissolve 7.3 g EDTA (BioRad, Catalog No. 161-0729) in 50 ml dH$_2$O. Store at room temperature.

15. 10% Tween: Mix 2 ml Tween 20 (Sigma, Catalog No. T2700) with 18 ml dH$_2$O. Store at 4°C.

16. 1% Tween: Mix 10 μl Tween 20 (Sigma, Catalog No. T2700) with 990 μl dH$_2$O. Store at 4°C.

17. 12× MES stock: Mix 0.70 g MES free acid monohydrate (Sigma, Catalog No. M5287), 1.9 g MES Sodium Salt (Sigma, Catalog No. M5057), 8 ml of H$_2$O and adjust volume to 10 ml. Adjust the pH to be between 6.5 and 6.7 using a pH meter or pH paper. Shield from light by wrapping bottle in foil. Store at 4°C. Replace if solution becomes visibly yellow or after 1 month.

18. 2× Hybridization buffer: Mix 8.3 ml of 12× MES Stock, 17.7 ml of 5 M NaCl (J.T. Baker, Catalog No. 3624-01), 4.0 ml of 0.5 M EDTA, 0.1 ml of 10% Tween 20 (vol/vol), 19.9 ml filtered dH$_2$O. Filter to remove dust particles that would interfere with scanning. Shield from light by wrapping bottle in foil. Store at 4°C. Replace as for MES stock.

19. Wash A: Mix 300 ml 20× SSPE (Sigma, Catalog No. S2015), 1 ml 10% Tween (vol/vol), 699 ml filtered dH$_2$O. Filter to remove dust particles that would interfere with scanning. Store at room temperature.

20. Wash B: Mix 150 ml 20× SSPE (Sigma, Catalog No. S2015), 1 ml 10% Tween (vol/vol), 849 ml dH$_2$O. Filter to remove dust particles that would interfere with scanning. Store at room temperature.

3. Methods

3.1. Yeast Deletion Strain Pool Construction

1. Allow at least 1 week to generate pooled aliquots of cells before beginning pooled growth experiments. Pooling is performed infrequently because pooled cells can be stored indefinitely at −80°C. *See* **Note 1** for information about strain background options.

2. Allow the frozen glycerol stocks for the strains of interest to thaw completely because cells may have settled prior to being frozen. Remove plates from the freezer in manageable numbers to avoid leaving all the plates at room temperature for an extended period.

3. To sterilize a 96-well pin tool, dip pin tool in water to rinse away any remaining cells, followed by two dips in 70% ethanol baths, then carefully flame the pin tool. Allow pin tool to cool for 1 min. Make certain that the level of the ethanol bath exceeds the level in the water bath to ensure all carry-over cells are flamed and removed. Change water frequently.

4. Insert the sterile 96-well pin tool into a thawed 96-well plate, swirl gently to capture settled cells and then transfer to a Nunc Omni Tray containing 50 ml of YPD-agar including 200 µg/ml G418. Allow pin to dwell on agar for 5–10 s. Repeat for the remaining plates, sterilizing the pin tool between transfers.

5. Grow colonies until they reach maximal size at 30°C *(2–3 days)*.

6. After colonies have reached full size, make note of any strains that are missing or appear as slow-growing colonies. Streak these strains out individually using standard yeast procedures to generate more cells.

7. Scrape the entire contents of all plates (in a laminar flow hood to avoid contamination) into a 50-ml conical centrifuge tube containing YPD liquid media + 200 µg/ml G418. Alternately, cells can be soaked off the plates by adding 12 ml YPD liquid media + 200 µg/ml G418 to each plate, soaking for 5 min, and gently agitating the plate to resuspend the cells. After cells are resuspended, pool liquid in a flask by transferring a fixed volume from each plate with a pipette.

8. For the slow-growing strains observed in **step 5**, add approximately three colony equivalents of cells using a sterile flat toothpick. This will help to ensure that these slow growing strains remain detectable by the end of a typical growth experiment.

9. Add glycerol to 15% or DMSO to 7% (vol/vol).

10. Measure the OD_{600} of the pool and adjust to a final concentration of 50 OD_{600}/ml (OD_{600} 1.0–2.2 × 10^7 cells/ml for diploid strains) with media containing 15% glycerol or 7% DMSO. If lower than 50 OD_{600}/ml, centrifuge the pool at 500 × g for 2 min. to pellet the cells, decant the supernatant, and resuspend to 50 OD_{600}/ml.

11. Aliquot 10–25 μl of pool into individually capped PCR tubes, and store at −80°C. Pooled aliquots can be stored indefinitely.

12. If desired, perform a hybridization test and create a pool supplement to boost the signal for strains that do not give good signal when hybridized (*see* **Note 2**).

3.2. Pooled Growth

1. Before beginning, *see* **Subheading 4** for information about growth equipment options (*see* **Note 3**), using a pre-screen growth step to allow the strains to recover from freezing (*see* **Note 4**), choosing the number of generations of growth (*see* **Note 5**), selecting the correct dose for drug conditions (*see* **Note 6**), choosing a sufficiently large culture volume and starting OD (*see* **Note 7**), selecting the correct control condition (*see* **Note 8**), and replicating samples (*see* **Note 9**).

2. Thaw a frozen aliquot of pooled cells on ice.

3. Immediately dilute the pool into media containing conditions of interest, including a control condition. For drug experiments, if the drug added is dissolved in a solvent, an equal amount of solvent should be added to the control culture. Choose culture volumes, starting OD, and final OD using the guidelines in **Note 7**, or use one of the two options given below.

    (a) *Option A:* Use a starting OD_{600} of 0.0625 in two 700-μl wells of a 48-well plate (*see* **Note 10**). Seal the plate with a plastic plate seal, rolling with a plate roller. If the condition requires aerobic growth (for example, non-fermentable carbon sources) poke a small hole in the membrane seal towards the side of each well to allow better aeration of the culture. Grow in a spectrophotometer at 30°C with shaking set to the highest setting. Cells can be grown in a single culture for five generations.

    (b) *Option B:* Use a starting OD_{600} of 0.0020 in a 50-ml culture in a 250-ml culture flask. Grow in a shaking incubator at 30°C and 250 rpm. Cells can be grown in a single culture for ten generations.

4. Grow cells for the desired number of generations, or until the OD_{600} = 2. If cells have reached the desired number of generations, proceed to the next step. Otherwise transfer the cells to a fresh culture as in the previous step.

5. Collect 1–2 OD_{600} of cells in a 1.5-ml tube. Spin the tubes at 500 g for 2 min in a microcentrifuge to pellet the cells, and remove the media.

6. If a starting cell sample (i.e., a "T0 time point") is needed, add 1–2 OD_{600} of pool from the freezer to a 1.5-ml tube. Spin the tube at max for 2 min in a microcentrifuge and remove the media.

7. If not proceeding to the next step immediately, store cell pellets at −20°C in a non-frost free freezer.

3.3. Purification and Amplification of Barcodes from Cell Samples

1. Purify genomic DNA from the cell pellets with the Zymo Research YeaStar kit using Protocol I (included with the kit), or another suitable method for purifying yeast genomic DNA. Elute the DNA with 300 μl 0.1× TE. Genomic DNA can be stored indefinitely at −20°C.

2. If desired, quantitate genomic DNA using a gel or a UV spectrophotometer. In practice, using a consistent volume of genomic DNA (we use 15 μl which contains ~0.1 μg) from the previous step without adjusting for variations in DNA concentration gives good results.

3. Set up two 60 μl PCR reactions for each sample, one for the uptags and one for the downtags (33 μl dH_2O, 6 μl 10× PCR buffer without $MgCl_2$, 3 μl 50 mM $MgCl_2$, 1.2 μl 10 mM dNTPs, 1.2 μl 50 μM Up or Down primer mix, 0.6 μl 5 U/μl Taq polymerase, ~0.1 μg genomic DNA in 15 μl).

4. Cycle as follows in a thermocycler with a heated lid: 94°C 3 min; repeat 30×: 94°C 30 s, 55°C 30 s, 72°C 30 s; 72°C 3 min, hold at 4°C. PCR products can be stored at −20°C.

5. If desired, check the resulting PCR products on a gel (*see* **Note 11**).

6. If not proceeding to the next step immediately, store PCR products at −20°C in a non-frost free freezer.

3.4. Array Hybridization and Scanning

1. These protocols use Affymetrix TAG4 arrays. *See* **Note 13** for more array platform options.

2. Set up a boiling water bath with a floating rack for 0.5-ml tubes and a slushy ice bucket. Set hybridization oven temperature to 42°C.

3. Fill the arrays with 90 μl 1× hybridization buffer.

4. Incubate at 42°C and 20 rpm for at least 10 min in the hybridization oven to pre-wet the array.

5. Prepare 60 μl hybridization mix per sample (45 μl 2× hybridization buffer, 0.3 μl B213 control oligonucleotide (1 fm/μl) (*see* **Note 14**), 7.2 μl of mixed oligonucleotides (12.5 pm/μl) (*see* **Note 15**), 1.8 μl 50× Denhardt's Solution) immediately before using and aliquot into 0.5-ml tubes suitable for boiling.

6. While arrays are equilibrating, add 20 μl of uptag PCR and 20 μl of downtag PCR to 60 μl of hybridization mix for a total volume of 100 μl.

7. Clip a tube-holder onto the lid of each tube to prevent the tubes from opening during boiling.

8. Boil each tube for 2 min and then set on ice for 2 min.

9. Spin tubes briefly to bring the samples to the bottom of the tubes.

10. Remove hybridization buffer from the arrays.

11. Add the samples to the arrays (90 μl per array).

12. Place a Tough-Spot over each of the two gaskets to prevent evaporation.

13. Hybridize for 3–16 h at 42°C and 20 rpm. It is best to keep the hybridization time constant for samples that are part of the same dataset.

14. Prime the fluidics station: put the tubings in place, empty the waste bottle, fill wash A, wash B, and water bottles, and run fluidics station protocol PRIME_450.

15. Prepare 600 μl biotin labeling mix per sample (180 μl 20× SSPE, 12 μl 50× Denhardt's Solution, 6 μl 1% Tween 20 (vol/vol), 1 μl 1 mg/ml streptavidin–phycoerythrin, 401 μl dH_2O). Prepare this mix immediately before using.

16. Aliquot 600 μl biotin labeling mix per chip into tubes.

17. Remove Tough-Spots from chips.

18. Remove hybridization mix and fill chips with 90 μl wash A.

19. Wrap chips that are waiting to be washed in aluminum foil.

20. Under Experiments: for each chip type in the sample name and the experiment name, enter the barcode.

21. Under Fluidics: enter the station, and which chip you will wash and label in which module using the "experiment" pull-down menu. Use the protocol "Genflex_TAG4_wash_protocol."

22. When the chips are ready for hybridization, check for air bubbles and wash again if necessary before you engage the wash block.

23. Clean the glass window on each array with isopropanol and a cotton swab or lint-free wipe.

24. Put Tough-Spots on the gaskets to prevent evaporation and put them in the scanner.

25. Scan at an emission wavelength of 560 nm.

26. When all fluidics operations are complete, put all cablings in Millipore water and run SHUTDOWN_450.

3.5. Analysis of Microarray Results

3.5.1. Outlier Masking

The Affymetrix TAG4 array contains at least five replicate features for each tag probe, dispersed across the array to make their performance as independent as possible *(35)*. This allows outlier features to be identified and discarded before calculating an average intensity value for each tag (*see* **Notes 12 and 16**).

1. For each feature on the array, examine the features within the surrounding 5 feature × 5 feature region. If at least 13 of the 25 probes in this region differ from their trimmed replicate mean (the mean of the three middle replicates, excluding the highest and lowest replicates) by more than 10%, consider this probe part of an outlier-dense region that is not suitable for data analysis.

2. Once these outlier-dense regions have been identified, pad them by also including any probes that are within a 5-probe radius, as defined by $((x_1 - x_2)^2 + (y_1 - y_2)^2)^{1/2} < 6$ where x_1, x_2, y_1, and y_2 are the x and y coordinates for the two features.

3. Also discard features for which (standard deviation of feature pixels/mean feature pixels) > 0.3. This unevenness is typically caused by bright pieces of debris on the chip that affect only a single probe, and are therefore not padded to include neighboring features. The standard deviation is included in the .cel file for Affymetrix arrays.

4. After identification and removal of outliers, intensity values are calculated for each tag by averaging all unmasked replicates.

3.5.2. Saturation Correction

Signal on the TAG4 array is not linearly related to tag concentration because of feature saturation *(35)*. If left uncorrected, this saturation will cause the degree of sensitivity or resistance to be underestimated for strains with brighter tags. Saturation is corrected by comparing the uptag and downtag ratios (*see* **Note 17**).

1. Using a pair of arrays that are not sample replicates, calculate $\ln(i_c - \mathrm{bg})/(i_t - \mathrm{bg})$ for each tag, where i_c is the control intensity, i_t is the treatment intensity, and bg is the background as estimated by taking the mean intensity of the unassigned tag probes.

2. Mark the ratios for any tags with minimum values less than 3× background as unusable (this mark is for saturation correction only; a more robust threshold for unusable tags will be re-calculated later for use in the final results).

3. Pair uptag and downtag ratios by strain. Ignore ratios for any strains with less than two usable tags.

4. For each strain, calculate the difference in average intensity for the two tags: $(i_{tu} + i_{cu})/2 - (i_{td} + i_{cd})/2$, where the subscript u indicates the uptag, and the subscript d indicates the downtag.

5. Sort ratios by the difference in average intensity as calculated above. Take a sliding window of 600 ratio pairs, sliding 100 pairs at a time. For each window fit a line to the uptag ratios (*x*-axis) versus the downtag ratios (*y*-axis) using least-squares and take the slope. Also calculate the mean of the differences in the average intensity for the window.

6. Fit a least-squares line to the intensity differences for all windows (*x*-axis), versus slopes for all windows (*y*-axis) and take the slope of this line.

7. Repeat **steps 4–6** using the reverse intensity difference: $(i_{td} + i_{cd})/2 - (i_{tu} + i_{cu})/2$, and taking the slope with the axes reversed: the downtag ratios on the *x*-axis, and the uptag ratios on the *y*-axis.

8. Average the slope calculated in **step 6** with the slope calculated in the repeat of **step 6**. This is the saturation correction factor S. A typical range for S is 0.0001–0.0005.

9. Adjust the raw intensity data using the following transformation: $f(i) = ie^{is}$.

10. To correct more than two arrays, calculate S for all possible pairs of arrays, then use the median of these values as the correction factor for all arrays in the set. Using a larger group of arrays will improve the accuracy of S.

3.5.3. Array Normalization

The uptags and downtags should be normalized separately, because they are amplified in separate PCR reactions, and the intensities of the individual PCR reactions will affect the intensities of their array. Normalize half of the unassigned probes with the uptags, and half with the downtags. These probes are normalized and saturation corrected with the other probes to keep their values comparable. Uptag background and downtag background are calculated separately, as normalization may have a different impact on the background level for each tag set. Normalize array data using either quantile normalization (option A) *(36)* or mean normalization (option B) (*see* **Notes 18** and **19**).

1. *Option A:* To quantile normalize a set of arrays:

2. For each set of tags (up and down), rank values obtained from each array in the order of increasing intensity.

3. For each rank, assign the tag at that rank for each array to the median of all values at that rank.

4. *Option B:* To mean normalize a set of arrays:

5. For each set of tags (up and down), divide by the mean.

6. Multiply each tag set by the mean across all arrays (this is for convenience only; it returns the tag intensities to approximately their original range).

3.5.4. Removing Unusable Tags

Tags with low intensity values in the control samples will give poor-quality results. An intensity value threshold for excluding these tags should be chosen by comparing the correlation of uptag and downtag ratios as a function of tag intensity (*see* **Note 20**).

1. Using any treatment-control pair, calculate log $2(i_c - \text{bg})/(i_t - \text{bg})$ for each tag, where i_c is the control intensity, i_t is the treatment intensity, and bg is the mean intensity of the unassigned tag probes for the appropriate tag set (uptag or downtag).

2. Pair uptag and downtag ratios by strain. Ignore ratios for any strains with only one tag (~200 strains only have an uptag).

3. For each pair, take the minimum intensity for the two tags in the two samples. Sort the ratio pairs by this minimum intensity.

4. Use a sliding window of size 50 on the ranked ratio pairs, starting with the lowest intensity pairs. Calculate the correlation of uptag and downtag ratios for pairs within the window. Also calculate the average of the minimum intensities calculated in the previous step.

5. Slide the window by 25 pairs, and repeat the previous step until all pairs have been traversed.

6. Plot the average minimum intensity versus the uptag–downtag correlation for all windows.

7. Chose an intensity threshold from this plot by eye, or by using a set formula. For example, the intensity value where the correlation first reaches 80% of its maximum level is a good cutoff.

8. Mark any tags that are below this cutoff in either of the samples as unusable for calculating log ratios. Mark any tags below this cutoff in any of the control samples as unusable for calculating p-values. The usability criteria are less strict for p-value calculation than for log 2 ratio calculation because log 2 ratios are a measure of how sensitive each strain is, whereas p-values are only a measure whether the strain is sensitive or not. It is possible for a strain that is above background in the control samples but below background in the treatment sample to be significantly sensitive, but without a usable treatment sample the degree of sensitivity cannot be measured.

3.5.5. Calculating p-Values and False Discovery Rates

This method works best for cases where a large set of common control arrays (more than ten) are used with one or two replicate arrays per experiment (*see* **Note 9**). A validation of this method based on a comparison to the number of false sensitivity calls in matched control–control comparisons has been described in detail *(15)*. If multiple replicates are available for the control and the treatment samples and a large control set is not available, other methods for calculating significance such as those used in the Significance Analysis of Microarrays (SAM) package *(37)* are preferable (*see* **Notes 21–27** for more information about interpreting results).

1. For each usable tag, calculate the mean of the controls (μ_c) and the standard deviation of the controls (σ_c).
2. For each strain, average all usable tags. If a strain has no useable tags, this strain should be excluded from the analysis.
3. For each strain, calculate a z-score: $(\mu_c - t)/\sigma_c$ where t is the treatment intensity for that strain. Strains that are sensitive will have positive scores, while strains that are resistant will have negative scores.
4. Calculate p-values from the z-scores by fitting a t-distribution with $n_c - 1$ degrees of freedom to all scores for the experiment, where n_c is the number of control arrays.
5. Calculate a corresponding false discovery rate (FDR) for each of the p-values as follows. Rank the p-values in ascending order. For the p-value with rank r, the corresponding FDR is $(pns)/r$ where n_s is the total number of strains in the analysis.

3.5.6. Calculating log 2 Ratios

Whereas p-values describe the level of confidence for calling strain sensitivity, the log 2 ratios give the best estimate of the actual level of sensitivity for each strain. For example, deletions in the same complex or linear pathway will often cause similar growth defects, and as a result these strains will often have similar log 2 ratios. However, sensitivity for these strains may be measured with different degrees of accuracy, and they will therefore not necessarily have similar p-values (*see* **Notes 21–27** for more information about interpreting results, and **Note 28** for sample ratio results).

1. For each tag, calculate log $2(\mu_c - bg)/(\mu_t - bg)$ where μ_c is the mean intensity for the control samples, μ_t is the mean intensity for the treatment samples, and bg is the mean intensity of the unassigned probes for the appropriate tag set.
2. For each strain, average the log 2 ratios for all usable tags to obtain a final sensitivity score.
3. Strains that are sensitive will have positive scores, while strains that are resistant will have negative scores. This score is proportional to the log 2 ratio of cells present in the control sample versus the treatment sample.

4. Notes

1. *Strain background options:* The yeast deletion collection contains four different sets of strains: homozygous diploids, heterozygous diploids, *MAT*a haploids, and *MAT*α haploids. In general it is preferable to work with the heterozygous and the homozygous collections because the diploid genome is

more robust against phenotypic effects caused by secondary site mutations. These mutations are typically heterozygous in the diploid collections, whereas for the haploid collections a secondary site mutation will affect the only available gene copy.

2. *Adjusting strains with low tag intensities:* To identify strains that are underrepresented in the pool, perform a test hybridization. It is best to grow the pool in the standard control condition before this test hybridization, because slow growing strains often drop out of the pool during this growth step (15% of the homozygous deletion strains and 3% of the heterozygous strains are slow-growing when compared to wild type *(1, 5)*). Once identified, the signal for these strains can be boosted by adding extra cells. Strains that are not detectable can be added in larger quantities to a supplementary pool to boost their signal. It is normal that 200 strains or more remain undetectable even after interventions of this type.

3. *Growth equipment options:* There are many possible ways to grow pooled collections of deletion mutants. We grow our pooled cultures in shaking spectrophotometers that are connected to pipetting robots, allowing the cells to be inoculated into new cultures or harvested robotically when they reach the desired optical density (OD). This greatly facilitates the process and increases reproducibility by making the growth step as consistent as possible, however it is by no means necessary for obtaining good results. Cells can be grown in a shaking spectrophotometer to facilitate OD tracking and then transferred and harvested manually, or cells can be grown in flasks with both OD readings and transfers performed manually. Cells could also be grown as a lawn on solid media, or in alternate growth equipment such as a chemostat.

4. *Using a pre-screen growth step to allow the strains to recover from freezing:* Experimental cultures can either be inoculated directly from frozen aliquots or from a starting culture that has been grown for a set number of generations before starting the experiment. Including a recovery step allows the cells to recover from freezing, and may improve the accuracy of the sensitivity results. The main disadvantage to including this step is that increasing the total growth time causes more slow-growing strains to drop out of the pool by the final time-point, resulting in a greater number of strains with unusable data.

5. *Choosing the number of generations of growth:* In general, a longer growth period can more sensitively detect subtle phenotypes, but results in the depletion of slow-growing strains from the population (in both the experimental and control condition), thus precluding the analysis of these strains. Therefore, the ideal growth period for a pooled culture will

vary. For example, it is generally preferable to grow heterozygous strains longer than homozygous strains because heterozygous growth phenotypes are generally more subtle. For resolution across a wide array of sensitivities, cells can also be collected at multiple time-points.

6. *Selecting the correct dose for drug conditions:* In general, a stronger treatment dose will reveal more sensitive strains by making the inhibition of slightly sensitive strains easier to detect. However, a higher dose will also increase the number of strains that do not grow at all, precluding the ranking of these strains' sensitivities. In general, we aim for treatment doses that cause a 10–20% decrease in growth rate when tested on a wild type. Standardizing the level of inhibition allows most experiments to be sampled after the same number of generations of growth. At 10–20% inhibition, optimal results for the heterozygous collection are usually obtained at the 20 generation time point, and optimal results for the homozygous collection are usually obtained at the five generation time point. In general, a more benign treatment will cause smaller changes in growth rate, and will therefore require a longer growth period to resolve differences between strains.

7. *Choosing a sufficiently large culture volume and starting OD:* Choosing an appropriate culture volume is critical for obtaining good results. At each inoculation step, the cells are diluted to their lowest numbers. Decreasing the average number of cells per strain at inoculation increases the effect of variation on the actual number of cells sampled for each strain. This variation is typically the most significant source of noise in the assay *(38)*. Using at least 300 cells per strain is recommended. The number of cells per strain at inoculation can be calculated as follows: Cells per strain = (Inoculation OD) × (culture volume (ml)) × (cells/ml at OD = 1)/(number of strains). A culture of diploid cells at OD_{600} 1.0 has approximately 2.2×10^7 cells/ml. If the culture vessel will not accommodate the volume needed to reach the desired cell numbers, multiple cultures can be grown in parallel and pooled at the end of the experiment. Increasing the number of inoculations will also increases noise, and should be minimized. Also keep in mind the specific needs of the growth setup when choosing a culture volume (for example, manual OD readings will require extra culture volume).

8. *Selecting the correct control condition:* Raw values of array intensity for a single sample are very poorly correlated with the abundance in absolute strain. For this reason, each experiment requires at least two samples – a control sample and a treatment sample. The intensity ratio between two samples is a good measure of the abundance ratio of the corresponding strain *(35,*

38). The appropriate control condition will vary depending on the experiment. To measure relative growth rates, different time-points should be compared. To measure the relative sensitivity of each strain to a treatment condition, cells should be grown in a treatment condition and a control condition for the same number of generations. This type of comparison is useful because it takes each strain's normal growth-rate into account, preventing strains that are slow-growing in the absence of treatment, from appearing falsely sensitive.

9. *Sample replicates:* Each experiment requires at least two arrays – one for a control sample, and one for a treatment sample. A single treatment-control pair gives good-quality data, especially for cases where the data set taken as a whole is more important than the accuracy of any single experiment. For example, in cases where the results of many experiments are used to build a gene network, the results will be relatively robust against false positives, and testing more conditions may improve the results more than increasing the number of experimental replicates. For cases where individual gene sensitivity predictions are important, replication may also be unnecessary if array experiments are confirmed with single-strain follow up tests. For more accurate data, two or more replicates of both the control and treatment conditions are recommended to allow for the identification and removal of rare outlier points. Increasing the number of replicates will also allow for more subtle phenotypes to be detected with statistical significance. We often use a large, shared set of control arrays (more than ten) for analyzing many different experimental arrays, each with only one replicate. This control set can be used to calculate the statistical significance of the final results, while minimizing the total number of experimental arrays needed. If using a large, shared control set it is important that the control arrays represent as diverse a set of samples as the treatment arrays (cells grown on different days, tags amplified in different runs, etc.) so that the variation in the control set will accurately represent the non-treatment related variation present in the treatment arrays. When not using a shared set of control arrays, it is best for each control-treatment pair to be processed together (cells grown on the same day, tags amplified in the same run, etc.) to minimize the non-treatment related differences between the two. As discussed in **Note 7**, the growth step is the main source of noise in the assay. For this reason, all replicate samples should be full biological replicates (grown separately, as opposed to separate hybridizations of the same growth sample). It is important to note that the uptag and downtag are not independent measures of strain sensitivity. The two tags will share the noise generated during the growth step, where any sampling error affects both tags simultaneously.

The uptag and downtag therefore can not be treated as independent sample replicates when using statistical analysis methods. All useable tags for each strain should therefore be averaged before using these methods, with only true experimental replicates treated as replicate samples.

10. *Comparing OD600 values between plates and cuvettes:* If growing cells in a shaking spectrophotometer, note that the OD_{600} measured for the plate will differ from the OD_{600} of the same culture measured in a cuvette due to differences in path length. Similarly, OD_{600} readings in a shaking spectrophotometer will also vary with differences in culture volume. All OD_{600} values given in this paper refer to those measured in a 1 ml, 1-cm path length cuvette.

11. *Checking for successful tag amplification:* To check for successful amplification, tagged PCR products can be separated on a gel. The desired product is ~60 bp. A second band is often seen because non-complementary tags can hybridize at their common primer regions, forming a partially single-stranded structure that migrates more slowly than the fully double-stranded tag products. Reactions that appear unusually dim (less than 5× the concentration of a typical reaction) will often give low-quality array data. These amplifications can be repeated using the leftover genomic DNA. It is common to have amplification in no-template control reactions because tag contamination is extremely difficult to prevent. This low-level of contamination will typically not impact results because the concentration of the experimental template will be much greater than the concentration of contamination in the experimental samples, preventing the contamination from having a serious impact on the results.

12. *Barcode sequences:* The deletion of a given gene will use the same two barcodes in all four strain backgrounds. Because of this barcode overlap, the homozygous deletion collection and the heterozygous deletion collection must be pooled and grown separately to ensure that each barcode corresponds to a unique strain within the pool. A full list of sequences is available *(38)*. It is an Affymetrix convention to list probe sequences for their arrays in 3′→5′ orientation because this is the direction of probe synthesis used in their manufacturing process. For this reason, the tag sequences listed in Affymetrix array files are the reverse of those listed elsewhere. This is a common source of confusion, so sequence orientation should be carefully checked when designing new barcoded strains or other samples for use with Affymetrix tag arrays.

13. *Array options:* These protocols use the Affymetrix tag arrays; however they can easily be adapted to the array platform of your choice. For alternative array examples *see* refs. *4, 12, 13, 39, 40.*

14. *Determining the correct B213 concentration:* It is common when diluting a new batch of B213 control oligo to have to test several concentrations to find a concentration that gives good results. The reason for this inconsistent performance from batch to batch is unknown. The B213 oligo binds to the border probes on the array, which are used by the Affymetrix software to align the grid that defines the probe borders. If the control spots are too dim, automated alignment will fail, and if the control probes are too bright, they can interfere with neighboring probes on the array. It is best to aim for an intensity that is bright enough for alignment to work consistently, but within the range of the tag probe intensities. If grid alignment fails because the border probes are too dim, it is possible to align the grid manually and recalculate the probe intensities using the Affymetrix software.

15. *Mixed oligos:* The mixed oligos are added to prevent re-annealing of the tagged product after melting. Non-complementary tags can stick together by their common primer sequences, forming a partially single-stranded product. Oligos complementary to the common primers are added in excess to prevent the formation of this product, making the tag sequences more accessible for hybridization.

16. *Re-hybridizing failed samples:* The PCR volume recommended here provides enough PCR for two hybridizations. If the first hybridization has problems such as an extremely uneven appearance (for example, due to large bubbles during washing or staining) or an unusually large amount of visible debris or damage, the remaining PCR can be hybridized to a new array. The replicate features can correct for minor damage, so in general, only damage that affects more than approximately one fifth of the array surface is a concern.

17. *Saturation correction:* The saturation correction function is derived from the data by exploiting the fact that the raw intensities for a strain's uptag and downtag often differ, despite the fact that each uptag–downtag pair is measuring the same true strain ratio. The saturation pattern can therefore be derived by comparing the difference in raw intensity to the difference in ratio for all uptag–downtag pairs. The data is then corrected to make the ratios independent of raw intensity. This transformation increases the correlation of uptag and downtag ratios. The correction algorithm requires at least two arrays, and works best with a pair of arrays that are not sample replicates. This is because all log ratios for a pair of sample replicates will be approximately zero, and saturation affects ratios close to zero the least. The level of saturation can vary from day to day because of differences in hybridization time, washing, and staining time. For this

reason it is best to derive the correction function for all arrays hybridized on a given day as a group, and avoid using pairs of arrays processed on different days for deriving the correction function.

18. *Normalizing array data:* Two alternate normalization methods are discussed here – mean normalization, and quantile normalization *(36)*. Quantile normalization is often preferred because it can normalize data effectively even when there is a non-linear relationship between the samples, but this method requires the assumption that the overall distribution of tag intensities is the same for all experiments being normalized. Mean normalization, though only capable of linear transformation of the data, does not require this assumption. In general, quantile normalization works best for experiments in which a reasonably small number of strains are affected by the condition tested. For experiments with a large number of sensitive strains, mean normalization often performs better. In particular, experiments with different growth periods (for example, the zero time-point and the 20 generation time-point) or from different starting pools (for example, homozygous pools constructed separately) should not be normalized together using quantile normalization because the tag intensities for these samples will have very different distributions *(38)*.

19. *Normalizing new arrays without renormalizing old data:* A slight variation on the normalization procedures described can help make data analysis more convenient by allowing new data to be normalized without having to re-normalize all existing data. When quantile normalizing a group of arrays, a standard curve is usually derived from the all arrays in the group. If you wish to keep the standard curve constant over time, it can be calculated from a starting set of arrays and then held constant for normalizing future arrays. This works well as long as all arrays normalized are of the same experimental type (same pooled cells, grown for the same number of generations). Similarly, when mean normalizing a set of arrays, the mean intensity across all arrays in the set is used to return the mean-normalized values to their original range. This mean can also be calculated from an initial set of arrays and then held constant.

20. *Removing unusable tags:* When tags become too dim to be usable, their correlation with their tag replicates degrades to zero. By comparing the correlation of uptag and downtag ratios to tag intensity, an effective cutoff can be chosen to remove these tags. In practice, this intensity cutoff remains relatively constant across experiments as long as the experimental setup (in particular the hybridization time and the scanner used) does not change, so once a threshold is chosen, it is not required to be recalculated with every experiment.

21. *Reasons expected hits may be absent:* If a strain for which a sensitive phenotype was expected is not called as significantly sensitive, a good first thing to check is what the maximum achievable score is for that strain based on the control experiments. For example, if a strain is already fairly close to the background in the controls, or if the control data for that strain is noisy, it may not be possible for the experimental sample to be dim enough to give a significant score. To calculate the maximum achievable score for each strain, create a mock experimental chip in which all of the tag values are set to background, which is the minimum possible experimental value for each strain. Calculate scores using this mock-experimental chip and the control set, and apply the same FDR cutoff used for the true experimental results. This will give the maximum score achievable for each strain. If a strain is not called as sensitive in this mock experiment, then the data for that strain can be considered inconclusive. It is also possible that the pooled growth phenotype for the strain of interest differs from its phenotype in another environment. Growth in a shaking spectrophotometer, growth in a flask, and growth on a plate may be slightly different because of differences in these growth environments. For example, a culture in a shaking spectrophotometer typically has a lower oxygen level than a standard flask culture. Pooled growth may also create differences in phenotype. A deletion strain grown in a pooled culture is predominantly surrounded by strains that are wild type at its deleted locus. Under certain conditions these surrounding strains may be able to compensate for the deficiencies caused by the deletion.

22. *Evaluating data quality and replicating sample agreement:* The most effective way to measure the quality of replicate samples is to measure the correlation of the log-transformed, normalized, and saturation-corrected intensity values, termed "processed raw values." This correlation should always be at least 0.98 for replicate samples. Processed raw values are a better indicator of sample quality than raw values because normalization and saturation correction remove artifacts that will artificially decrease replicate correlation, and log transformation adjusts the values so that a given deviation from the diagonal will have a consistent impact on the final log ratios, regardless of tag intensity. Lower correlation may indicate a sampling error problem, which can be corrected with increased culture volume as described in **Note 7**. Uptag and downtag ratios for a given sample will be positively correlated; however it is important to note that low correlation for uptag and downtag ratios does not necessarily indicate poor sample quality. In experimental samples where few strains are sensitive, most of the ratios are centered at zero, causing low correlation. In perfect data, the uptag–downtag log 2 ratio correlation for a control–control pair would be

zero, but sampling error causes slight strain representation differences between replicates, and the two tags correlate in measuring this noise. Replicate samples for which few strains are sensitive will also have low log 2 ratio correlation for the same reason. Because of this property, log 2 ratio correlation is a poor measure of replicate sample agreement.

23. *Comparing results across different treatment conditions:* The same score in two different experiments does not indicate the same level of inhibition relative to the control culture. To understand why, note that for any given condition, the phenotype for the majority of gene-deletion mutants is not significantly different from that of the wild-type control strain, and the median score for the experiment can therefore be assumed to be approximately that of the wild-type strain. This median score is also typically close to zero, because the final sample always contains the same number of cells no matter how slow growing a culture is. Therefore, for an experiment where the treatment culture grew 20% slower than the control culture, strains with a log 2 ratio of zero also grew approximately 20% slower than in the control culture. Similarly, in a case where the treatment culture grew 15% slower than the control culture, strains with a log 2 ratio of zero grew approximately 15% slower than in the control culture. Therefore, when comparing treatment inhibition for the same strain across multiple treatments, the growth rate of each pooled culture needs to be taken into account. Theoretically it should be possible to adjust log 2 ratios by multiplying the scores for each experiment by the average level of inhibition for each culture, however this idea has not been tested in practice.

24. *Reduced data quality for slow growing strains:* The data for strains with lower-than-average representation in the pool is noisier due to increased sampling error. Strains that are slow-growing in the control condition are especially prone to this problem because they often have low representation in the pool *(5)*. Data from these strains should be carefully confirmed. It is also common for slow-growing strains to appear slightly resistant to treatments that cause a large decrease in growth rate. While the reason for this effect is unknown, it is thought that slow growth may be less limiting in a condition where the pool as a whole is growing more slowly, causing these strains to be slowed proportionately less than others and thereby appear resistant.

25. *Frequently sensitive strains:* Some gene deletions are known to cause sensitivity in a large number treatment conditions *(13, 15)*. While the sensitivity of these deletions is genuine, strains specifically sensitive to the condition of interest may

be better candidates for follow-up experiments than genes known to mediate sensitivity to a broad range of conditions. Common gene families include: ergosterol biosynthesis, aromatic amino-acid biosynthesis, genes related to transcription, and genes with a variety of membrane functions such as ER to Golgi transport, vesicle mediated transport, and vacuolar targeting *(5, 15)*. The frequency of sensitivity for each strain in a large number of published experiments can be used as a reference when evaluating results *(15)*.

26. *Interference from neighboring gene deletions:* Gene function can be disrupted by deletion of a neighboring gene, either by partial deletion of the ORF due to overlap, disruption of promoter regions, or by generation of secondary site mutations during homologous recombination. It is common to see pairs of adjacent genes that are both sensitive to the same conditions. The true gene responsible for the phenotype can be determined by adding each of the two genes back to each of the two deletion strains on a low copy-number plasmid. The gene that is responsible for the phenotype should complement the deletion in both of these strains.

27. *Interpreting data for drug-target identification:* In addition to the deletion strain corresponding to the drug target, there may be other heterozygous deletions that are sensitive. For example, a gene involved in coping with the stress caused by the drug may also be sensitive due to reduced protein level. Because of this, there is no set way for determining which, if any, of the sensitive strains corresponds to the drug target from the heterozygous profiling data alone. Other data types are often helpful for narrowing the candidates. Synthetic lethality data can aid in interpreting the results *(12, 13, 39, 41, 42)*. Homozygous deletions corresponding to synthetic lethal partners of the drug target should be more sensitive to the drug because the drug mimics a deletion (or partial deletion) of the drug target. Synthetic lethality data is available for many complete deletions and also for essential genes in the form of promoter replacement alleles *(43)*. and 3′ untranslated region disruption alleles *(44)*. both of which reduce gene dosage. Both of these data types are useful for clarifying heterozygous profiling results. Genetic interaction data can be found in the BioGRID database *(45)*. Genome-wide overexpression data is also useful, as drug resistance often results from overexpression of the gene target *(17)*. In addition, the corresponding homozygous deletion data (if a full deletion of the gene of interest is viable) should be examined. If both the heterozygote and the homozygote are sensitive the gene is unlikely to encode the drug target because a strain that lacks the drug target should not be affected by the drug.

28. *Sample results:* **Fig. 6** shows sample data for an experiment in which the amino acid lysine was omitted. This data shows the typical level of agreement for raw data from multiple experiments, and the typical transformation of the raw data caused

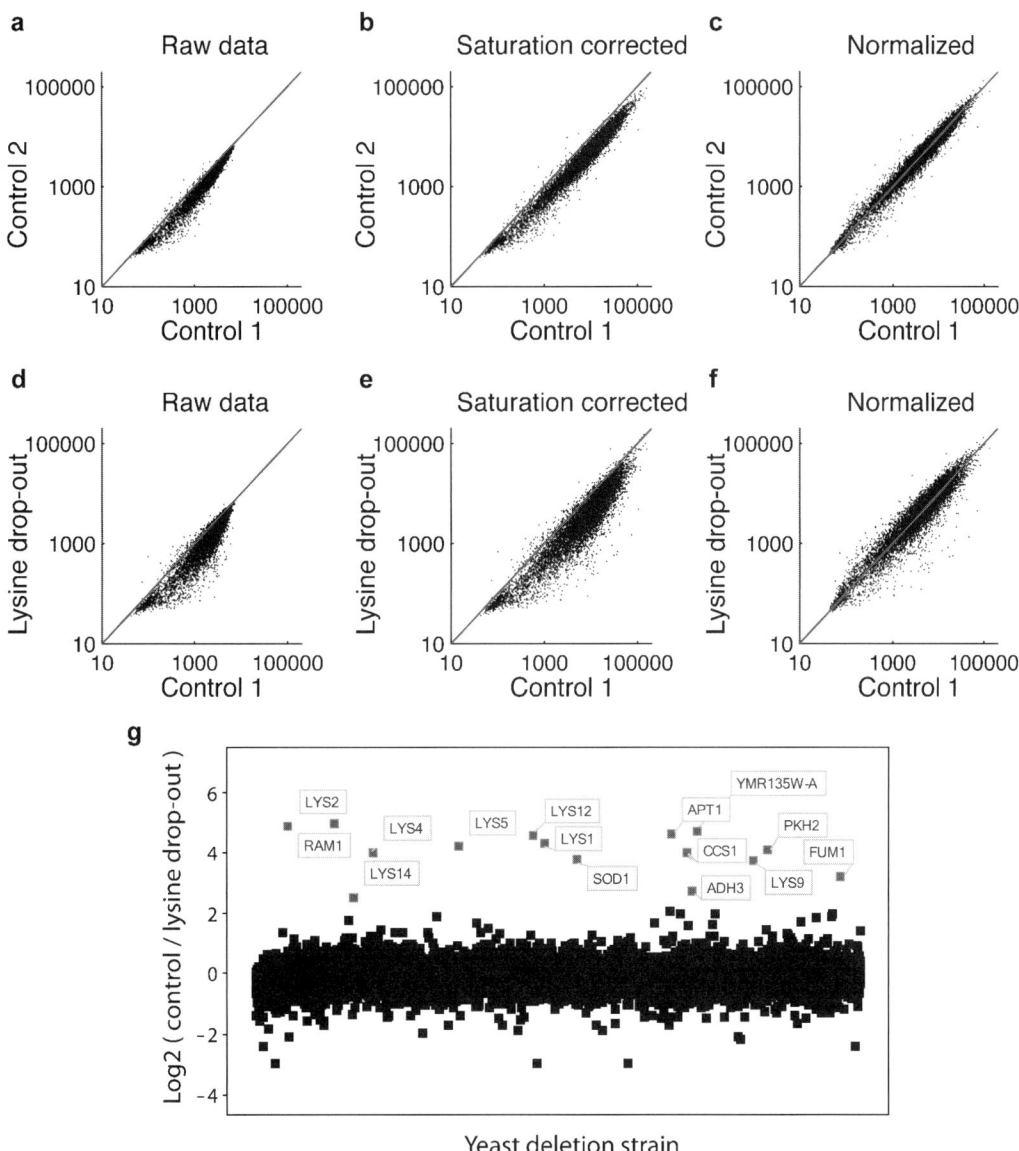

Fig. 6. Sample data showing data analysis steps and expected results. Sample data is shown for a control–control pair, and for a lysine drop-out treatment sample compared to a control. (**a**) Raw tag intensities for the control–control pair (replicate array features have been averaged for each tag). (**b**) Data after saturation correction. Note that the range of intensities increases. (**c**) Data after quantile normalization. Note that the data is now centered along the diagonal. (**d–f**) The same steps are shown for the lysine drop-out treatment versus a control. (**g**) log 2 intensity ratios for the lysine drop out experiment. Positive values indicate sensitivity in the treatment condition, negative values indicate resistance. As expected, genes known to be involved in lysine biosynthesis are overrepresented among the sensitive strains ($p = 1.13 \times 10^{-14}$). Modified from ref. 38.

by saturation correction and quantile normalization. Data is shown for both a pair of controls and a control-treatment pair. As expected, the raw data for the pair of control samples is more similar than the raw data for the control-treatment pair. Final log 2 ratios are shown for the lysine drop-out condition versus the control condition. As expected, the majority of sensitive strains have known roles in lysine biosynthesis. Note that the log 2 ratios are similar for these strains, suggesting that they have a similar level of sensitivity.

Acknowledgments

Work In the Giaever and Nislow labs is supported by the NHGRI, the Canadian Foundation for Innovation, and the CIHR (MOP-81340 to GG) and (MOP-84305 to CN).

References

1. Giaever, G. et al. Functional profiling of the *Saccharomyces cerevisiae* genome. *Nature* **418**, 387–91 (2002).
2. Winzeler, E.A. et al. Functional characterization of the *S. cerevisiae* genome by gene deletion and parallel analysis. *Science* **285**, 901–6 (1999).
3. Shoemaker, D.D., Lashkari, D.A., Morris, D., Mittmann, M. & Davis, R.W. Quantitative phenotypic analysis of yeast deletion mutants using a highly parallel molecular bar-coding strategy. *Nat Genet* **14**, 450–6 (1996).
4. Birrell, G.W. et al. Transcriptional response of *Saccharomyces cerevisiae* to DNA-damaging agents does not identify the genes that protect against these agents. *Proc Natl Acad Sci USA* **99**, 8778–83 (2002).
5. Deutschbauer, A.M. et al. Mechanisms of haploinsufficiency revealed by genome-wide profiling in yeast. *Genetics* **169**, 1915–25 (2005).
6. Giaever, G. et al. Chemogenomic profiling: identifying the functional interactions of small molecules in yeast. *Proc Natl Acad Sci USA* **101**, 793–8 (2004).
7. Giaever, G. et al. Genomic profiling of drug sensitivities via induced haploinsufficiency. *Nat Genet* **21**, 278–83 (1999).
8. Kastenmayer, J.P. et al. Functional genomics of genes with small open reading frames (sORFs) in S. cerevisiae. *Genome Res* **16**, 365–73 (2006).
9. Lee, W. et al. Genome-wide requirements for resistance to functionally distinct DNA-damaging agents. *PLoS Genet* **1**, e24 (2005).
10. Lum, P.Y. et al. Discovering modes of action for therapeutic compounds using a genome-wide screen of yeast heterozygotes. *Cell* **116**, 121–37 (2004).
11. Ooi, S.L., Shoemaker, D.D. & Boeke, J.D. A DNA microarray-based genetic screen for non-homologous end-joining mutants in Saccharomyces cerevisiae. *Science* **294**, 2552–6 (2001).
12. Parsons, A.B. et al. Integration of chemical-genetic and genetic interaction data links bioactive compounds to cellular target pathways. *Nat Biotechnol* **22**, 62–9 (2004).
13. Parsons, A.B. et al. Exploring the mode-of-action of bioactive compounds by chemical-genetic profiling in yeast. *Cell* **126**, 611–25 (2006).
14. Steinmetz, L.M. et al. Systematic screen for human disease genes in yeast. *Nat Genet* **31**, 400–4 (2002).
15. Hillenmeyer, M.E. et al. The chemical genomic portrait of yeast: uncovering a phenotype for all genes. *Science* **320**, 362–5 (2008).
16. Hoon, S. et al. An integrated platform of genome-wide assays reveals small molecule bioactivities. *Nat Chem Biol* **4**(8), 498–506 (2008).
17. Ericson, E. et al. Off-target effects of psychoactive drugs revealed by genome-wide assays in yeast. *PLoS Genet* **4**(8), e1000151 (2008).

18. Workman, C.T. et al. A systems approach to mapping DNA damage response pathways. *Science* **312**, 1054–9 (2006).
19. Jensen, L.J., Jensen, T.S., de Lichtenberg, U., Brunak, S. & Bork, P. Co-evolution of transcriptional and post-translational cell-cycle regulation. *Nature* **443**, 594–7 (2006).
20. Pollack, J.R. et al. Microarray analysis reveals a major direct role of DNA copy number alteration in the transcriptional program of human breast tumors. *Proc Natl Acad Sci USA* **99**, 12963–8 (2002).
21. Groh, J.L., Luo, Q., Ballard, J.D. & Krumholz, L.R. A method adapting microarray technology for signature-tagged mutagenesis of Desulfovibrio desulfuricans G20 and Shewanella oneidensis MR-1 in anaerobic sediment survival experiments. *Appl Environ Microbiol* **71**, 7064–74 (2005).
22. Karlyshev, A.V. et al. Application of high-density array-based signature-tagged mutagenesis to discover novel Yersinia virulence-associated genes. *Infect Immun* **69**, 7810–9 (2001).
23. Berns, K. et al. A large-scale RNAi screen in human cells identifies new components of the p53 pathway. *Nature* **428**, 431–7 (2004).
24. Brummelkamp, T.R. et al. An shRNA barcode screen provides insight into cancer cell vulnerability to MDM2 inhibitors. *Nat Chem Biol* **2**, 202–6 (2006).
25. Fischer, K.D. et al. Defective T-cell receptor signalling and positive selection of Vav-deficient CD4 + CD8 + thymocytes. *Nature* **374**, 474–7 (1995).
26. Fraser, A. RNA interference: human genes hit the big screen. *Nature* **428**, 375–8 (2004).
27. Kolfschoten, I.G. et al. A genetic screen identifies PITX1 as a suppressor of RAS activity and tumorigenicity. *Cell* **121**, 849–58 (2005).
28. Ngo, V.N. et al. A loss-of-function RNA interference screen for molecular targets in cancer. *Nature* **441**, 106–10 (2006).
29. Ngo, V.N. et al. A loss-of-function RNA interference screen for molecular targets in cancer. *Nature* **441**, 106–10 (2006).
30. Westbrook, T.F. et al. A genetic screen for candidate tumor suppressors identifies REST. *Cell* **121**, 837–48 (2005).
31. Akhras, M.S. et al. PathogenMip assay: a multiplex pathogen detection assay. *PLoS ONE* **2**, e223 (2007).
32. Clayton, D.G. et al. Population structure, differential bias and genomic control in a large-scale, case-control association study. *Nat Genet* **37**, 1243–6 (2005).
33. Hardenbol, P. et al. Multiplexed genotyping with sequence-tagged molecular inversion probes. *Nat Biotechnol* **21**, 673–8 (2003).
34. Hardenbol, P. et al. Highly multiplexed molecular inversion probe genotyping: over 10,000 targeted SNPs genotyped in a single tube assay. *Genome Res* **15**, 269–75 (2005).
35. Pierce, S.E. et al. A unique and universal molecular barcode array. *Nat Methods* **3**, 601–3 (2006).
36. Bolstad, B.M., Irizarry, R.A., Astrand, M. & Speed, T.P. A comparison of normalization methods for high density oligonucleotide array data based on variance and bias. *Bioinformatics* **19**, 185–93 (2003).
37. Tusher, V.G., Tibshirani, R. & Chu, G. Significance analysis of microarrays applied to the ionizing radiation response. *Proc Natl Acad Sci USA* **98**, 5116–21 (2001).
38. Pierce, S.E., Davis, R.W., Nislow, C., & Giaever, G. Genome-wide analysis of barcoded *S. cerevisiae* gene-deletion mutants in pooled cultures. *Nat Protoc* **2**, 2958–74 (2007).
39. Pan, X. et al. A robust toolkit for functional profiling of the yeast genome. *Mol Cell* **16**, 487–96 (2004).
40. Yuan, D.S. et al. Improved microarray methods for profiling the Yeast Knockout strain collection. *Nucleic Acids Res* **33**, e103 (2005).
41. Tong, A.H. et al. Systematic genetic analysis with ordered arrays of yeast deletion mutants. *Science* **294**, 2364–8 (2001).
42. Tong, A.H. et al. Global mapping of the yeast genetic interaction network. *Science* **303**, 808–13 (2004).
43. Davierwala, A.P. et al. The synthetic genetic interaction spectrum of essential genes. *Nat Genet* **37**, 1147–52 (2005).
44. Schuldiner, M. et al. Exploration of the function and organization of the yeast early secretory pathway through an epistatic miniarray profile. *Cell* **123**, 507–19 (2005).
45. Stark, C. et al. BioGRID: a general repository for interaction datasets. *Nucleic Acids Res* **34**, D535–9 (2006).

Chapter 8

Identification of Inhibitors of Chromatin Modifying Enzymes Using the Yeast Phenotypic Screens

Benjamin Newcomb and Antonio Bedalov

Summary

A multitude of enzymes that modify histones and remodel nucleosomes are required for the formation, maintenance, and propagation of the transcriptionally repressed chromatin state in eukaryotes. Robust phenotypic screens in yeast *S. cerevisiae* have proved instrumental in identifying these activities and for providing mechanistic insights into epigenetic regulation. These phenotypic assays, amenable for high throughput small molecule screening, enable identification and characterization of inhibitors of chromatin modifying enzymes largely bypassing traditional biochemical approaches.

Key words: Chromatin, Chemical genetics, Histone deacetylase, Chromatin modifying enzyme, Epigenetics

1. Introduction

1.1. Understanding Chromatin Biology Through Chemical Genetics

1.1.1. Chromatin Structure and Known Chemical Modifiers

Portions of the genome in eukaryotes is maintained in a transcriptionally inactive, or silenced state as a result of the local chromatin structure. The formation and the maintenance of the silenced state is an active process requiring a multitude of enzymes that act on DNA (methylation) and histones (acetylation, phosphorylation, and methylation) *(1)*. Following cell division the silenced state of chromatin is passed onto the daughter cells thus forming a basis for the epigenetic propagation of cellular memory. Faithful transmission of the epigenetic state from mother to daughter plays a key role in many cellular processes in eukaryotes such as mating in yeast *(2)* or development in multicellular organisms *(3)*. Epigenetic mechanisms also play an important role in the pathogenesis of many human neoplasms *(4)*. The importance of epigenetic regulation in cancer is underscored by the observations that tumor suppressor genes are often silenced rather than mutated and that many dominant oncogenes require epigenetic regulators for their activity. These epigenetic underpinnings of cancer can be exploited as a therapeutic strategy for two reasons. First, since silenced copies of tumor suppressor genes do not harbor genetic mutations, their reactivation in the context of malignant cells may suppress growth or induce death. Second, while transcription factors have traditionally been considered poor drug targets, the enzymatic activities required for their function (e.g., histone acetyl transferases HAT, histone deacetylases HDAC) can be inhibited pharmacologically. Together, these observations point to epigenetic regulation as a major new therapeutic area for cancer.

At the present time our ability to pharmacologically influence epigenetics in cancer cells, and to use this as therapy, is limited by the scarcity of effective small molecule inhibitors of enzymes that control epigenetic states. Classic HDAC inhibitors (e.g., SAHA) and DNA demethylating agents (e.g., deoxy-5-azacytidine) are the only two classes of chromatin modifying drugs in clinical use. This highlights the need to develop new drugs that target other enzymes involved in the establishment and maintenance of epigenetic states. Traditional approaches to identify enzyme inhibitors rely on high throughput biochemical screens. However, the enzymatic activities and proteins required for epigenetic regulation are extremely well conserved among eukaryotes, which makes drug discovery possible in vivo using model organisms.

Yeast is an attractive model system because of its rapid growth rate, ease of genetic manipulation, and because many yeast strains have already been developed to study epigenetics. Using a cell-based screen for compounds that can abrogate silencing at telomeres in yeast we have identified splitomicin, the first inhibitor of Sir2, a major nuclear NAD-dependent histone deacetylase and epigenetic regulator in yeast *(5)* and a founding member of a broadly conserved class of enzymes, sirtuins *(6)*. Conditional inactivation of Sir2 with splitomicin and its analogues has proved valuable in dissecting chromatin biology in yeast *(5, 7, 8)* and mammalian cells *(9)*, and in evaluating inhibition of sirtuins as a therapeutic strategy in cancer *(10)*. Our success in identifying Sir2 inhibitors through phenotypic screens for epigenetic regulators in yeast, suggests that the same strategy can be used for the identification of inhibitors of other enzymes required for propagation of epigenetic memory. In the following sections we provide an overview of silencing in yeast, the enzymatic activities required for efficient silencing, and a description of the silencing assays available. Additionally, we provide a detailed high throughput screening protocol for identifying compounds that disrupt telomeric silencing, a description of the methods employed for characterization of the hits, and an overview of the strategies for identifying the molecular targets of the compounds.

1.1.2. Yeast Silent Chromatin and Enzymatic Activities

Silent chromatin occurs at three distinct sites in the yeast genome: silent mating-type loci (HML and HMR), telomeres and at ribosomal RNA genes (rDNA) *(11)*. The formation of silent chromatin, best understood at the silent mating type loci and telomeres, depends on DNA elements or silencers. These silencers are located in close proximity to the genes they regulate and contain binding sites for several DNA binding proteins including Rap1, Abf1, and the origin recognition complex. These DNA binding proteins, through protein–protein interactions, recruit the SIR (silent information regulator) complex (Sir2–4). Once nucleated at the silencers, the SIR complex is thought to spread along the chromatin through the binding of Sir3 and Sir4 to the hypoacetylated NH_2-terminal tails of histones H3 and H4. The NH_2-terminal tails of histones H3 and H4 are kept in the hypoacetylated state through the action of Sir2 and this activity is critical for the formation of silent chromatin. Sir2 deacetylase activity is also required for silent chromatin formation at rDNA. However, at rDNA Sir2 is part of a different protein complex, which does not include Sir3 or Sir4.

The formation of silent chromatin leads to transcriptional repression at the silent mating types and at subtelomeric regions. The mating type of the yeast cell is determined by the mating type information (MATa or MATα) present at the mating locus (MAT) on chromosome III *(12)*. The ability of a and α cells to respond to mating pheromones and mate depends on expression of the haploid-specific and MATa or MATα specific genes controlled by MATa or MATα transcription factors. In addition to being present at the active MAT locus, a copy of the MATα gene exists in a silent state at the silent HML locus and a copy of MATa exists at the silent HMR locus on the same chromosome *(12)*. Normal diploid a/α state is defined by the coexpression of MATa and MATα information from the active MAT loci. However, the loss of silencing at the HMR and HML loci can create a pseudo diploid a/α state in a haploid cell by allowing the expression of MATa or MATα from the silent mating type loci. Accordingly, loss of silencing at the HML and HMR loci creates a mating defect and cells become insensitive to mating pheromones. The response of MATa cells to the alpha factor can be used as an assay for scoring derepressions at the silent HMR locus.

Silencing at the rDNA locus has two important effects. It represses transcription of the ribosomal RNA gene or an inserted reporter gene *(13)*, and it suppresses recombination between the tandem copies of rRNA genes *(14)*.

Beside the NAD-dependent deacetylase activity of Sir2, which is required for the silenced state at all silenced sites in the yeast genome *(15)*, other nuclear processes including DNA replication *(16)*, nucleosome assembly *(17)* and remodeling machinery *(18)* and several other histone modifying enzymes also participate in silent chromatin formation and maintenance. Intriguingly, the main role for several histone modifying enzymes, such as Dot1 *(19)*, Set1 *(20, 21)*, and Sas2 *(22–24)*, which are required for efficient silencing at telomeres, appears to be in limiting heterochromatin spreading. In their absence, the SIR complex extends beyond its normal boundaries and the redistribution of the SIR complexes weakens silencing at the native sites. While detailed mechanistic insight into how each of these and other enzymes contribute to the epigenetic memory is still lacking, it is clear that silencing of the reporter genes at telomeres is a very sensitive readout of the local chromatin state.

1.2. Silencing Assays in Yeast

Robust phenotypic assays have been developed for examining each of the silenced loci in the yeast genome *(25)*. In the following sections we will discuss the assays that are suitable for high throughput small molecule screening. **Table 1** lists the strains used to assay silencing at the telomeres, silent mating loci, and rDNA.

1.2.1. Silencing at the Telomeres

When a URA3 gene is introduced in the vicinity of a telomere its transcription is repressed *(26)*. However, the transcriptional repression of a telomeric URA3 gene is overcome in synthetic medium lacking uracil (C-Ura). Low uracil growth conditions up regulate the transcription factor Ppr1, which transactivates URA3 and other genes required for uracil biosynthesis *(28)*. Deletion of PPR1 can be used to modulate the strength of URA3 transcription. In the absence of PPR1, URA3 expression is largely dictated by the local chromatin state and is insensitive to the lack of uracil in cells. Accordingly, in the S288C strain background, the Δppr1

Table 1
Yeast strains used in silencing assays

Assay	Relevant genotype	Selection criteria	Strain
Telomeric silencing	ppr1- ΔURA3-TEL (VII-L)	C -Ura media	UCC2210 *(26)*
	URA3-TEL (VII-L)	C +5-FOA media	UCC3503 *(26)*
	ADE2-TEL (V-R)	Colony color	UCC3503 *(26)*
HML silencing	Mat a	Sensitivity to α-factor	BY4741
HMR silencing	hmrΔ::TRP1/HMR	C -Trp media	YJB959 *(27)*
rDNA silencing	RDN1::Ty1-mURA3	C -Ura or C +5-FOA	S6 *(13)*

URA3-TEL (VII-L) reporter strain will grow poorly in media without uracil. In the Δppr1 strain, perturbed silencing at the telomeres permits sufficient URA3 transcription for growth in a medium lacking uracil. In addition to positive selection, the URA3 telomeric reporter strain can also be employed in a negative selection screen using 5-Flouroorotic Acid (5-FOA). When silencing is perturbed the URA3 gene products convert the nontoxic 5-FOA into 5-Flurouracil, which is highly toxic.

Other telomeric reporter strains have been developed and include the ADE2-TEL (V-R) strain *(26)*. The ADE2 telomeric reporter strain relies on a straightforward assay based on ADE2 expression and colony color. Strains expressing a wild type copy of ADE2 produce white colonies and strains with silenced or deleted ADE2 form red colonies. The ADE2-TEL (V-R) strain stochastically switches the ADE2 gene from being transcriptionally repressed to being transcriptionally active and this process leads to the variegated expression of the ADE2-TEL (V-R) gene. As the colonies grow, the transcriptional state of the telomere is passed on to daughter cells and red and white sectors appear in the colony (*see* **Notes 1** and **2**). The amount of sectoring in the colony is directly related to the transcriptional state of the ADE2 gene, and if silencing is perturbed colonies appear with less red sectoring (*see* **Note 3**). In addition to the URA3 and ADE2 reporter strains, other telomeric reporters, such as strains utilizing HIS3, or TRP1 reporters, are also available (reviewed in *(25)*) and can be easily adapted to screens for small molecules with antisilencing properties.

1.2.2. Silencing at HMR and HML Loci

Silencing at the HMR locus is assayed using a TRP1 reporter gene integrated at the HMR locus. In a cell with intact silencing, the TRP1 gene will not be expressed and the cell will be unable to grow in media lacking tryptophan (C −Trp). When silencing at the HMR locus is disrupted, the TRP1 gene will be expressed and the strain will grow in C −Trp media.

Testing silencing at the HML locus relies on the mating pheromone α-factor. When two haploid yeast cells, of opposite mating type, come in close proximity to each other they secrete cell type specific mating pheromones *(2)*. MATa cells secrete a-factor, a modified hydrophobic peptide, and MATα cells secrete α-factor, an unmodified soluble peptide. Typically α-factor is used in experiments utilizing a mating pheromone because it can be dissolved in liquid media and it is easily obtained commercially. α-Factor induces MATa cells to arrest in the G_1 phase of the cell cycle and undergo morphological changes, known as shmooing, in preparation for cell and nuclear fusion. MATa cells with aberrant silencing at the HML locus are insensitive to α-factor and do not arrest in G_1 or shmoo.

1.2.3. Silencing at the rDNA

Yeast ribosomal DNA (rDNA) consists of a 9.1 kb region of DNA, containing a 5S rRNA gene and a 35S pre-rRNA gene, which is repeated 100–200 times along chromosome XII. Silencing at the rDNA is assayed with a strain that utilizes a URA3 reporter, with a minimal TRP1 promoter (mURA3), integrated at the rDNA repeats *(13)*. Use of the mURA3 instead of native URA3 improves URA3 silencing and permits scoring by positive and negative selection. In addition to silencing, recombination at the rDNA locus can be measured through the loss rate of an ADE2 gene integrated into the rDNA *(29)*.

1.3. Screening Strategy

1.3.1. Primary Screen

The use of a telomeric URA3 reporter as a primary screen has several advantages over other available reporter systems. Primarily, silencing at the telomere is a very sensitive readout of the chromatin state (*see* **Note 4**) that is influenced by the products of many genes. Furthermore, the loss of silencing-induced activation of the URA3 reporter can be scored as growth in medium lacking uracil, or as toxicity in medium containing 5-FOA. The design of a high throughput screen utilizing selection for growth has an advantage in that it requires that the compound inhibits a relevant target at concentrations that do not perturb other cellular processes (*see* **Note 5**).

1.3.2. Secondary Assays

Compounds identified in the primary screen that promote growth in C –Ura, or demonstrate toxicity in 5-FOA are subjected to secondary screens with the goal of identifying compounds that specifically inhibit silencing. This is carried out by comparing toxicity of compounds in medium with and without 5-FOA, and by examining the ability of compounds to stimulate growth in medium lacking uracil. A dose response curve using the same concentration of drug employed in the primary screen and several twofold dilutions is carried out in complete medium, medium containing 5-FOA and medium lacking uracil. The concentration of an ideal antisilencing drug that inhibits growth by 50% (IC_{50}) is expected to be at least 2–4-fold lower in medium containing 5-FOA relative to the IC_{50} in C-medium (**Fig. 1**). The dose response curve in medium lacking uracil is expected to identify the concentration of drug that activates the reporter and stimulates growth. Growth inhibition at the higher drug concentrations for compounds that activate a telomeric reporter may indicate inhibition of an essential activity that is important for telomeric silencing, or indicate inhibition of an unrelated essential target (off target activity).

In order to demonstrate that a compound interferes specifically with the state of chromatin, and does not solely promote transactivation of the URA reporter, silencing of other reporters at telomeres and at loci other than the telomeres should be characterized. For this purpose, the sectoring assay using a strain containing ADE2 at the telomere can be used. The presence of

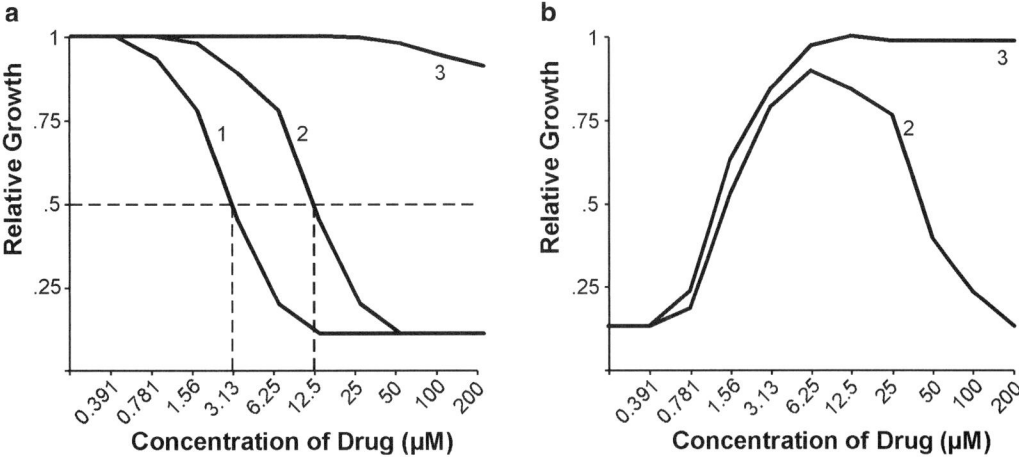

Fig. 1. (**a**) Example dose response curves for a compound identified in a primary screen. *Curve 1* represents the dose response in C +5-FOA. The IC_{50} for the compound in C +5-FOA is 3.13 µM. *Curve 3* represents the dose response in C media, expected if the compound used for *Curve 1* is highly specific and nontoxic (e.g., highly specific Sir2 inhibitor with no off target activity). *Curve 2* represents the dose response curve in C medium that would be expected if the compound used in *Curve 1* had off target activity or activity against an essential enzyme involved with silencing (IC_{50} in C medium is 12.5). Fourfold greater sensitivity in C +5-FOA relative to C medium suggests activation of the URA3 reporter. (**a**) *Curve 3* represents the dose response in C-Ura medium for a highly specific compound that inhibits silencing and that has no off target (e.g., highly specific Sir2 inhibitor with no off target activity). *Curve 2* shows the expected dose response curve in C-Ura medium for a compound that at higher concentrations exhibits off target activity or that inhibits an enzyme that is, in addition to silencing, required for normal cell growth.

heterochromatin at sites other than the telomeres can be examined by monitoring activation of a TRP1 reporter integrated at the HMR locus or the responsiveness of MATa cells to α-factor can be used to assay silencing of the endogenous HML locus. Furthermore, both silencing and recombination rates at rDNA can be assayed using available strains. A URA3 reporter strain has been developed for assaying silencing, and strains have been developed with ADE2 integrated at the rDNA to determine the recombination rate by measuring the loss of the ADE2 gene.

1.4. Characterizing Hits

1.4.1. Identification of Molecular Targets

Identification of the targets of compounds identified through cell based screens presents one of the major challenges of chemical genetics. Traditional target identification strategies rely on biochemical affinity methods such as purification of the target protein using a drug affinity matrix. One of the major drawbacks of this approach is the need for ligand modification, which is often time consuming, and can lead to loss of drug activity. Furthermore, this approach is limited to compounds that have a relatively high affinity to their corresponding target. Beside the affinity purification methods, the large body of preexisting knowledge of yeast epigenetic regulation and the facile genetics offer additional means for target identification. The ideal drug, upon binding the target protein, mimics either a loss of function or a gain of function mutation in the corresponding gene. This premise serves as a basis for identifying drug targets by *drug-mutant* matching approaches. An antisilencing drug that inhibits a specific chromatin modifying enzyme is expected to *(1)* recreate the effect of a point mutation in the corresponding enzyme that abrogates its enzymatic activity in vitro, *(2)* recreate the same global or local chromatin alterations, *(3)* change the global transcriptional profile, and *(4)* replicate the synthetic interaction profiles of their corresponding loss of function mutants.

1.4.2. In vitro Inhibition of Known Enzymatic Activities Important for Silencing

Several of the genes that affect telomeric silencing in yeast encode proteins that have defined in vitro enzymatic activities. While the published assays, such as enzymatic assays for Dot1, Set1, Sir2, and Sas2 described in references *(19, 20, 23, 30, 31)*, may not be suitable for high throughput screening, the number of compounds that need to be characterized in this step is expected to be limited, which makes the in vitro testing for enzyme inhibition the most straightforward approach for evaluating whether the antisilencing compounds inhibit these known enzymatic activities.

1.4.3. Histone Modifications in Drug Treated Cells and Silencing Mutants

A deficiency of histone modifying enzymes is expected to lead to local (i.e., telomeric) or global alterations in specific histone post-translational modifications (PTM). The ability of compounds to alter histone PTM can be evaluated using a panel of antibodies specific to different modifications of histone proteins using Western blots, for global, or chromatin immunoprecipitation for local telomere alterations respectively. Antibodies targeting more than 40 different histone modifications (methylation, acetylation of different lysine residues) are commercially available.

1.4.4. Matching of Transcriptional Effects

Silencing in yeast occurs at three locations: telomeres, silent mating loci and rDNA. While the loss of SIR2 affects silencing at all three sites, the effect of inactivating other chromatin modifying enzymes may be restricted to specific locations. In addition to known sites affected by silencing, proteins required for silencing also affect transcription, directly or indirectly, at other locations in the genome. As a result, the global transcriptional changes in drug treated wild type cells are expected to be highly correlated with the changes observed in cells with mutations in proteins required for transcription at various loci. The gene which, when deleted, creates a transcriptional profile most highly correlated with the transcription profile created by drug treated wild type cells should be further analyzed as potentially encoding for a drug target. Large numbers of yeast mutants have been subjected to transcriptional profiling *(32, 33)*, and the data are readily available.

The definite proof that the drug affects a putative target will rely on:

(a) Inhibition of the targets' *in vitro* enzymatic activity,

(b) Demonstrating specific binding of the drug to its target if no biochemical activity has been ascribed to the protein,

(c) Obtaining a mutant protein through in vivo screening for alleles that confer drug resistance and have alterations in drug binding or are resistant to enzyme inhibition in vitro.

2. Materials

2.1. Positive Selection Screen

1. Strain UCC2210 Mat α ade2d::hisG his3d200 leu2d0 lys2d0 met15d0 trp1d63 ura3d0 adh4::URA3-TEL (VII-L) ppr1::HIS3 (*see* **Note 6**)

2. YEPD medium (For 2 L of media): 20 g bacto yeast extract (cat. no. DF0127-17-9, Fisher), 40 g bacto peptone, 40 g dextrose.

3. Complete synthetic (C) medium (for 2 L of media): 2.9 g yeast nitrogen base without amino acids and ammonium sulfate, 10 g ammonium sulfate, 40 g dextrose, and amino acids (omit Uracil for C –URA media) according to **Table 2**.

4. Flat bottom, clear 96-well microplate.

5. Compound library and a positive control, such as splitomicin (*see* **Note 7**), known to abrogate silencing in yeast.

6. 30°C incubator.

7. Microplate reader (e.g., Molecular Devices VERSAMAX Tunable Microplate Reader).

8. Distributing cells and compounds to 96-well plates is greatly simplified with the use of an automated 96-channel pipettor (e.g., a PlateMate or similar programmable device).

2.2. Negative Selection Screen

1. Strain UCC3503 Mat α ura3-52 lys2-801 ade2-101 leu2-d1 his3-d200 adh4::URA3-TEL (VII-L) ADE2-TEL (V-R)

2. C +5-FOA media is C media as described in **Subheading 2.1**, **item 3** with 2 g 5-FOA (Zymoresearch) added to 2 L of C media.

3. See required materials for Positive Selection (**Subheading 2.1**).

2.3. Dose Response Assay

1. Strain UCC3503 Mat α ura3-52 lys2-801 ade2-101 leu2-d1 his3-d200 adh4::URA3-TEL (VII-L) ADE2-TEL (V-R) or Strain UCC2210 Mat α ade2d::hisG his3d200 leu2d0 lys2d0 met15d0 trp1d63 ura3d0 adh4::URA3-TEL (VII-L) ppr1::HIS3.

Table 2
Amino acids used for preparation of synthetic complete media

Amino acid	Grams	Cat. No.	Vendor
Adenine (6-aminopurine)	1	A-8626	Sigma
L-arginine	1	BP370-100	Fisher
L-aspartic acid	5	BP374-100	Fisher
L-glutamic acid	5	BP378-100	Fisher
L-histidine	1	BP382-100	Fisher
L-isoleucine	4	BP384-100	Fisher
L-leucine	4	BP385-100	Fisher
L-lysine	3	L9037	Sigma
L-methionine	1	BP388-100	Fisher
L-phenylalanine	2.5	BP391-100	Fisher
L-serine	20	BP393-100	Fisher
L-threonine	10	BP394-100	Fisher
L-tryptophan	4	BP395-100	Fisher
L-tyrosine	3	BP396-100	Fisher
Uracil	1	U-1128	Sigma
L-valine	7.5	BP397-100	Fisher

2. C media, C −URA media, and C +5-FOA media as described above.
3. See required materials for Positive Selection (**Subheading 2.1**).

2.4. α-Factor Resistance Assay

1. Strain BY4741 Mat a his3 leu2 met15 ura3.
2. C media (described in **Subheading 2.1**).
3. α-Factor (Zymoresearch).
4. See required materials for Positive Selection (**Subheading 2.1**).

2.5. HMR Silencing Assay

1. Strain YJB959 MATa ura3-1 ade2-1 his3-11 leu 2-3,112 can 1-100 trp1-1 HMR::TRP1
2. C media prepared without tryptophan (*see* **Subheading 2.1**, **item 3**)
3. See required materials for Positive Selection (**Subheading 2.1**).

2.6. Red/White Colony Formation

1. Strain UCC3503 Mat α ura3-52 lys2-801 ade2-101 leu2-1 his3-200 adh4::URA3-TEL (VII-L) ADE2-TEL (V-R)

2.7. rDNA Silencing

1. Strain S26 Mat α his3 200 leu2 1 ura3-167 RDN1::Ty1-mURA3.
2. C-Ura media, or +5-FOA media as described above in **Subheading 2.2, item 2**.
3. See required materials for Positive Selection (**Subheading 2.1**).

2. YEPD agar (For 2 L of media): 20 g bacto yeast extract, 40 g bacto peptone, 40 g dextrose, 40 g bacto agar.
3. YEPD agar plates containing drug at the IC_{50} concentration and plates without drug.

3. Methods

3.1. Primary Screens

3.1.1. Positive Selection Screen

1. Grow a culture of the assay strain in YEPD media to saturation (approximately 2×10^8 cells/mL).
2. Harvest cells by centrifugation and wash once with water.
3. Resuspend washed cell pellet in C-media prepared without uracil (referred to as C –URA) to a density of 5×10^4 cells/mL.
4. Deliver 135 μL of cell suspension to each well of a 96-well plate.
5. The compounds to be assayed should be diluted with water to 10× the desired assay concentration. Generally stock concentrations of compounds should be between 50 and 150 μM, solubility permitting.
6. Deliver 15 μL of the compound stocks to the yeast cells.
7. The total volume of each well will be 150 μL, with cells at a density of 5×10^4 and the compounds to be assayed at final concentrations in the range of 5–15 μM (*see* **Note 7**).
8. Incubate the plates at 30°C for 24–48 h (*see* **Note 8**).
9. Score growth by reading the optical density (OD) of the wells at 660 nm on a microplate reader (Molecular Devices VERSAMAX Tunable Microplate Reader) (*see* **Notes 7 and 9**).

3.1.2. Negative Selection Screen

Conducting a negative selection screen in 5-FOA media is carried out in the same fashion as the positive selection screen, but it is necessary to use a strain that is wild type for PPR1. Incubate plates for 24–48 h at 30°C (*see* **Notes 7, 8, and 10**).

3.2. Secondary Screens

3.2.1. Dose Response Assay

1. Prepare cells as in the positive selection screen (**Subheading 3.1.1**), and resuspend in the appropriate media (*see* **Note 11**).
2. Deliver 135 μL of cell suspension to each well of a 96-well plate.

3. Prepare 1:2 serial dilutions of the compound of interest, spanning a range of concentrations above and below the concentration used in the primary screen.

4. Deliver 15 μL of compound dilutions to cells (*see* **Note 7**).

5. Incubate at 30°C for 24–48 h and read the optical density (*see* **Notes 9** and **10**).

3.2.2. α-Factor Resistance (HML Silencing)

1. Prepare cells as in the positive selection screen (**Subheading 3.1.1**) and resuspend in C media containing 5 μM α-Factor.

2. Distribute 135 μL of cell suspension into each well of a 96-well plate and add 15 μL of the compounds of interest at the desired concentration (*see* **Notes 7, 12,** and **13**).

3. Incubate the plate(s) at 30°C for 12–24 h, and read the optical density at 660 nm.

3.2.3. HMR Silencing Assay

1. Prepare cells as in the positive selection screen (**Subheading 3.1.1**) and resuspend in C media prepared without tryptophan.

2. Distribute 135 μL of cells to each well of a 96-well plate, and deliver 15 μL of the compounds of interest (*see* **Note 7** and **12**).

3. Incubate the plate(s) at 30°C for 12–36 h, and read the optical density at 660 nm.

3.2.4. Red/White Colony Formation Assay

The Red/White colony formation assay can be carried out using 96-well plate format. Grow cells in YEPD liquid to saturation.

1. Dilute and plate cells for singles (*see* **Note 14**) on YEPD agar and YEPD agar containing the drug of interest.

2. Grow colonies at 30°C for 24–48 h and transfer plates to 4°C for 72 h to develop red pigmentation (*see* **Notes 1, 2, 3,** and **7**).

3.2.5. rDNA Silencing Assay

1. The S26 strain (Mat α his3 200 leu2 1 ura3-167 RDN1::Ty1-mURA3) can be used in a positive selection screen in medium lacking uracil as described in **Subheading 3.1.1**, or in a negative selection screen in medium containing 5-FOA as described in **Subheading 3.1.2**.

4. Notes

1. Red-pigmented colonies do not grow as robustly as normal white colonies. The use of YEPD media instead of C media will help minimize the disparity.

2. The degree of Red/White sectoring is not uniform among colonies of the same strain.

3. Development of the red pigmentation is greatly improved by placing plates at 4°C for several days after colonies have formed.

4. We have noted that derepression of telomeric reporters can be achieved by at least fivefold lower concentrations of the Sir2 inhibitor splitomicin, than those required for derepression of the reporters at the silent mating loci.

5. The choice of a positive or negative selection strategy for the primary screen should be based on the characteristics of the chemical libraries being assayed. If the library is enriched for toxic compounds, (e.g., 20–25% of compounds in the National Cancer Institute repository exhibit toxicity), a primary screen using positive selection, which assays for stimulation of growth in medium lacking uracil, is expected to be more efficient than a screen for toxicity in 5-FOA medium. Libraries that are not enriched for toxic compounds can be efficiently screened for toxicity in 5-FOA as a primary screen followed by dose response evaluations of toxicity in C medium and growth enhancement in medium lacking uracil as secondary screens.

6. Deletion of *PDR1*, *PDR3* which encode transcription factors that activate pleotrophic drug resistance genes, and *ERG6* required for ergosterol biosynthesis may increase the intracellular drug concentrations by impairing drug efflux and enhancing membrane permeability.

7. The use of splitomicin is helpful to gauge the performance of the assay and to determine the stopping point of the assay. At early timepoints, such as 18–36 h, splitomicin will permit cells to grow in C-Ura media and untreated cells will exhibit no growth. At later timepoints (48 h and longer), the stochastic nature of silencing and the resulting spontaneous derepression of the *URA3* reporter in a small fraction of cells will permit growth in some, though not all, negative control wells. Thus, scoring at later time points allows for greater sensitivity but may lead to an increased rate of false positive hits. In order to minimize false positives, growth should be scored at a timepoint when wells treated with splitomicin have an OD_{660} of 0.25-0.4 (on a Molecular Devices VERSAMAX Tunable Microplate Reader) and no growth is observed in control wells. Splitomicin can be used as a positive control at a concentration of 10 μM in **Subheadings 3.1.1**, **3.1.2**, **3.2.2**, **3.2.3**, **3.2.4**, and **3.2.5**. For the dose response assays in **Subheading 3.2.1**, a range of splitomicin concentrations from 0.1 to 50 μM should be analyzed. Colonies growing on agar containing 10 μM splitomicin appear white. Splitomicin should be diluted in the media immediately prior to use as it rapidly hydrolyzes in aqueous solutions (half life 1–8 h depending on pH).

8. Place plates inside a clean plastic bag, or seal with parafilm to prevent evaporation during the incubation period.
9. In a positive selection screen, an OD between 0.25 and 0.4 indicates growth.
10. In a negative selection screen (growth in 5-FOA), >75% growth inhibition should be scored as a hit (growth inhibition = $1 - (OD_{treated}/OD_{untreated})$).
11. For the dose response assays, resuspend cells in C media, C-Ura media and C +5-FOA media, and conduct the dose response assay in each type of media.
12. When assaying the compounds for silencing at the mating loci and ADE-TEL (V-R) locus, use the concentration of drug determined to best stimulate growth in C -Ura media in **Subheading 3.2.1**. When assaying compounds for rDNA silencing use the concentration from the dose response curve in C-Ura media that best stimulates growth, for a positive selection screen; or the IC_{50} concentration in 5-FOA, for a negative selection screen.
13. In addition to growth, loss of silencing is also expected to abrogate the ability of MATa cells to shmoo, which can be scored using a microscope.
14. For 96-well format, plate 5–10 cells per well.

Acknowledgments

This work was supported by the National Institute of Health (CA129132-01A1) and the Leukamia and Lymphoma Society grants to A. Bedalov.

References

1. Kouzarides, T. (2007) Chromatin modifications and their function *Cell* **128**, 693–705.
2. Marsh, L., Neiman, A. M., and Herskowitz, I. (1991) Signal transduction during pheromone response in yeast *Annual Review of Cell Biology* **7**, 699–728.
3. Schuettengruber, B., Chourrout, D., Vervoort, M., Leblanc, B., and Cavalli, G. (2007) Genome regulation by polycomb and trithorax proteins. *Cell* **128**, 735–45.
4. Jones, P. A., and Baylin, S. B. (2007) The epigenomics of cancer *Cell* **128**, 683–92.
5. Bedalov, A., Gatbonton, T., Irvine, W. P., Gottschling, D. E., and Simon, J. A. (2001) Identification of a small molecule inhibitor of Sir2p. *Proc Natl Acad Sci U S A* **98**, 15113–8.
6. Sauve, A. A., Wolberger, C., Schramm, V. L., and Boeke, J. D. (2006) The Biochemistry of Sirtuins. *Annu Rev Biochem* **75**, 435–65.
7. Bedalov, A., Hirao, M., Posakony, J., Nelson, M., and Simon, J. (2003) NAD-dependent deacetylase Hst1p controls biosynthesis and cellular NAD levels in Saccharomyces cerevisiae. *Mol Cell Biol* **23**.
8. Chang, C. R., Wu, C. S., Hom, Y., and Gartenberg, M. R. (2005) Targeting of cohesin by transcriptionally silent chromatin. *Genes Dev* **19**, 3031–42.

9. Pruitt, K., Zinn, R. L., Ohm, J. E., McGarvey, K. M., Kang, S. H., Watkins, D. N., Herman, J. G., and Baylin, S. B. (2006) Inhibition of SIRT1 reactivates silenced cancer genes without loss of promoter DNA hypermethylation. *PLoS Genet* **2**, e40.

10. Heltweg, B., Gatbonton, T., Schuler, A. D., Posakony, J., Li, H., Goehle, S., Kollipara, R., DePinho, R. A., Gu, Y., Simon, J. A., and Bedalov, A. (2006) Antitumor activity of a small molecule inhibitor of human Sir2 enzymes. *Cancer Res* **66**, 4368–77.

11. Rusche, L. N., Kirchmaier, A. L., and Rine, J. (2003) The establishment, inheritance, and function of silenced chromatin in Saccharomyces cerevisiae. *Annu Rev Biochem* **72**, 481–516.

12. Herskowitz, I. (1988) Life cycle of the budding yeast Saccharomyces cerevisiae. *Microbiol Rev* **52**, 536–53.

13. Smith, J. S., and Boeke, J. D. (1997) An unusual form of transcriptional silencing in yeast ribosomal DNA. *Genes Dev* **11**, 241–54.

14. Gottlieb, S., and Esposito, R. E. (1989) A new role for a yeast transcriptional silencer gene, SIR2, in regulation of recombination in ribosomal DNA. *Cell* **56**, 771–6.

15. Imai, S., Armstrong, C. M., Kaeberlein, M., and Guarente, L. (2000) Transcriptional silencing and longevity protein Sir2 is an NAD-dependent histone deacetylase. *Nature* **403**, 795–800.

16. Suter, B., Tong, A., Chang, M., Yu, L., Brown, G. W., Boone, C., and Rine, J. (2004) The origin recognition complex links replication, sister chromatid cohesion and transcriptional silencing in Saccharomyces cerevisiae. *Genetics* **167**, 579–91.

17. Sharp, J. A., Fouts, E. T., Krawitz, D. C., and Kaufman, P. D. (2001) Yeast histone deposition protein Asf1p requires Hir proteins and PCNA for heterochromatic silencing. *Curr Biol* **11**, 463–73.

18. Dror, V., and Winston, F. (2004) The Swi/Snf chromatin remodeling complex is required for ribosomal DNA and telomeric silencing in Saccharomyces cerevisiae. *Mol Cell Biol* **24**, 8227–35.

19. van Leeuwen, F., Gafken, P. R., and Gottschling, D. E. (2002) Dot1p modulates silencing in yeast by methylation of the nucleosome core. *Cell* **109**, 745–56.

20. Krogan, N. J., Dover, J., Khorrami, S., Greenblatt, J. F., Schneider, J., Johnston, M., and Shilatifard, A. (2002) COMPASS, a histone H3 (Lysine 4) methyltransferase required for telomeric silencing of gene expression. *J Biol Chem* **277**, 10753–5.

21. Venkatasubrahmanyam, S., Hwang, W. W., Meneghini, M. D., Tong, A. H., and Madhani, H. D. (2007) Genome-wide, as opposed to local, antisilencing is mediated redundantly by the euchromatic factors Set1 and H2A.Z. *Proc Natl Acad Sci U S A* **104**, 16609–14.

22. Reifsnyder, C., Lowell, J., Clarke, A., and Pillus, L. (1996) Yeast SAS silencing genes and human genes associated with AML and HIV-1 Tat interactions are homologous with acetyltransferases. *Nat Genet* **14**, 42–9.

23. Sutton, A., Shia, W. J., Band, D., Kaufman, P. D., Osada, S., Workman, J. L., and Sternglanz, R. (2003) Sas4 and Sas5 are required for the histone acetyltransferase activity of Sas2 in the SAS complex. *J Biol Chem* **278**, 16887–92.

24. Kimura, A., Umehara, T., and Horikoshi, M. (2002) Chromosomal gradient of histone acetylation established by Sas2p and Sir2p functions as a shield against gene silencing. *Nat Genet* **32**, 370–7.

25. van Leeuwen, F., and Gottschling, D. E. (2002) Assays for gene silencing in yeast. *Methods Enzymol* **350**, 165–86.

26. Singer, M. S., Kahana, A., Wolf, A. J., Meisinger, L. L., Peterson, S. E., Goggin, C., Mahowald, M., and Gottschling, D. E. (1998) Identification of high-copy disruptors of telomeric silencing in Saccharomyces cerevisiae. *Genetics* **150**, 613–32.

27. Buck, S. W., and Shore, D. (1995) Action of a RAP1 carboxy-terminal silencing domain reveals an underlying competition between HMR and telomeres in yeast. *Genes Dev* **9**, 370–84.

28. Aparicio, O. M., and Gottschling, D. E. (1994) Overcoming telomeric silencing: a trans-activator competes to establish gene expression in a cell cycle-dependent way. *Genes Dev* **8**, 1133–46.

29. Kaeberlein, M., McVey, M., and Guarente, L. (1999) The SIR2/3/4 complex and SIR2 alone promote longevity in *Saccharomyces cerevisiae* by two different mechanisms. *Genes Dev* **13**, 2570–80.

30. Smith, J. S., Avalos, J., Celic, I., Muhammad, S., Wolberger, C., and Boeke, J. D. (2002) SIR2 family of NAD(+)-dependent protein deacetylases. *Methods Enzymol* **353**, 282–300.

31. Alcasabas, A. A., Osborn, A. J., Bachant, J., Hu, F., Werler, P. J., Bousset, K., Furuya, K., Diffley, J. F., Carr, A. M., and Elledge, S. J. (2001) Mrc1 transduces signals of DNA replication stress to activate Rad53. *Nat Cell Biol* **3**, 958–65.

32. Mnaimneh, S., Davierwala, A. P., Haynes, J., Moffat, J., Peng, W. T., Zhang, W., Yang, X., Pootoolal, J., Chua, G., Lopez, A., Trochesset, M., Morse, D., Krogan, N. J., Hiley, S. L., Li, Z., Morris, Q., Grigull, J., Mitsakakis, N., Roberts, C. J., Greenblatt, J. F., Boone, C.,

Kaiser, C. A., Andrews, B. J., and Hughes, T. R. (2004) Exploration of essential gene functions via titratable promoter alleles. *Cell* **118**, 31–44.

33. Hughes, T. R., Marton, M. J., Jones, A. R., Roberts, C. J., Stoughton, R., Armour, C. D., Bennett, H. A., Coffey, E., Dai, H., He, Y. D., Kidd, M. J., King, A. M., Meyer, M. R., Slade, D., Lum, P. Y., Stepaniants, S. B., Shoemaker, D. D., Gachotte, D., Chakraburtty, K., Simon, J., Bard, M., and Friend, S. H. (2000) Functional discovery via a compendium of expression profiles. *Cell* **102**, 109–26.

Chapter 9

Exploiting Yeast Genetics to Inform Therapeutic Strategies for Huntington's Disease

Flaviano Giorgini and Paul J. Muchowski

Summary

Huntington's disease (HD) is a devastating neurodegenerative disorder that is inherited in an autosomal dominant fashion and is caused by a polyglutamine expansion in the protein huntingtin (htt). In recent years, modeling of various aspects of HD in the yeast *Saccharomyces cerevisiae* has provided insight into the conserved mechanisms of mutant htt toxicity in eukaryotic cells. The high degree of conservation of cellular and molecular processes between yeast and mammalian cells have made it a valuable system for studying basic mechanisms underlying human disease. Yeast models of HD recapitulate conserved disease-relevant phenotypes and can be used for drug discovery efforts as well as to gain mechanistic and genetic insights into candidate drugs. Here we provide a detailed overview of yeast models of mutant htt misfolding and toxicity and the molecular and phenotypic characterization of these models. We also review how these models identified novel therapeutic targets and compounds for HD and discuss the benefits and limitations of this model genetic system. Finally, we discuss how yeast may be used to provide further insight into the molecular and cellular mechanisms underlying HD and treatment strategies for this devastating disorder.

Key words: Huntington's disease, Huntingtin, Yeast models, Protein misfolding, Neurodegeneration

1. Introduction

Although the Huntington's disease (HD) gene was cloned over a decade ago *(1)*, no therapeutic options are available for HD patients. HD is caused by an expansion of a polyglutamine (polyQ) stretch in the protein huntingtin (htt), which leads to misfolding and aggregation of mutant htt. Many studies support the hypothesis that mutant htt causes HD through a gain-of-function mechanism. Models of HD have been developed by expressing mutant htt or mutant htt fragments in mice, mammalian cells, *Drosophila*, *Caenorhabditis elegans*, and the baker's yeast *Saccharomyces cerevisiae*.

Yeast has been used for centuries to produce foods such as bread and wine *(2)*. Over the past 40 years, studies in yeast have provided critical insight into basic cellular processes that are conserved in higher eukaryotes, including cell division, replication, metabolism, protein folding, and intracellular transport *(3)*. The conserved cellular mechanisms elucidated in these organisms, such as cell-cycle regulation, have played a major role in understanding cancer and other complex diseases *(4, 5)*. More recently, yeast has been used to model various aspects of protein misfolding disorders, such as HD and Parkinson's disease (PD), and these models have been used for high-throughput screens to identify candidate therapeutic targets and candidate drugs.

Yeast is an ideal model organism for biological research, owing to its power and ease of use for genetic studies. Classical genetics is easily performed in *S. cerevisiae* because it can exist in both haploid and diploid states and because it is easy to mate haploid strains and sporulate diploid strains. Yeast is perhaps the best-characterized eukaryotic organism. Its ~14 Mb genome was the first eukaryotic genome to be fully sequenced *(6)*, and *Saccharomyces* genes and proteins are extensively annotated in several genomic and proteomic databases. Other useful properties of yeast include rapid growth on defined media, the ease of replica plating and mutant isolation, and importantly, a highly versatile DNA transformation system *(7)*. In addition, genes of interest can easily be knocked out or altered by homologous recombination *(7)*.

These features, along with the ease of mutagenesis and the availability of several mutant collections and open reading frame (ORF) libraries, allow for rapid isolation of genetic modifiers of specific processes or phenotypes. In addition, molecular genetic manipulations, such as DNA transformation, targeted disruption of specific genes, generation of point mutations in cloned genes, and overexpression of proteins of interest can be performed in a matter of days, as compared to months or years in other model organisms. Development of arrayed libraries for systematic deletion or overexpression of most yeast genes has permitted rapid genomic screening and identification of novel gene interactions and drug targets.

Here we review how studies in yeast have allowed dissection of the basic underlying mechanisms of mutant htt toxicity and have aided in the identification of novel small molecules and drug targets with therapeutic potential for HD.

2. Huntington's Disease

HD is a fatal neurodegenerative disorder that predominantly affects brain cells of the striatum and cortex, resulting in progressive chorea, rigidity, and dementia *(8)*. The prevalence of HD is 5–10 in 100,000 people in the United States and United Kingdom *(9)*. In adults, the median age of onset is the late forties or early fifties, but the juvenile form can appear as early as 2 years of age *(10)*.

The mutation that causes HD is an expansion of CAG repeats in the HD gene and is inherited in an autosomal dominant manner *(1)*. In the general population, the number of CAG repeats ranges from 4 to 35. Expansions of 36–39 CAG repeats increase the risk of developing HD, while expansions of 40 or more CAG repeats are fully penetrant *(11, 12)*. The length of the CAG repeat correlates inversely with the age of onset *(13)*, with expansions of 70 CAG repeats or longer inevitably leading to juvenile-onset HD *(1, 14)*. Although it is inversely correlated with the number of CAG repeats, the age of onset is highly variable when controlling for repeat length. Indeed, work with HD kindreds containing over 18,000 individuals found that approximately 40% of variation in age of onset in HD patients is due to genetic modifiers *(15)*, suggesting that many therapeutic targets may be available.

Pathologically, HD is characterized by a 10–20% loss of brain mass. The atrophy occurs in many brain regions but is most prominent in the striatum *(16)*. Medium spiny neurons, which account for about 95% of striatal cells, are the most affected cells. These cells, which express the γ-aminobutyric acid (GABA) receptor, are inhibitory projection neurons that carry the output of the striatum to the globus pallidus and the substantia nigra *(16)*. The striatal pathology probably underlies the involuntary movements and several other symptoms of the disease.

The major neuropathological hallmark in HD is brain lesions composed of intranuclear and cytoplasmic inclusions that contain mutant htt. The CAG repeat in the HD gene encodes a polyQ stretch at the N-terminal end of htt. A protein fragment encoding the first exon of htt with an expanded polyQ can spontaneously form amyloid fibers in vitro *(17)*, and aggregates containing mutant htt are present in the brains of mice expressing mutant htt *(18)* and in brains of HD patients *(19)*.

The length of the polyQ expansion in htt correlates directly with its aggregation kinetics in vitro and with severity of the disease. The 17-amino acid amino-terminus of htt, which lies directly upstream of the polyglutamine tract, forms an amphipathic α-helical domain that can bind membranes and be targeted to vesicles and the endoplasmic reticulum *(20, 21)*. This domain appears to modulate the toxicity and aggregation of mutant htt, and it disrupts calcium homeostasis in glutamate-challenged PC12 cells *(20)*. Immediately after the polyglutamine tract, htt contains two proline-rich regions; in mutant htt, these regions have been implicated in aggregation, cellular toxicity, sequestration of trafficking proteins, and membrane interactions *(22–28)*. Further downstream of the proline-rich region, htt contains several HEAT domains, each approximately 40 amino acids long, that form hydrophobic α-helices and are likely to be involved in protein interactions *(29)*. The structure of htt led to the prediction that wild-type htt protein is a multifunctional scaffolding protein *(30, 31)*.

Despite the many putative roles of htt in the cell, the pathogenesis of HD appears to reflect predominantly a gain-of-function mechanism due to expansion of the polyQ tract rather than a loss of wild-type htt function *(32)*. Several molecular mechanisms have been implicated in the pathogenesis of HD, including transcriptional dysregulation, perturbations of kynurenine pathway metabolites which result in excitotoxicity, oxidative stress, impaired energy metabolism, defective vesicle trafficking in axons, and impairment of ubiquitination and proteasomal function.

3. Using Yeast to Model Neurodegenerative Diseases

Yeast have been used to model features of several neurodegenerative disorders, including HD, PD, Alzheimer's disease (AD), Friedreich's ataxia, and amyotrophic lateral sclerosis *(33, 34)*. Additionally, studies in yeast using both yeast prions and mammalian prion-related protein have provided insight into the mechanism of prion action and disease *(35, 36)*. While models of disease in higher eukaryotes (fruit flies, nematodes, and mice) are extremely powerful tools for analysis of genetic and pharmacological modification of phenotype/pathology, yeast models provide a rapid and facile alternative to the higher cost and slower pace of these models. Yeast models of protein misfolding allow both genetic and pharmacological screens with a rapidity and ease not possible in fly and worm models, and can be used to identify conserved cellular mechanisms critical to toxicity and candidate therapeutic targets *(37–39)*. Ultimately, candidate modifier genes or compounds identified by screens in yeast must be validated in mammalian systems.

Aspects of neurodegenerative diseases have been modeled in yeast by analyzing the function of yeast homologs of human disease genes or by analyzing the phenotypes caused by expressing human disease genes in yeast. The first approach has been used to model aspects of Friedreich's ataxia, amyotrophic lateral sclerosis, and prion disease. Studies of the yeast homologs of the human genes implicated in Friedreich's ataxia (*YHF1*) and amyotrophic lateral sclerosis (*SOD1*) have helped elucidate pathogenic mechanisms underlying those diseases *(33, 40)*. In the second approach, genes of interest are expressed in yeast, and this has proven successful for modeling aspects of HD, PD, and AD. These models recapitulate many disease-relevant phenotypes and have provided mechanistic insight into the pathology of these diseases.

Obviously, there are limitations to studying mechanisms of protein misfolding in yeast, as there are with any model system. First, genes involved in neurodegeneration may not be present in the yeast genome. Nonetheless, several promising therapeutic targets and compounds relevant to HD have been identified in yeast, indicating that targeting evolutionarily conserved genes and pathways in yeast is a valid approach for drug discovery. Second, cellular toxicity in yeast may not be related to that involved in neurodegeneration. However, yeast can undergo apoptosis-like cell death in response to several stimuli, and several yeast orthologs of crucial apoptotic regulators exist *(41)*. Intriguingly, expression of a toxic mutant htt fragment in yeast produces hallmarks of apoptosis, including mitochondrial fragmentation and caspase activation *(42)*. In addition, expression of wild-type α-synuclein or the inherited mutants (A30P, A53T) in a yeast model of PD triggers several markers of apoptosis, and deletion of a yeast metacaspase gene suppresses many of these apoptosis-like phenotypes *(43)*. In summary, several thorough and insightful studies have shown that yeast can be used to model pathogenic mechanisms involved in neurodegenerative diseases.

4. Overview of Yeast Models of HD and Phenotypic Analyses

Several yeast models of HD have been developed to study the folding and behavior of mutant htt within a eukaryotic cell *(44–47)* (**Table 1**). These studies have supported and provided further insight into observations made with mutant htt in mammalian cells. Expression of an amino-terminal fragment of htt containing the polyQ region fused to green fluorescent protein (GFP) in yeast results in a polyQ length–dependent aggregation and formation of cytoplasmic and nuclear inclusions that can easily be detected in living cells by fluorescence microscopy *(42, 44, 46, 47)*. Genetic impairment of the ubiquitin/proteasome degradation pathway via lesions in the genes encoding the ubiquitin/proteasome pathway components *UBA1*, *DOA3*, and *SEN3* does

Table 1
Studies in which yeast were used to model aspects of HD

Model	IBs[a]	Aggregates[b]	Localization	Toxicity
Krobitsch et al. (2000)	+	+	Cytoplasmic	−
Muchowski et al. (2000)	+	+	Cytoplasmic	−
Hughes et al. (2001)	+	n/a	Cytoplasmic/nuclear[c]	−
Meriin et al. (2002)	+	+	Cytoplasmic/nuclear[d]	+

IBs inclusion bodies
[a]"+" indicates polyglutamine length-dependent formation of IBs based upon microscopy
[b]"+" indicates polyglutamine length-dependent aggregation confirmed by biochemical approaches
[c]SV40 nuclear localization signal used to target htt constructs to the nucleus
[d]Sokolov et al., 2006

not alter the formation of mutant htt inclusions function in yeast, suggesting that ubiquitination is not required for their formation *(44)*. A simple filter-retention assay has also been used to isolate detergent-insoluble aggregates formed by polyQ proteins with expanded repeats in the disease-causing range *(45)*.

In several of these models, modulation of chaperone expression alters the formation of mutant htt aggregates or inclusions. Co-expression of Hsp40 (Ydj1) or Hsp70 (Ssa1 or Ssb1) with mutant htt (53Q) in yeast inhibited formation of these detergent-insoluble aggregates and resulted in the formation of detergent-soluble aggregates *(45)*. In addition, Hsp40 and Hsp70 family members co-immunoprecipitated with the mutant htt fragment but not with the wild-type control. In another study, co-expression of Hsp40 and Hsp70 family members (Sis1 and Ssa1, respectively) with mutant htt fragments containing longer polyQ expansions (72Q and 103Q) resulted in a larger number of smaller inclusions *(44)*. Overexpression of the chaperone Hsp104 increased the number of inclusions containing mutant htt, while deletion of Hsp104 eliminated their formation and resulted in diffuse GFP fluorescence, most likely due to the indirect effect of curing the yeast prion [RNQ+] (see below). Depletion of the chaperonin TRiC increases aggregation of a mutant htt fragment in yeast *(48)*. Therefore, as in mammalian cells, expression of mutant htt fragments in yeast results in polyQ-length dependent aggregation that can be modulated by chaperone expression.

Perhaps the most-studied and best-characterized yeast model of HD, the Meriin model *(46)*, exhibits polyQ length–dependent toxicity. Expression of an htt fragment with a polyQ length in the pathogenic range (Htt103Q) is toxic in yeast, while expression of the same construct containing a nonexpanded polyQ tract (Htt25Q) shows no negative effect on growth *(46)*. This model has allowed genetic characterization of conserved cellular pathways required for mutant htt toxicity in yeast. Aggregation of mutant htt (103Q) was suppressed when the activities of certain chaperone proteins were impaired, including Hsp104, Sis1, and Ydj1, supporting observations made in the models discussed above. Interestingly, the polyQ length–dependent toxicity in the Meriin model requires the presence of the yeast prion Rnq1 in its prion conformation, as well as other putative prion proteins *(39, 46)*.

This model is unique among the yeast models in that it exhibits polyQ length-dependent toxicity, and we have focused much effort on understanding how protein misfolding/aggregation in this model leads to cellular toxicity. We learned that both cis and trans factors are critical for mutant htt toxicity in yeast. First, we found that several aggregation prone proteins play a role in mutant htt toxicity and affect the modulation of aggregation *(27, 39)*. Several of these proteins are known or predicted yeast prion proteins. We and others have also observed that the flanking amino acid sequences are required for modulating toxic conformations of these htt fragments in yeast *(25, 27)*. This conversion of benign species to toxic species (and vice-versa) can operate in cis or in trans *(27)*. Several factors downstream of mutant htt misfolding/aggregation may contribute to toxicity in this yeast model of HD, including perturbations in the kynurenine pathway *(39)*, defects in endocytosis *(49, 50)*, transcriptional dysregulation *(47)*, active transport along microtubules *(51)*, apoptotic-like events *(42)*, increased levels of reactive oxygen species (ROS) *(39, 52)*, and mitochondrial dysfunction *(42, 52)*. Our work dissecting the central role of the kynurenine pathway in mutant htt toxicity in yeast is described at length below.

Using the Meriin model, we performed a genomic screen for null mutations that suppress the toxicity of Htt103Q *(39)*. One of the most potent suppressors was a deletion of the gene *BNA4*, which encodes the yeast homolog of kynurenine 3-monooxygenase (KMO), which functions in the kynurenine pathway of tryptophan degradation. Intriguingly, this pathway is activated in HD patients and in animal models of HD *(53)*. Several tryptophan metabolites are neuroactive, including 3-hydroxykynurenine (3-HK) and quinolinic acid (QUIN), whose neurotoxic effects are mediated by mechanisms that are likely to include the generation of reactive oxygen species (ROS). Kynurenine pathway metabolites and enzymes are well conserved from yeast to humans, and the genetics of the pathway have been extensively characterized in yeast *(54)*.

Since deletion of KMO is predicted to eliminate the synthesis of QUIN and 3-HK, the identification of *bna4* as a suppressor led us to hypothesize that upregulation of the QUIN branch of the kynurenine pathway could contribute to toxicity in yeast, as observed in HD patients and in mouse models of HD. If so, increased levels of QUIN and 3-HK could contribute to Htt103Q-dependent toxicity, and elimination of these metabolites by deletion of KMO activity could be suppress the toxicity.

To test this hypothesis, we measured 3-HK and QUIN levels in wild-type yeast expressing Htt25Q or Htt103Q and in the *bna4* suppressor strain expressing Htt103Q. Excitingly, 3-HK and QUIN levels were significantly higher in wild-type cells expressing the toxic Htt103Q construct than in controls expressing nontoxic Htt25Q, as in HD patients and in the mouse models of HD. As predicted, deletion of *BNA4* eliminated 3-HK and QUIN and reduced Htt103Q-mediated toxicity. This was the first evidence for a direct relationship between lower levels of endogenous 3-HK and QUIN and reduced toxicity of mutant htt-dependent toxicity in a genetic model of HD.

We also measured ROS levels using two oxidation-sensitive dyes: dihydroethidium and dihydrorhodamine-123. ROS levels were approximately eightfold higher in wild-type yeast expressing Htt103Q, than in Htt25Q controls; however, the levels of ROS in *bna4Δ* yeast expressing Htt103Q were not significantly elevated. Thus, there is a direct correlation between the levels of 3-HK and QUIN, the levels of ROS, and the levels of cellular toxicity in this yeast model of HD. Interestingly, KMO is an outer mitochondrial membrane protein and mitochondrial dysfunction has been implicated in HD pathology and the generation of ROS. These findings suggest that expression of Htt103Q leads to upregulation of the kynurenine pathway, causing cellular toxicity through a ROS-mediated mechanism.

We also found that pharmacological inhibition of KMO in yeast with the compound Ro 61-8048 decreased levels of 3-HK, decreased generation of ROS, and ameliorated the Htt103Q growth impairment. Ro 61-8048 is currently in trials with HD model mice, and preliminary results indicate that though this compound does not readily cross the blood brain barrier, it improves several outcome measures in these mice (unpublished observations).

Many of the above observations made in yeast have since been validated in other models of HD, including flies, mammalian cells, and mice, highlighting the value of yeast as a tool for studying conserved pathogenic mechanisms in HD and human disorders in general. Such confirmation suggests that these models can be reliably used to screen for both genes and compounds that modulate the toxicity and aggregation of mutant htt.

5. Using Yeast Models of HD for Genomic Screens and for Identifying Therapeutic Compounds

Yeast models of HD have enabled us to perform genomic screens to identify gene deletions that either enhance or suppress mutant htt toxicity in yeast *(37, 39)* (**Fig. 1**). These models have also allowed for screening and individual testing of small molecules to identify novel candidate therapeutic compounds which have been subsequently validated in *Drosophila* and mammalian cells *(38, 55)*. The genetic screens have taken advantage of a collection of yeast gene deletion strains (YGDS) developed by the Saccharomyces Genome Deletion Project *(56, 57)*. The YGDS collection contains 4,850 viable mutant haploid strains, each lacking a single gene, and has been used to identify new genes and pathways in several biological processes.

To identify gene deletions that enhance mutant htt toxicity, we used a nontoxic exon 1 mutant htt fragment construct to screen the YGDS collection *(37)* (**Fig. 1**). This enhancer screen identified gene deletions that produced polyQ length-dependent

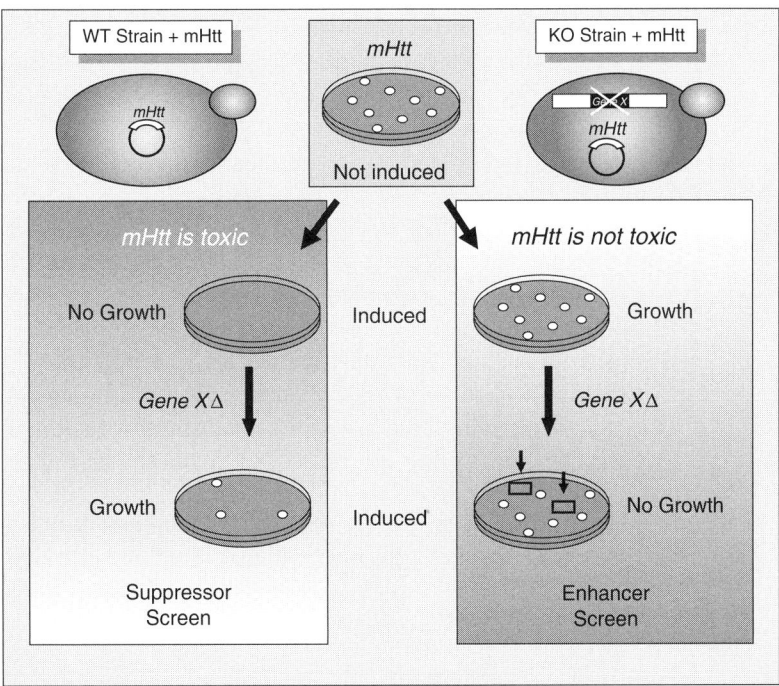

Fig. 1. Overview of loss-of-function screens for suppressors or enhancers of toxicity of a mutant htt fragment (mHtt) in yeast. A plasmid containing mHtt under the control of an inducible promoter is transformed into wild-type yeast cells and the YKO deletion set. If mHtt is toxic to wild-type yeast cells, then the YKO set can be screened for gene deletions (i.e., Gene X) that restore growth or suppress toxicity. If mHtt is not toxic to wild-type cells, the YKO set can be screened for gene deletions that enhance the toxicity of mHtt.

synthetic lethality in combination with the mutant htt fragment. The YGDS strains were transformed with the mutant htt construct, and 52 gene deletions that enhanced toxicity of the mutant htt fragment were isolated. Forty (77%) of these enhancers are genes with known or predicted functions; 35% belong to the functionally related categories of protein folding, response to cell stress, and ubiquitin-dependent protein catabolism; the rest were dispersed among several diverse functional categories. Among the suppressors were two Hsp40 homologs (Apj1 and Hlj1), suggesting that Hsp40 chaperones are necessary for suppression of polyQ toxicity. Interestingly, Hsp40 chaperones cooperate with Hsp70 partners, which modulate pathogenesis in mouse models of polyglutamine disorders *(58, 59)*.

We also screened the YGDS for loss-of-function suppressors of toxicity in the Meriin model *(39, 46)* (**Fig. 1**). The primary goal was to identify targets that, when genetically inhibited (deleted), reduce the toxicity of a mutant htt fragment and would presumably produce a similar phenotype if inhibited pharmacologically. We also sought to identify cellular pathways required for this toxicity. Of the genes identified, 25% encode proteins that cluster into the functionally related categories of vesicular transport, vacuolar protein sorting, and vacuolar import; 25% encode proteins involved directly in transcription or in the establishment/maintenance of chromatin architecture; and ~21% encode known yeast prions or proteins containing Q/N-rich regions that may mediate prion-like aggregation.

The genes encoding proteins involved in vesicular transport and vacuolar function are of particular interest because induction of autophagy in fly and mouse models of HD reduces the toxicity of expanded polyglutamine *(60)*. Thus, in yeast, Htt103Q might be targeted for degradation in the vacuole, and the autophagic machinery might become saturated or impaired as Htt103Q accumulates in cells. We hypothesize that perturbation of protein sorting to the vacuole, a phenotype found in many of these gene deletion strains, may allow for resumption of normal autophagy, perhaps by excluding Htt103Q from the vacuole. Consistent with this model, deletion of the gene encoding the vacuolar protease Cps1 enhances the toxicity of a mutant htt fragment *(37)*. For both the enhancer and suppressor screens described above, over half of the suppressors have mammalian homologs *(61)* In the case of the suppressors, ~87% are known to be expressed in the brain.

Recently, a high-throughput screen of 16,000 compounds was performed to identify small-molecule inhibitors of polyQ aggregation in yeast expressing a GFP-tagged mutant htt fragment *(38)*. This screen yielded nine small molecules that ameliorated growth and/or fluorescence. Upon further testing, one of these compounds (C2-8) suppressed polyQ aggregation in a PC12 cell–based model and in neurons of cultured brain slices from HD model mice; it also ameliorated neurodegeneration in a *Drosophila* model of HD. Intriguingly, C2-8 is a structural analog of the aforementioned KMO inhibitor that ameliorates Htt103Q toxicity. This finding suggests that genetic and chemical-genetic approaches can converge at the same candidate targets. Moreover, C2-8 readily crosses the blood-brain barrier in mice and improves both motor performance and neuronal atrophy in a mouse model of HD *(62)*. As predicted by the aggregation assay used to identify C2-8, the volume of aggregate in striatal neurons was reduced.

A yeast model has also been used to validate green tea (−)-epigallocatechin-gallate (EGCG) as a therapeutic compound for HD *(55)*. A library of ~5,000 natural substances was screened with a membrane filer retardation assay. One purified natural compound (EGCG) inhibited aggregation of mutant htt exon 1. When tested in yeast expressing a toxic mutant htt fragment (Htt72Q), EGCG improved growth and reduced the number of cells carrying Htt72Q inclusion bodies. Further tests showed that EGCG reduces neurodegeneration and improves motor impairment in a fly model of HD. These studies highlight the potential for yeast-based approaches to screen for candidate therapeutic compounds for treatment of neurodegenerative disorders.

In summary, yeast models of HD have been very useful for identifying genetic modifiers of mutant htt toxicity and compounds that modulate the toxicity and aggregation of mutant htt. This work has led to further validation studies of KMO inhibitors, EGCG, and C2-8 as potential therapies for HD. Thus, despite their simplicity, yeast models are an important tool for identifying and characterizing therapeutics targets and compounds for HD.

6. Conclusions and Future Directions

Within the last decade, several models in *S. cerevisiae* have been developed that recapitulate important features characterized by HD. The success of genetic and compound screens will likely lead to further exploitation of this model in the hopes of informing HD therapeutics in the future. A number of genetic approaches can be implemented with these models that have not yet been attempted, including overexpression screens with ORF libraries. Random clone libraries have been used for overexpression studies in yeast, but only recently has the Yeast ORF Collection (Open Biosystems) become available, an array of ~5,500 overexpression constructs covering the vast majority of the yeast genome. Systematic screening of the collection is efficient as the clones are arrayed in 96-well microtiter plates. Isolation of overexpression suppressors of mutant htt toxicity would identify candidate ORFs that, when overexpressed, protect against cellular toxicity in yeast.

So far, only nonessential genes have been tested in loss-of-function screens. Therefore many important targets have not been screened, and approaches need to be implemented to identify essential modifier genes in yeast.

Critically, candidate modifiers or compounds identified in yeast need to be validated in more physiologically relevant models of neurodegeneration, such as mammalian cell and mouse models of HD. Nonetheless, yeast will undoubtedly continue to be useful for advancing our knowledge of the underlying cellular mechanisms in HD and many other human diseases, even in cases in which it is initially difficult to conceive of the utility of this simple, yet powerful organism. The promising candidate targets and compounds for HD identified in yeast thus far have provided further support for the role of yeast as "living test tubes" in studies of human disease.

References

1. A novel gene containing a trinucleotide repeat that is expanded and unstable on Huntington's disease chromosomes. The Huntington's Disease Collaborative Research Group. Cell 1993;72(6):971–83.
2. Botstein D, Fink GR. Yeast: an experimental organism for modern biology. Science 1988; 240(4858):1439–43.
3. Fields S, Johnston M. Cell biology. Whither model organism research. Science 2005; 307(5717):1885–6.
4. Hartwell LH. Nobel Lecture. Yeast and cancer. Biosci Rep 2002;22(3–4):373–94.
5. Nurse PM. Nobel Lecture. Cyclin dependent kinases and cell cycle control. Biosci Rep 2002;22(5–6):487–99.
6. Goffeau A, Barrell BG, Bussey H, et al. Life with 6000 genes. Science 1996;274(5287):546,63–7.
7. Sherman F. Getting started with yeast. In: Guthrie CaF, G.R. (Eds), Methods Enzymol 1991;194:3–21.
8. Martin JB, Gusella JF. Huntington's disease. Pathogenesis and management. N Engl J Med 1986;315(20):1267–76.
9. Harper PS. The epidemiology of Huntington's disease. In: Bates G, Harper PS, Jones L, eds. Huntington's Disease. Oxford, UK: Oxford University Press; 2002:159–97.
10. Kremer B. Clinical neurology of Huntington's disease. In: Bates G, Harper PS, Jones L, eds. Huntington's disease. Oxford: Oxford University Press; 2002:28–61.
11. Rubinsztein DC, Leggo J, Coles R, et al. Phenotypic characterization of individuals with 30-40 CAG repeats in the Huntington disease (HD) gene reveals HD cases with 36 repeats and apparently normal elderly individuals with 36-39 repeats. Am J Hum Genet 1996;59(1):16–22.
12. Myers RH, Marans KS, MacDonald ME. Genetic Instabilities and Hereditary Neurological Diseases. In: Wells RD, Warren ST, eds. San Diego: Academic Press; 1998:301–23.
13. Duyao M, Ambrose C, Myers R, et al. Trinucleotide repeat length instability and age of onset in Huntington's disease. Nat Genet 1993;4(4):387–92.
14. Telenius H, Kremer HP, Theilmann J, et al. Molecular analysis of juvenile Huntington disease: the major influence on (CAG)n repeat length is the sex of the affected parent. Hum Mol Genet 1993;2(10):1535–40.
15. Wexler NS, Lorimer J, Porter J, et al. Venezuelan kindreds reveal that genetic and environmental factors modulate Huntington's disease age of onset. Proc Natl Acad Sci U S A 2004;101(10):3498–503.
16. Gutekunst CA, Norflus F, Hersch SM. The neuropathology of Huntington's disease. In: Bates G, Harper PS, Jones L, eds. Huntington's disease. Oxford: Oxford University Press; 2002:251–75.

17. Scherzinger E, Lurz R, Turmaine M, et al. Huntingtin-encoded polyglutamine expansions form amyloid-like protein aggregates in vitro and in vivo. Cell 1997;90(3):549–58.
18. Davies SW, Turmaine M, Cozens BA, et al. Formation of neuronal intranuclear inclusions underlies the neurological dysfunction in mice transgenic for the HD mutation. Cell 1997;90(3):537–48.
19. DiFiglia M, Sapp E, Chase KO, et al. Aggregation of huntingtin in neuronal intranuclear inclusions and dystropic neurites in brain. Science 1997;277(5334):1990–3.
20. Rockabrand E, Slepko N, Pantalone A, et al. The first 17 amino acids of Huntingtin modulate its sub-cellular localization, aggregation and effects on calcium homeostasis. Hum Mol Genet 2007;16(1):61–77.
21. Atwal RS, Xia J, Pinchev D, Taylor J, Epand RM, Truant R. Huntingtin has a membrane association signal that can modulate huntingtin aggregation, nuclear entry and toxicity. Hum Mol Genet 2007;16(21):2600–15.
22. Khoshnan A, Ko J, Patterson PH. Effects of intracellular expression of anti-huntingtin antibodies of various specificities on mutant huntingtin aggregation and toxicity. Proc Natl Acad Sci U S A 2002;99(2):1002–7.
23. Qin ZH, Wang Y, Sapp E, et al. Huntingtin bodies sequester vesicle-associated proteins by a polyproline-dependent interaction. J Neurosci 2004;24(1):269–81.
24. Bhattacharyya A, Thakur AK, Chellgren VM, et al. Oligoproline effects on polyglutamine conformation and aggregation. J Mol Biol 2006;355(3):524–35.
25. Dehay B, Bertolotti A. Critical role of the proline-rich region in Huntingtin for aggregation and cytotoxicity in yeast. J Biol Chem 2006;281(47):35608–15.
26. Suopanki J, Gotz C, Lutsch G, et al. Interaction of huntingtin fragments with brain membranes–clues to early dysfunction in Huntington's disease. J Neurochem 2006;96(3):870–84.
27. Duennwald ML, Jagadish S, Muchowski PJ, Lindquist S. Flanking sequences profoundly alter polyglutamine toxicity in yeast. Proc Natl Acad Sci U S A 2006;103(29):11045–50.
28. Duennwald ML, Jagadish S, Giorgini F, Muchowski PJ, Lindquist S. A network of protein interactions determines polyglutamine toxicity. Proc Natl Acad Sci U S A 2006;103(29):11051–6.
29. Andrade MA, Bork P. HEAT repeats in the Huntington's disease protein. Nat Genet 1995;11(2):115–6.
30. Harjes P, Wanker EE. The hunt for huntingtin function: interaction partners tell many different stories. Trends Biochem Sci 2003;28(8):425–33.
31. Bates GP. History of genetic disease: the molecular genetics of Huntington disease – a history. Nat Rev Genet 2005;6(10):766–73.
32. Li SH, Li XJ. Huntingtin-protein interactions and the pathogenesis of Huntington's disease. Trends Genet 2004;20(3):146–54.
33. Outeiro TF, Muchowski PJ. Molecular genetics approaches in yeast to study amyloid diseases. J Mol Neurosci 2004;23(1–2):49–60.
34. Outeiro TF, Giorgini F. Yeast as a drug discovery platform in Huntington's and Parkinson's diseases. Biotechnol J 2006;1(3):258–69.
35. Uptain SM, Lindquist S. Prions as protein-based genetic elements. Annu Rev Microbiol 2002;56:703–41.
36. Chien P, Weissman JS, DePace AH. Emerging principles of conformation-based prion inheritance. Annu Rev Biochem 2004;73:617–56.
37. Willingham S, Outeiro TF, DeVit MJ, Lindquist SL, Muchowski PJ. Yeast genes that enhance the toxicity of a mutant huntingtin fragment or alpha-synuclein. Science 2003;302(5651):1769–72.
38. Zhang X, Smith DL, Meriin AB, et al. A potent small molecule inhibits polyglutamine aggregation in Huntington's disease neurons and suppresses neurodegeneration in vivo. Proc Natl Acad Sci U S A 2005;102(3):892–7.
39. Giorgini F, Guidetti P, Nguyen Q, Bennett SC, Muchowski PJ. A genomic screen in yeast implicates kynurenine 3-monooxygenase as a therapeutic target for Huntington disease. Nat Genet 2005;37(5):526–31.
40. Puccio H, Koenig M. Recent advances in the molecular pathogenesis of Friedreich ataxia. Hum Mol Genet 2000;9(6):887–92.
41. Madeo F, Herker E, Wissing S, Jungwirth H, Eisenberg T, Frohlich KU. Apoptosis in yeast. Curr Opin Microbiol 2004;7(6):655–60.
42. Sokolov S, Pozniakovsky A, Bocharova N, Knorre D, Severin F. Expression of an expanded polyglutamine domain in yeast causes death with apoptotic markers. Biochim Biophys Acta 2006;1757(5–6):660–6.
43. Flower TR, Chesnokova LS, Froelich CA, Dixon C, Witt SN. Heat shock prevents alpha-synuclein-induced apoptosis in a yeast model of Parkinson's disease. J Mol Biol 2005;351(5):1081–100.
44. Krobitsch S, Lindquist S. Aggregation of huntingtin in yeast varies with the length of the polyglutamine expansion and the expression

of chaperone proteins. Proc Natl Acad Sci U S A 2000;97(4):1589–94.
45. Muchowski PJ, Schaffar G, Sittler A, Wanker EE, Hayer-Hartl MK, Hartl FU. Hsp70 and hsp40 chaperones can inhibit self-assembly of polyglutamine proteins into amyloid-like fibrils. Proc Natl Acad Sci U S A 2000;97(14):7841–6.
46. Meriin AB, Zhang X, He X, Newnam GP, Chernoff YO, Sherman MY. Huntington toxicity in yeast model depends on polyglutamine aggregation mediated by a prion-like protein Rnq1. J Cell Biol 2002;157(6):997–1004.
47. Hughes RE, Lo RS, Davis C, et al. Altered transcription in yeast expressing expanded polyglutamine. Proc Natl Acad Sci U S A 2001;98(23):13201–6.
48. Tam S, Geller R, Spiess C, Frydman J. The chaperonin TRiC controls polyglutamine aggregation and toxicity through subunit-specific interactions. Nat Cell Biol 2006;8(10):1155–62.
49. Meriin AB, Zhang X, Miliaras NB, et al. Aggregation of expanded polyglutamine domain in yeast leads to defects in endocytosis. Mol Cell Biol 2003;23(21):7554–65.
50. Meriin AB, Zhang X, Alexandrov IM, et al. Endocytosis machinery is involved in aggregation of proteins with expanded polyglutamine domains. Faseb J 2007;21(8):1915–25.
51. Muchowski PJ, Ning K, D'Souza-Schorey C, Fields S. Requirement of an intact microtubule cytoskeleton for aggregation and inclusion body formation by a mutant huntingtin fragment. Proc Natl Acad Sci U S A 2002;99(2):727–32.
52. Solans A, Zambrano A, Rodriguez M, Barrientos A. Cytotoxicity of a mutant huntingtin fragment in yeast involves early alterations in mitochondrial OXPHOS complexes II and III. Hum Mol Genet 2006;15(20):3063–81.
53. Schwarcz R. The kynurenine pathway of tryptophan degradation as a drug target. Curr Opin Pharmacol 2004;4(1):12–7.
54. Panozzo C, Nawara M, Suski C, et al. Aerobic and anaerobic NAD+ metabolism in *Saccharomyces cerevisiae*. FEBS Lett 2002;517(1–3):97–102.
55. Ehrnhoefer DE, Duennwald M, Markovic P, et al. Green tea (-)-epigallocatechin-gallate modulates early events in huntingtin misfolding and reduces toxicity in Huntington's disease models. Hum Mol Genet 2006;15(18):2743–51.
56. Giaever G, Chu AM, Ni L, et al. Functional profiling of the *Saccharomyces cerevisiae* genome. Nature 2002;418(6896):387–91.
57. Winzeler EA, Shoemaker DD, Astromoff A, et al. Functional characterization of the *S. cerevisiae* genome by gene deletion and parallel analysis. Science 1999;285(5429):901–6.
58. Hansson O, Nylandsted J, Castilho RF, Leist M, Jaattela M, Brundin P. Overexpression of heat shock protein 70 in R6/2 Huntington's disease mice has only modest effects on disease progression. Brain Res 2003;970(1–2):47–57.
59. Hay DG, Sathasivam K, Tobaben S, et al. Progressive decrease in chaperone protein levels in a mouse model of Huntington's disease and induction of stress proteins as a therapeutic approach. Hum Mol Genet 2004;13(13):1389–405.
60. Ravikumar B, Vacher C, Berger Z, et al. Inhibition of mTOR induces autophagy and reduces toxicity of polyglutamine expansions in fly and mouse models of Huntington disease. Nat Genet 2004;36(6):585–95.
61. Botstein D, Chervitz SA, Cherry JM. Yeast as a model organism. Science 1997;277(5330):1259–60.
62. Chopra V, Fox JH, Lieberman G, et al. A small-molecule therapeutic lead for Huntington's disease: preclinical pharmacology and efficacy of C2-8 in the R6/2 transgenic mouse. Proc Natl Acad Sci U S A 2007;104(42):16685–89.

Chapter 10

Global Proteomic Analysis of *Saccharomyces cerevisiae* Identifies Molecular Pathways of Histone Modifications

Jessica Jackson and Ali Shilatifard

Summary

The very long DNA of the eukaryotic cells must remain functional when packaged into the cell nucleus. Although we know very little about this process, it is clear at this time that chromatin and its post-translational modifications play a pivotal role. Yeast *Saccharomyces cerevisiae* provides a powerful genetic and biochemical model system for deciphering the molecular machinery involved in chromatin modification and transcriptional regulation. In this chapter, we describe a novel method, the Global Proteomic analysis in *S. cerevisiae* (GPS), for the global analysis of the molecular machinery required for proper histone modifications. Since many of the molecular machineries involved in chromatin biology are highly conserved from yeast to humans, GPS has proven to be an outstanding method for the identification of the molecular pathways involved in chromatin modifications.

Key words: Chromatin, Histones, Histone modifications, Transcription, RNA polymerase II, COMPASS, Histone methylations, Histone acetylations, Methylase, GPS

1. Introduction

Chromatin is an array of nucleosomes containing 146 bp of DNA wrapped twice around an octamer composed of two copies of each of the core histone proteins H2A, H2B, H3, and H4 *(1–3)*. Linker histones (e.g., H1) and other non-histone proteins can further pack nucleosomes to form higher order chromatin structures. Core histones are highly evolutionarily conserved proteins, with a flexible amino-terminal tail and a globular domain *(2)*. Structural studies have demonstrated that interactions between the globular domains of each of the core histones form the nucleosome scaffold *(3)*. These studies also confirmed that histone amino-terminal tails protrude outward from the nucleosome and can influence nucleosome–nucleosome interactions as well as interactions between nucleosomes and regulatory factors *(3)*. Histones can be modified by a variety of post-translational modifications such as acetylation, phosphorylation, ubiquitination, methylation, sumoylation, and ADP ribosylation *(4–7)*. However, several novel post-translational modifications of histones have been identified recently that occur within the core region of histones *(8–11)*. Modifications within the histone core can alter DNA–histone interactions within and between nucleosomes and thus affect higher order chromatin structures. Nucleosomes have now emerged as an active participant in the regulation of many pathways associated through chromosomal DNA. In this regard, the covalent modifications of histones can provide a combinatorial effect which can then be translated by nuclear proteins to influence a multitude of cellular processes such as transcription, replication, DNA repair, and cell cycle progression *(12)*.

Two protein families, the Trithorax group (Trx), which is the Drosophila homolog of human mixed-lineage leukemia (MLL), and the Polycomb group (Pc), are essential in the regulation of gene expression throughout development *(13–15)*. Chromosomal translocations in the *MLL* gene result in hematological malignancies, including acute myeloid and lymphoid leukemia *(13–16)*. Although these cytogenic abnormalities were discovered over 25 years ago *(14)*, little is known about the biochemical functions of *MLL*, its protein complexes, its translocation partners, and, particularly, why translocations result in leukemia. The best understood target genes for MLL are the clustered homeotic *Hox* genes. *Hox* gene products are involved in the specification of cell fate. Genetic studies have demonstrated that the Trx and Pc groups of proteins cooperate to antagonistically regulate *Hox* gene expression. The Trx group positively regulates gene expression, while the Pc group of proteins is essential for the maintenance of the silenced state of homeotic genes.

Many members of the Trx and Pc groups of proteins contain a 130–140-amino acid motif, termed the SET domain *(17, 18)*. The SET domain is found in a variety of chromatin-associated proteins. We identified the *Saccharomyces cerevisiae* Set1 protein as the homolog of MLL and isolated the yeast Set1-containing protein complex. This complex is named COMPASS (Complex of Proteins Associated with Set1) *(19)*. COMPASS was the first enzyme complex identified that catalyzes the mono-, di-, and trimethylation of H3K4 *(19–22)*. Following the identification of yeast COMPASS, the MLL protein was also isolated in a COMPASS-like complex capable of catalyzing the methylation of H3K4 *(23–26)*. Recent biochemical studies demonstrated that the Pc group of proteins is also histone-methyltransferase-capable of methylating lysine 27 of histone H3 (H3K27) *(27–29)*. Together, these biochemical studies suggest the existence of a fundamental role for the regulation of the patterns of histone methylations in development.

Large-scale alterations in the histone modification landscape in chromatin accompany gene expression changes associated with differentiation and cancer. Histone modifications play a major role in the reprogramming of the genome. Histone H3K4 methylation is a hallmark of transcriptionally active genes *(5)*. It antagonistically functions with H3K27 methylation patterns to regulate stem cell self-renewal and differentiation. Understanding how these modifications contribute to the differential regulation of the over 30,000 human genes for proper differentiation and development remains a major challenge. Much emphasis has focused on identifying the modifications and their locations within the genome; however, very little is known about the machinery that catalyzes the addition or the removal of the modifications throughout development.

To better define the molecular machinery and regulatory pathways involved in the post-translational modifications of histone H3K4 methylation in yeast, we devised a proteomic screen we call Global Proteomic analysis in *S. cerevisiae* (GPS) *(30)* (**Fig. 1**). In GPS, we prepare extracts from each of the nonessential yeast gene deletion mutants. These extracts will then be analyzed for defects in modifications of the histones by sodium

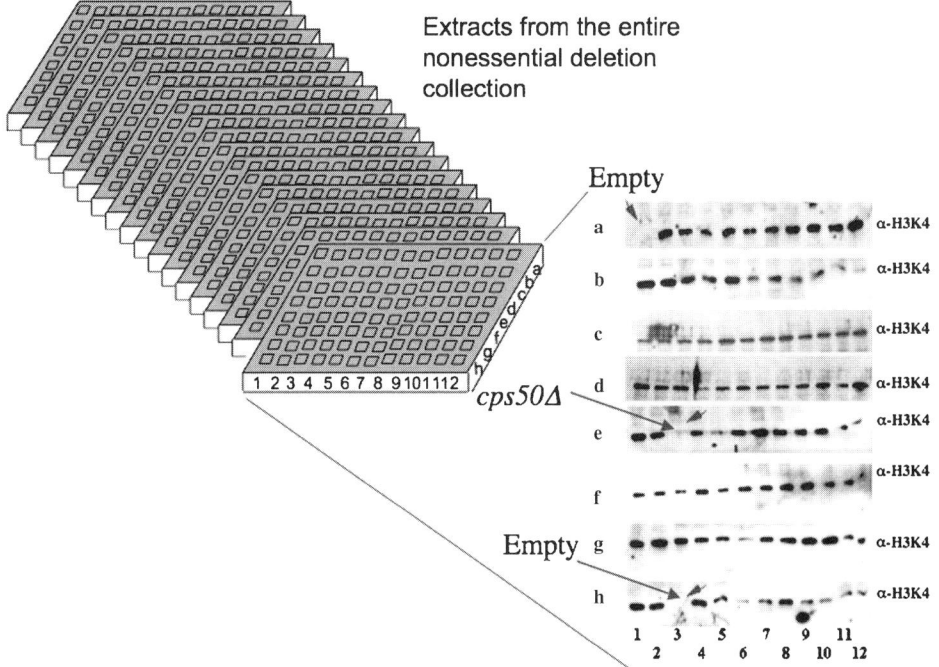

Fig. 1. Global proteomic analysis of *S. cerevisiae* (GPS). Schematic representation of GPS employing a polyclonal antibody specific to lysine 4 methylated histone H3. As described in detail in the text, cell extracts from the entire yeast deletion collection were made, and these extracts were transferred to 96-well plates. Each extract was applied to a 16% SDS/PAGE and subjected to Western analysis using antibodies towards methylated histone H3. In this figure, the Western data from a plate containing a known component of COMPASS are shown. Position E3 represents extracts from a strain deleted for the CPS50 subunit of COMPASS. *Arrows* at position A1 and H3 indicate empty wells as plate markers.

dodecyl sulfate polyacrylamide gel electrophoresis (SDS/PAGE) and Western blotting. Antibodies specific to H3K4-methylated histones will be used as probes. Employing GPS, we have been able to identify factors that are required for proper histone H3K4 methylation by COMPASS.

2. Materials

2.1. Growth of Nonessential Gene Deletion Mutants

1. The YPD medium used for growing the yeast gene deletion mutants is composed of 1% yeast extract, 2% proteo-peptone, and 2% dextrose (Sigma catalog # G8270). The dextrose was autoclaved separately from the remaining ingredients and was added immediately after it was removed from the autoclave while the solutions were still hot. Nunc™ brand OmniTray, single-well containers were used when making agar plates. For each plate, 2% agar was added to the YP media before autoclaving, and then combined with the dextrose.

2. The cells from which the protein extracts were made were inoculated using disposable, 96-well pinning devices (Genetix catalog # X5054) onto Nunc™ brand OmniTrays (Fisher Scientific catalog # 12-565-450) containing solid YPD media from frozen stocks in which the yeast gene deletion mutants were kept in 96-well plates (Fisher Scientific catalog # 07-200-90). The plates were allowed to grow for 48 h at 30 °C. From there, the cells were then inoculated into 5 ml of liquid YPD contained in sterile, polypropylene, 15-ml conical tubes, (MidSci catalog # 3018Y) and grown for 24 h at 30 °C.

2.2. Preparation of the Extracts

1. Cells were collected by centrifugation, transferred to 1.5-ml microcentrifuge tubes (SciMart catalog # GS-8509-N), and resuspended in 200 µl of lysis buffer (0.25 M Sucrose, 60 mM potassium chloride, 14 mM sodium chloride, 5 mM magnesium chloride, 1 mM calcium chloride, and 0.8% Triton X-100), and then 200 µl of 0.5 mm glass beads (BioSpec catalog # 11079105) was added.

2. After lysis and centrifugation, the supernatants were removed and the cell pellets were transferred, with water, to flexible 96-well PCR plates (Fisher Scientific catalog # 05-500-48) that were seated in Costar nonsterile 96-well flat-bottom culture plates (Fisher Scientific catalog # 07-200-90).

3. 4× Laemmli loading buffer (1 ml water, 2 ml, 1M Tris-HCl, pH 6.8, 3.2 ml 100% glycerol, 0.64 g SDS (Fisher Scientific catalog # BP166-500), 1.6 ml 2-mercaptoethanol (Sigma catalog # M7154), and 1 mg bromophenol blue) were added

2.3. Analysis of Extracts

to the pellets of the lysed cells, in addition to water, and then heated to 95 °C in an MJ thermocycler for 5 min.

1. 16% SDS gels were used to separate the extracts by PAGE, using a 1:10 dilution of the electrophoresis buffer (0.25 M Tris, 2 M glycine, 10% SDS).

2. The resolved proteins were blotted to GE™ brand 0.45-μm nitrocellulose membranes (ISC Bioexpress catalog # F-3190-30X3) that were submerged in Western transfer buffer (30 mM Tris, 0.15 M glycine, 20% methanol).

3. Nitrocellulose-bound proteins were probed with mono- and polyclonal antibodies specific for numerous histone modifications. Antibodies were diluted into a 1:10 dilution of Tris-buffered saline (TBS) pH 7.4 (0.1 M Tris, 1.5 M sodium chloride, 7 mM calcium chloride, 4 mM magnesium chloride). Washing of the membranes was performed in a 1:10 dilution of TBST pH 7.4 (0.1 M Tris, 1.5 M sodium chloride, 7 mM calcium chloride, 4 mM magnesium chloride, 0.5% Tween).

4. Blots were visualized by the exposure of the membranes to 14 × 17 in. IsoMax film (SciMart catalog # GX-331417) through the use of enhanced chemiluminescence (Perkin Elmer catalog # NEL104001EA).

3. Methods

1. Through the aid of a 96-well pinning device, frozen stocks from each of the approximately 4,800 nonessential yeast gene deletion mutants were inoculated onto an agar plate containing YPD, and allowed to grow for 48 h at 30 °C. The cells were then inoculated into 96 conical tubes (15 ml) containing 5 ml YPD, and grown for 24 h at 30 °C with continuous rotary movement.

2. The cells were then centrifuged at 500 g for 5 min. Using 1 ml of water, each cell pellet was washed and transferred to 1.5-ml microcentrifuge tubes. The tubes were centrifuged at 2,000 rpm for 5 min, the water was removed, and the cell pellets were resuspended in 200 μl of lysis buffer (0.25 M sucrose, 60 mM potassium chloride, 14 mM sodium chloride, 5 mM magnesium chloride, 1 mM calcium chloride, 0.8% Triton X-100), along with 200 μl of 0.5 mm glass beads. The tubes were then vigorously vortexed for 25 min at 4 °C in order to lyse the cells.

3. Using a 16G needle, cell lysates were obtained by puncturing a small hole at the bottom of the 1.5-ml tube. The tube was

then placed in a 15-ml conical tube to collect the contents. The 1.5-ml/15-ml tube construct was centrifuged for 10 min at 4 °C, which formed a white pellet and cloudy supernatant. The supernatant was removed and the pellet was resuspended in 100 µl of water and transferred to a 96-well PCR plate, to which 50 µl of 4× Laemmli loading buffer (1 ml water, 2 ml 1 M Tris, pH 6.8, 3.2 ml 100% glycerol, 0.64 g SDS, 1.6 ml 2-mercaptoethanol, 1 mg bromophenol blue) was added. Aluminum sealing film was placed on top of the 96-well plate, and the extracts were briefly vortexed before being heated at 95 °C for 5 min.

4. Approximately 10 µl of each extract was loaded onto a 16% SDS gel and subjected to PAGE, where the proteins were resolved at 50 mA for 1 h in an electrophoresis buffer. The resolved proteins contained in the gel were blotted onto nitrocellulose membranes submerged in a Western transfer buffer, and ran at 400 mA for 1 h. Protein-containing membranes were blocked in a milk solution consisting of non-fat dry milk and water for 30 min.

5. Antibodies specific for a variety of histone modifications were diluted in TBS and incubated with the membranes overnight at 4 °C with constant shaking. Following the removal of the antibody, the blots were subjected to three 10-min washes with TBST, after which a secondary antibody was applied and allowed to incubate with the membranes for 30 min. Three additional washes were performed in TBST. The membranes were exposed to 14 × 17 in. IsoMax film with exposure times ranging from 2 to 30 min. Through a chemiluminescent reaction, specific alterations to the histones were visualized on the film.

4. Notes

For COMPASS to trimethylate histone H3 in vivo, histone H2B must be monoubiquitinated by the Rad6/Bre1 complex *(5, 30–32)*. To identify other factors and the molecular machinery required for proper H2B monoubiquitination and, therefore, H3K4 methylation, we screened the entire yeast collection via GPS *(30)*. Our screen revealed that several other gene products, including the subunit of the Paf1 complex (**Fig. 2**) *(33, 34)* and the subunits of the Bur1/Bur2 kinase (**Fig. 3**) *(35)*, are required for proper H2B monoubiquitination and, therefore, H3K4 methylation. Furthermore, using polyclonal antibodies specifically recognizing mono-, di-, and trimethylated H3K4, we were able to demonstrate that the Cps40 subunit of COMPASS is required for proper H3K4 trimethylation by the enzyme (**Fig. 4**) *(36)*.

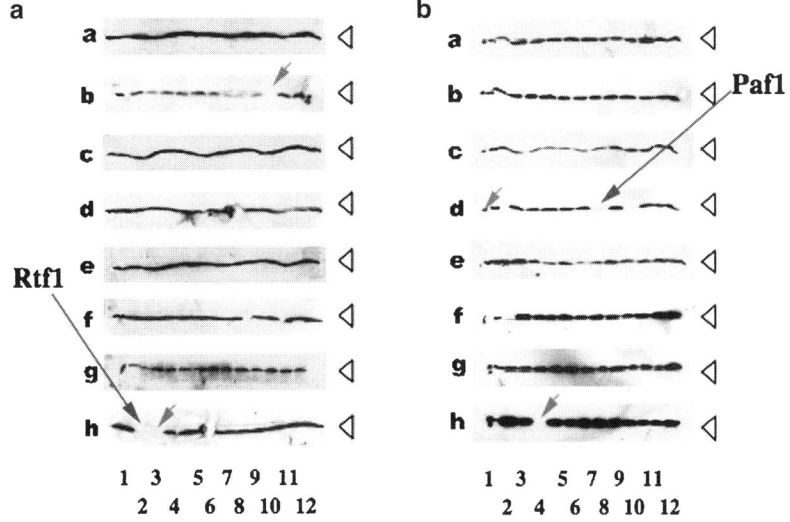

Fig. 2. Surveying the *S. cerevisiae* genome with GPS identified the RNA polymerase II elongation complex, Paf1, required for the methylation of lysine 4 of histone H3. (**a, b**) Employing GPS, extracts of *S. cerevisiae* mutants missing one of the approximately 4,800 nonessential genes were tested for the presence of Lys4-methylated histone H3. Strains lacking either (**a**) Rtf1 or (**b**) Paf1 were defective for this histone modification. *Short arrows* at position b10 in (**a**), d1 in (**b**), and h3 in both, indicate empty wells as plate markers.

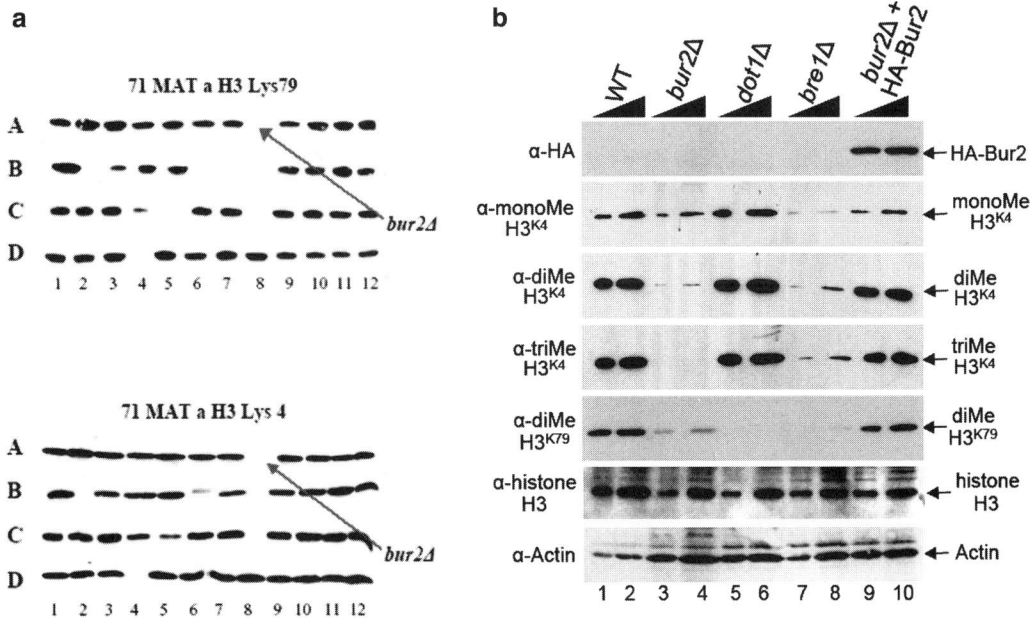

Fig. 3. Deletion of *BUR2* results in a reduction of the levels of histone H3 lysine 4 and lysine 79 methylation. (**a**) Extracts prepared from plates containing single deletion mutants from the nonessential deletion consortium were subjected to SDS-PAGE and Western analysis using antibodies directed against dimethylated lysine 4 and lysine 79 of histone H3. *Arrows* indicate the lane containing the *BUR2* deletion mutant. Several other sites that are null for histone methylation represent slow-growing strains or plate markers. (**b**) Deletion of *BUR2* results in a decrease in the di- and trimethylation of H3K4 (*lanes 3* and *4*). After a *CEN-URA3* plasmid containing HA-tagged Bur2 was introduced into the *BUR2* deletion mutant, histone H3K4 methylation was restored to levels comparable to wild-type (*lanes 9* and *10*).

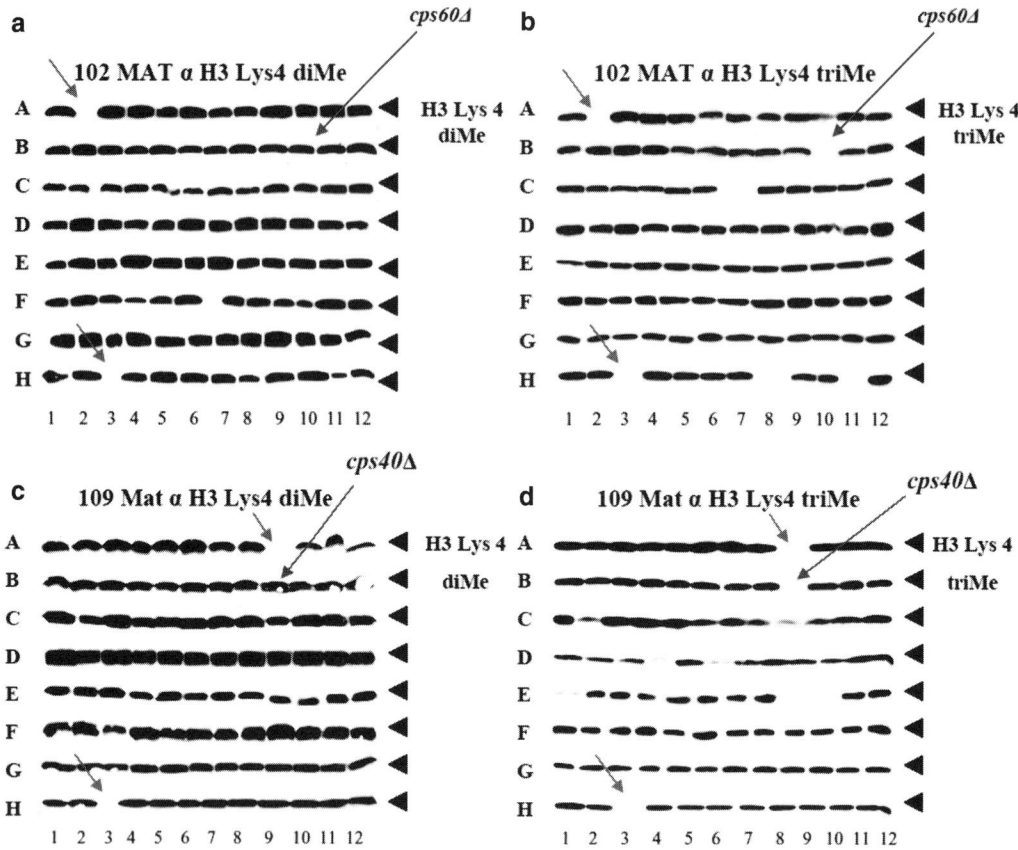

Fig. 4. Multiple subunits of COMPASS are required for histone H3 lysine 4 trimethylation. (**a**, **b**) Extracts were made from each of the nonessential gene deletion mutants of the *S. cerevisiae* genome. In GPS, the extracts from the collection were resolved by SDS-PAGE and tested for the presence of (**a**) dimethylated or (**b**) trimethylated lysine 4 of histone H3. *Short arrows* at positions A2 and H3 indicate empty wells as place markers. *Long arrows* indicate the position of the *CPS60* deletion mutant at plate location B10. (**c**, **d**) GPS results tested for the presence of (**c**) dimethylated or (**d**) trimethylated lysine 4 of histone H3. *Long arrows* indicate the position of the *CPS40* deletion mutant at plate location B9.

The human homolog of Cps40, the Ash2 protein, also functions similarly within mammalian H3K4 methylase complexes, indicating that the H3K4 methylation machinery is highly conserved from yeast to humans *(37)*. Via GPS, we also determined that the Ctk complex, which is required for RNA Polymerase II CTD phosphorylation and, therefore, the rate of transcription elongation, can also regulate the pattern of H3K4 monomethylation (**Fig. 5**) *(38)*. GPS has been instrumental in defining the molecular pathway required for proper H3K4 methylation by COMPASS (**Fig. 6**).

In addition to GPS being powerful in defining the molecular pathway for H3K4 methylation, the employment of an H3K56 acetyl specific antibody was also able to identify the enzymatic machinery responsible for this histone acetylation (**Fig. 7**) *(10)*. Work by other laboratories has also resulted in this same observation *(39–42)*.

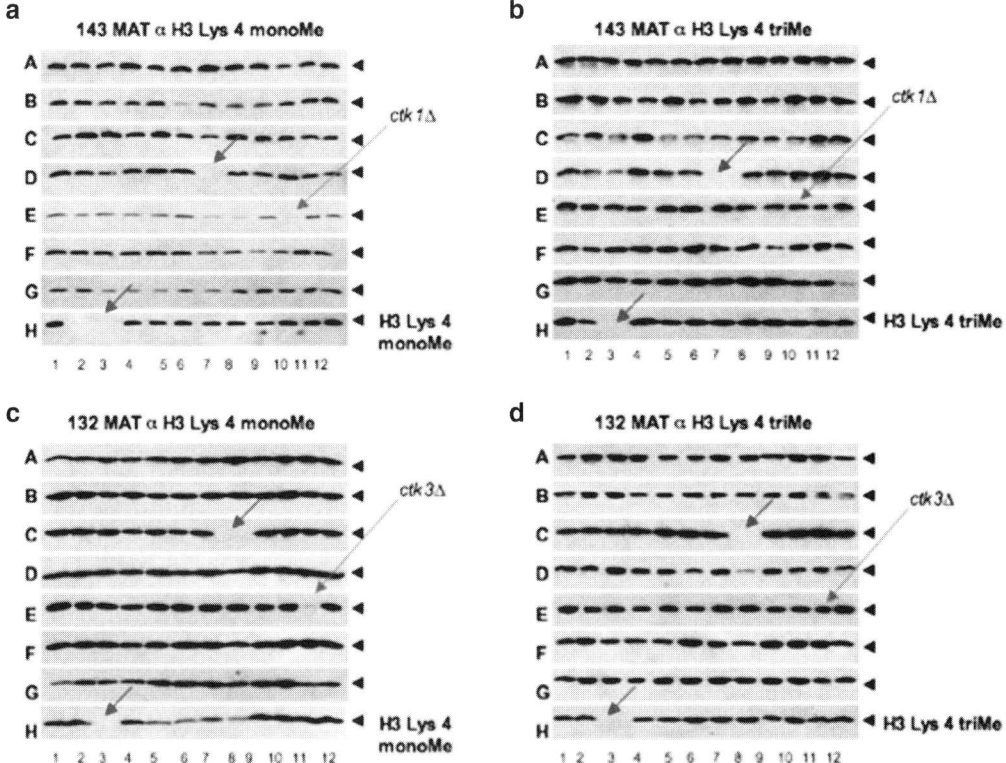

Fig. 5. Loss of CTD kinase-1 complex members affects histone H3K4 monomethylation. (**a–d**) GPS screen demonstrating that the deletion of either CTK1 or CTK3 results in reduced levels of H3K4 monomethylation and (**a** and **c**) but not H3K4 trimethylation (**b** and **d**).

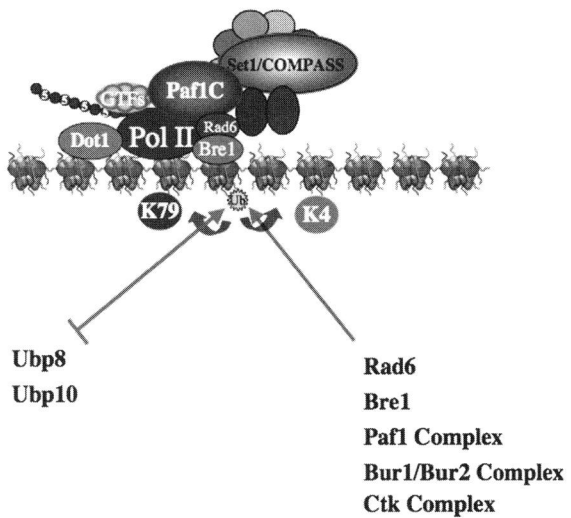

Fig. 6. Schematic representation of the GPS identified molecular pathway required for proper histone H3 methylation on lysines 4 and 79. Employing GPS we identified in yeast initially that the Rad6/Bre1 complex is required for proper monoubiquitination of histone H2B on lysine 123. Histone H2B monoubiquitination signals for the activation of histone H3 methylation on lysines 4 and 79 by COMPASS and Dot1p, respectively. This mechanism is known as histone crosstalk. GPS has also uncovered the role of other factors such as the Paf1 complex and the Bur1/Bur2 complex in proper histone H2B monoubiquitination. The Paf1 complex can also serve as a platform for the association of COMPASS and other histone methyltransferases with the elongating form of RNA Polymerase II (Pol II). The monoubiquitinated histone H2B is deubiquitinated by the action of deubiquitinating enzymes such as Ubp8 and Ubp10. The enzymatic machinery involved in the histone crosstalk pathway is highly conserved from yeast to humans.

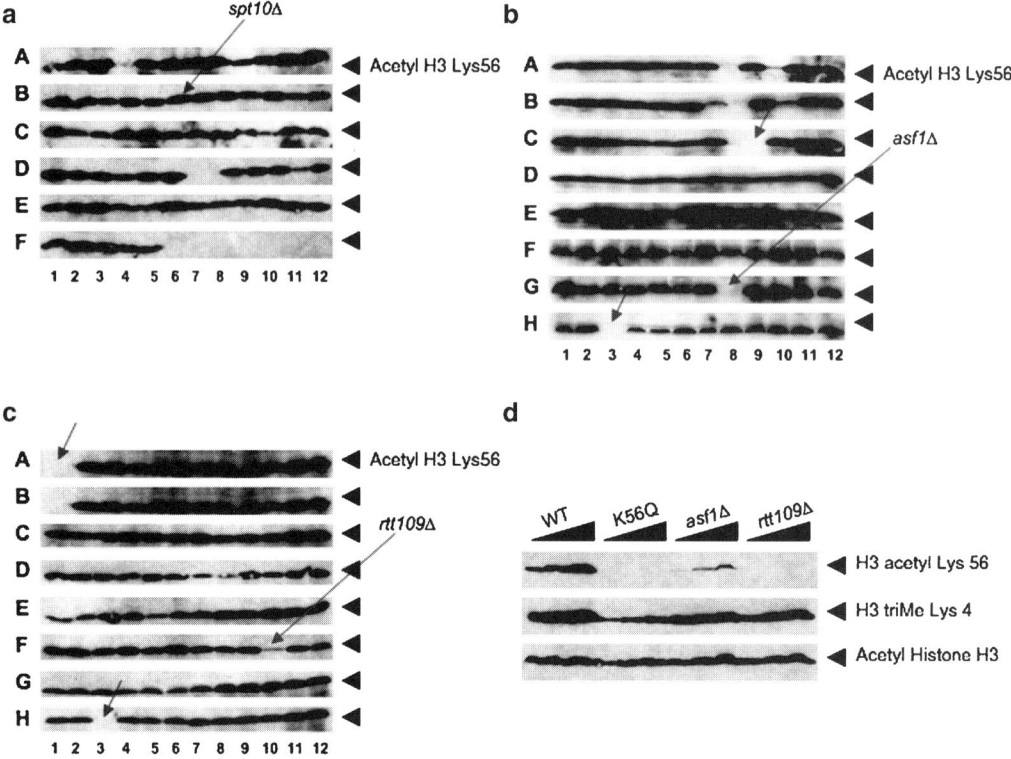

Fig. 7. Global proteomic analyses in defining factors required for proper H3K56 acetylation. (**a–c**) Total cellular extracts were prepared from each of the nonessential yeast gene deletion mutants of the *S. cerevisiae* genome. Each extract was subjected to SDS-PAGE, Western blotted, and probed with the anti-acetyl H3 Lys56 antibody. *Long arrows* indicate the positions of (**a**) *spt10Δ* in which H3K56 acetylation is present, (**b**) *asf1Δ*, and (**c**) *rtt109Δ* where H3K56 acetylation is absent. *Short arrows* indicate empty wells as plate markers. (**d**) Extracts from the *rtt109Δ* strain were titrated and analyzed by SDS-PAGE to determine the level of the H3K56 loss in the absence of Rtt109. Western blots probed with the appropriate antibodies demonstrated a loss in H3K56 acetylation. H3 Lys4 trimethylation remained unaffected and anti-acetyl Histone H3 was used as a loading control.

GPS has most certainly been a very successful method for identifying the molecular pathway required for proper H3K4 methylation in yeast (**Fig. 6**). The information obtained from this screen has also proven to be very useful as a template for identifying factors involved in H3K4 methylation in higher eukaryotic organisms. This method is a testimony to the great value of the *S. cerevisiae* deletion collection. One disadvantage of GPS is the requirement for an antiserum sensitive enough to detect the relatively small amounts of protein made available by our method. To surmount this problem, we are now developing methods that do not require an antibody specific for the given post-translational modifications. Having said this, GPS can also be used for defining the molecular machinery required for other post-translational modification pathways such as phosphorylation, acetylation, and glycosylation when antibodies capable of identifying such modifications on proteins in yeast *S. cerevisiae* are available.

References

1. Kornberg, R. D. 1974. Chromatin structure: a repeating unit of histones and DNA. Science 184:868–71.
2. Kornberg, R. D., and Y. Lorch. 1999. Twenty-five years of the nucleosome, fundamental particle of the eukaryote chromosome. Cell 98:285–94.
3. Luger, K., A. W. Mader, R. K. Richmond, D. F. Sargent, and T. J. Richmond. 1997. Crystal structure of the nucleosome core particle at 2.8 A resolution. Nature 389:251–60.
4. Berger, S. L. 2007. The complex language of chromatin regulation during transcription. Nature 447:407–12.
5. Shilatifard, A. 2006. Chromatin modifications by methylation and ubiquitination: implications in the regulation of gene expression. Annu Rev Biochem 75:243–69.
6. Workman, J. L., and R. E. Kingston. 1998. Alteration of nucleosome structure as a mechanism of transcriptional regulation. Annu Rev Biochem 67:545–79.
7. Wu, J., and M. Grunstein. 2000. 25 years after the nucleosome model: chromatin modifications. Trends Biochem Sci 25:619–23.
8. Lacoste, N., R. T. Utley, J. M. Hunter, G. G. Poirier, and J. Cote. 2002. Disruptor of telomeric silencing-1 is a chromatin-specific histone H3 methyltransferase. J Biol Chem 277:30421–4.
9. Ng, H. H., Q. Feng, H. Wang, H. Erdjument-Bromage, P. Tempst, Y. Zhang, and K. Struhl. 2002. Lysine methylation within the globular domain of histone H3 by Dot1 is important for telomeric silencing and Sir protein association. Genes Dev 16:1518–27.
10. Schneider, J., P. Bajwa, F. C. Johnson, S. R. Bhaumik, and A. Shilatifard. 2006. Rtt109 is required for proper H3K56 acetylation: a chromatin mark associated with the elongating RNA polymerase II. J Biol Chem 281:37270–4.
11. van Leeuwen, F., P. R. Gafken, and D. E. Gottschling. 2002. Dot1p modulates silencing in yeast by methylation of the nucleosome core. Cell 109:745–56.
12. Bhaumik, S. R., E. Smith, and A. Shilatifard. 2007. Covalent modifications of histones during development and disease pathogenesis. Nat Struct Mol Biol 14:1008–16.
13. Hess, J. L. 2004. MLL: a histone methyltransferase disrupted in leukemia. Trends Mol Med 10:500–7.
14. Rowley, J. D. 1998. The critical role of chromosome translocations in human leukemias. Annu Rev Genet 32:495–519.
15. Tenney, K., and A. Shilatifard. 2005. A COMPASS in the voyage of defining the role of trithorax/MLL-containing complexes: linking leukemogenesis to covalent modifications of chromatin. J Cell Biochem 95:429–36.
16. Tenney, K., M. Gerber, A. Ilvarsonn, J. Schneider, M. Gause, D. Dorsett, J. C. Eissenberg, and A. Shilatifard. 2006. Drosophila Rtf1 functions in histone methylation, gene expression, and Notch signaling. Proc Natl Acad Sci U S A 103:11970–4.
17. Jones, R. S., and W. M. Gelbart. 1993. The Drosophila Polycomb-group gene enhancer of zeste contains a region with sequence similarity to trithorax. Mol Cell Biol 13:6357–66.
18. Tschiersch, B., A. Hofmann, V. Krauss, R. Dorn, G. Korge, and G. Reuter. 1994. The protein encoded by the Drosophila position-effect variegation suppressor gene Su(var)3–9 combines domains of antagonistic regulators of homeotic gene complexes. EMBO J 13:3822–31.
19. Miller, T., N. J. Krogan, J. Dover, H. Erdjument-Bromage, P. Tempst, M. Johnston, J. F. Greenblatt, and A. Shilatifard. 2001. COMPASS: a complex of proteins associated with a trithorax-related SET domain protein. Proc Natl Acad Sci U S A 98:12902–7.
20. Krogan, N. J., J. Dover, S. Khorrami, J. F. Greenblatt, J. Schneider, M. Johnston, and A. Shilatifard. 2002. COMPASS, a histone H3 (lysine 4) methyltransferase required for telomeric silencing of gene expression. J Biol Chem 277:10753–5.
21. Nagy, P. L., J. Griesenbeck, R. D. Kornberg, and M. L. Cleary. 2002. A trithorax-group complex purified from *Saccharomyces cerevisiae* is required for methylation of histone H3. Proc Natl Acad Sci U S A 99:90–4.
22. Roguev, A., D. Schaft, A. Shevchenko, W. W. Pijnappel, M. Wilm, R. Aasland, and A. F. Stewart. 2001. The *Saccharomyces cerevisiae* Set1 complex includes an Ash2 homologue and methylates histone 3 lysine 4. EMBO J 20:7137–48.
23. Cho, Y. W., T. Hong, S. Hong, H. Guo, H. Yu, D. Kim, T. Guszczynski, G. R. Dressler, T. D. Copeland, M. Kalkum, and K. Ge. 2007. PTIP associates with MLL3- and MLL4-containing histone H3 lysine 4 methyltransferase complex. J Biol Chem 282(28):20395–406.
24. Hughes, C. M., O. Rozenblatt-Rosen, T. A. Milne, T. D. Copeland, S. S. Levine, J. C. Lee, D. N. Hayes, K. S. Shanmugam, A. Bhattacharjee, C. A. Biondi, G. F. Kay, N. K. Hayward, J. L. Hess, and M. Meyerson. 2004.

Menin associates with a trithorax family histone methyltransferase complex and with the hoxc8 locus. Mol Cell 13:587–97.

25. Lee, J. H., and D. G. Skalnik. 2005. CpG-binding protein (CXXC finger protein 1) is a component of the mammalian Set1 histone H3-Lys4 methyltransferase complex, the analogue of the yeast Set1/COMPASS complex. J Biol Chem 280:41725–31.

26. Lee, J. H., C. M. Tate, J. S. You, and D. G. Skalnik. 2007. Identification and characterization of the human Set1B histone H3-Lys4 methyltransferase complex. J Biol Chem 282:13419–28.

27. Barski, A., S. Cuddapah, K. Cui, T. Y. Roh, D. E. Schones, Z. Wang, G. Wei, I. Chepelev, and K. Zhao. 2007. High-resolution profiling of histone methylations in the human genome. Cell 129:823–37.

28. Cao, R., L. Wang, H. Wang, L. Xia, H. Erdjument-Bromage, P. Tempst, R. S. Jones, and Y. Zhang. 2002. Role of histone H3 lysine 27 methylation in Polycomb-group silencing. Science 298:1039–43.

29. Lee, T. I., R. G. Jenner, L. A. Boyer, M. G. Guenther, S. S. Levine, R. M. Kumar, B. Chevalier, S. E. Johnstone, M. F. Cole, K. Isono, H. Koseki, T. Fuchikami, K. Abe, H. L. Murray, J. P. Zucker, B. Yuan, G. W. Bell, E. Herbolsheimer, N. M. Hannett, K. Sun, D. T. Odom, A. P. Otte, T. L. Volkert, D. P. Bartel, D. A. Melton, D. K. Gifford, R. Jaenisch, and R. A. Young. 2006. Control of developmental regulators by Polycomb in human embryonic stem cells. Cell 125:301–13.

30. Dover, J., J. Schneider, M. A. Tawiah-Boateng, A. Wood, K. Dean, M. Johnston, and A. Shilatifard. 2002. Methylation of histone H3 by COMPASS requires ubiquitination of histone H2B by Rad6. J Biol Chem 277:28368–71.

31. Sun, Z. W., and C. D. Allis. 2002. Ubiquitination of histone H2B regulates H3 methylation and gene silencing in yeast. Nature 418:104–8.

32. Wood, A., N. J. Krogan, J. Dover, J. Schneider, J. Heidt, M. A. Boateng, K. Dean, A. Golshani, Y. Zhang, J. F. Greenblatt, M. Johnston, and A. Shilatifard. 2003. Bre1, an E3 ubiquitin ligase required for recruitment and substrate selection of Rad6 at a promoter. Mol Cell 11:267–74.

33. Krogan, N. J., J. Dover, A. Wood, J. Schneider, J. Heidt, M. A. Boateng, K. Dean, O. W. Ryan, A. Golshani, M. Johnston, J. F. Greenblatt, and A. Shilatifard. 2003. The Paf1 complex is required for histone H3 methylation by COMPASS and Dot1p: linking transcriptional elongation to histone methylation. Mol Cell 11:721–9.

34. Wood, A., J. Schneider, J. Dover, M. Johnston, and A. Shilatifard. 2003. The Paf1 complex is essential for histone monoubiquitination by the Rad6-Bre1 complex, which signals for histone methylation by COMPASS and Dot1p. J Biol Chem 278:34739–42.

35. Wood, A., J. Schneider, J. Dover, M. Johnston, and A. Shilatifard. 2005. The Bur1/Bur2 complex is required for histone H2B monoubiquitination by Rad6/Bre1 and histone methylation by COMPASS. Mol Cell 20:589–99.

36. Schneider, J., A. Wood, J. S. Lee, R. Schuster, J. Dueker, C. Maguire, S. K. Swanson, L. Florens, M. P. Washburn, and A. Shilatifard. 2005. Molecular regulation of histone H3 trimethylation by COMPASS and the regulation of gene expression. Mol Cell 19:849–56.

37. Steward, M. M., J. S. Lee, A. O'Donovan, M. Wyatt, B. E. Bernstein, and A. Shilatifard. 2006. Molecular regulation of H3K4 trimethylation by ASH2L, a shared subunit of MLL complexes. Nat Struct Mol Biol 13:852–4.

38. Wood, A., A. Shukla, J. Schneider, J. S. Lee, J. D. Stanton, T. Dzuiba, S. K. Swanson, L. Florens, M. P. Washburn, J. Wyrick, S. R. Bhaumik, and A. Shilatifard. 2007. Ctk complex-mediated regulation of histone methylation by COMPASS. Mol Cell Biol 27:709–20.

39. Collins, S. R., K. M. Miller, N. L. Maas, A. Roguev, J. Fillingham, C. S. Chu, M. Schuldiner, M. Gebbia, J. Recht, M. Shales, H. Ding, H. Xu, J. Han, K. Ingvarsdottir, B. Cheng, B. Andrews, C. Boone, S. L. Berger, P. Hieter, Z. Zhang, G. W. Brown, C. J. Ingles, A. Emili, C. D. Allis, D. P. Toczyski, J. S. Weissman, J. F. Greenblatt, and N. J. Krogan. 2007. Functional dissection of protein complexes involved in yeast chromosome biology using a genetic interaction map. Nature 446:806–10.

40. Driscoll, R., A. Hudson, and S. P. Jackson. 2007. Yeast Rtt109 promotes genome stability by acetylating histone H3 on lysine 56. Science 315:649–52.

41. Han, J., H. Zhou, B. Horazdovsky, K. Zhang, R. M. Xu, and Z. Zhang. 2007. Rtt109 acetylates histone H3 lysine 56 and functions in DNA replication. Science 315:653–5.

42. Tsubota, T., C. E. Berndsen, J. A. Erkmann, C. L. Smith, L. Yang, M. A. Freitas, J. M. Denu, and P. D. Kaufman. 2007. Histone H3-K56 acetylation is catalyzed by histone chaperone-dependent complexes. Mol Cell 25:703–12.

Chapter 11

Systematic Characterization of the Protein Interaction Network and Protein Complexes in *Saccharomyces cerevisiae* Using Tandem Affinity Purification and Mass Spectrometry

Mohan Babu, Nevan J. Krogan, Donald E. Awrey, Andrew Emili, and Jack F. Greenblatt

Summary

Defining protein complexes is a vital aspect of cell biology because cellular processes are often carried out by stable protein complexes and their characterization often provides insights into their function. Accurate identification of the interacting proteins in macromolecular complexes is easiest after purification to near homogeneity. To this end, the tandem affinity purification (TAP) system with subsequent protein identification by high-throughput mass spectrometry was developed *(1, 2)* to systematically characterize native protein complexes and transient protein interactions under near-physiological conditions. The TAP tag containing two adjacent affinity purification tags (calmodulin-binding peptide and *Staphylococcus aureus* protein A) separated by a tobacco etch virus (TEV) protease cleavage site is fused with the open reading frame of interest. Using homologous recombination, a fusion library was constructed for the yeast *Saccharomyces cerevisiae (3)* in which the carboxy-terminal end of each predicted open reading frame is individually tagged in the chromosome so that the resulting fusion proteins are expressed under the control of their natural promoters *(3)*. In this chapter, an optimized protocol for systematic protein purification and subsequent mass spectrometry-based protein identification is described in detail for the protein complexes of *S. cerevisiae (4–6)*.

Key words: *Saccharomyces cerevisiae*, TAP tagging, Affinity purification, Mass spectrometry, LC-MS, MALDI-TOF, Protein complex

1. Introduction

Since protein interactions are important for most cellular processes, and multi-subunit protein machines actively participate in most or all such processes *(5–7)*, it is of immense significance to systematically characterize protein–protein interactions (PPIs) in various organisms. In the budding yeast *Saccharomyces cerevisiae*, large-scale PPI networks were first generated by systematically using the yeast two-hybrid technique *(8, 9)*. Subsequently, the tandem affinity purification (TAP) approach was developed for the purification of native yeast protein complexes *(2, 4)*. In this approach, TAP-tagged proteins are expressed under normal physiological conditions, purified $\sim 10^6$-fold via a two-step enrichment procedure, and then characterized by mass spectrometry. Because it is relatively simple to fuse affinity tags with target proteins, this approach has also been successfully applied to many other evolutionarily diverse organisms, including *Caenorhabditis elegans*, *Drosophila melanogaster*, *Escherichia coli*, mammalian cells, and plants *(10–15)*. Exploring PPI networks on a large scale provides not only additional information about well-characterized proteins but also a rational framework for elucidating the biological functions of uncharacterized proteins based on the concept of "guilt by association" *(16)*.

The TAP tag consists of two affinity purification tags, a calmodulin-binding peptide (CBP) and *Staphylococcus aureus* protein A, separated by a tobacco etch virus (TEV) protease cleavage site *(2)*, to allow highly selective two-stage protein enrichment and to help reduce the number of nonspecifically bound proteins (**Fig. 1a**). Each tagged protein is purified first by binding to beads containing immobilized IgG and subsequently by binding to beads containing immobilized calmodulin. To facilitate protein analyses on a global scale, Ghaemmaghami et al. *(3)* used homologous recombination to integrate a TAP-tag in-frame immediately after the stop codon of each predicted open reading frame in its natural chromosomal location in *S. cerevisiae*. The TAP-tagged fusion proteins are then expressed under the control of their natural promoters, and the abundance of each TAP-tagged fusion protein was assessed by quantitative Western blotting using an antibody that binds to the protein A component of the TAP tag *(3)*. The entire collection of TAP-tagged yeast strains has been made commercially available for academic use (http://www.openbiosystems.com/GeneExpression/Yeast/TAP/).

We purified a large number of the TAP-tagged bait proteins from 2 L yeast cultures under native conditions *(6)*. To increase the interactome coverage and confidence, the identities of the co-purifying proteins (preys) were determined using two complementary, highly sensitive mass spectrometry methods: gel-free liquid chromatography-tandem mass spectrometry (LC-MS/MS), and gel-based peptide mass fingerprinting using matrix-assisted laser desorption/ionization time-of-flight mass spectrometry (MALDI-TOF MS). We attempted to purify 4,562 different soluble proteins, and an extensive PPI network containing high-confidence protein interactions generated from this

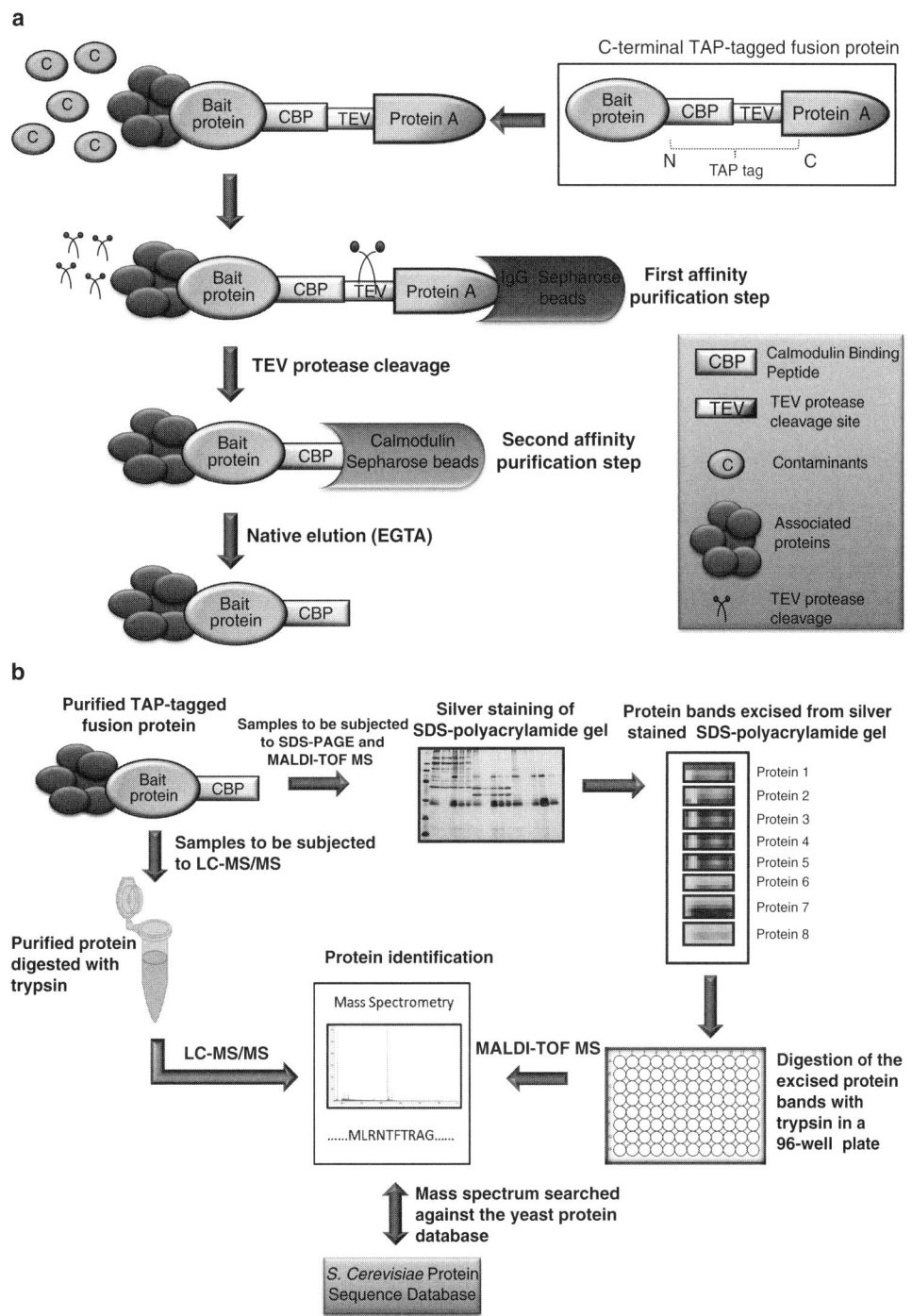

Fig. 1. Schematic representation of the procedures involved in the identification of protein complexes using the tandem affinity purification (TAP) approach followed by mass spectrometry analyses. The TAP-tag structure and an overview of the TAP strategy are shown in (**a**). Identification of the co-purifying proteins is performed using two complementary mass spectrometry methods (**b**). First, the purified TAP-tagged fusion protein and associated proteins are separated by SDS-PAGE followed by silver staining. The excised protein bands are digested with trypsin, and the resulting tryptic peptides are analyzed by peptide mass finger printing using MALDI-TOF MS. A yeast protein database is then searched using the Knexus software. Second, the purified TAP-tagged fusion proteins and associated proteins are digested with trypsin and analyzed by gel-free liquid chromatography-tandem mass spectrometry (LC-MS/MS). Searches are performed with the tandem mass spectra of the complete *S. cerevisiae* protein database using the SEQUEST computer algorithm to generate probable identifications of the proteins.

study was published in 2006 *(6)* at the same time as Gavin et al. *(5)*. These authors also used the TAP method and MALDI-TOF MS to elucidate a similarly extensive set of yeast PPI. More recently, the protein interaction data generated from these two large-scale studies were amalgamated into a single, more reliable collection of experimentally based PPIs using a novel purification enrichment (PE) scoring system *(17)*, and this larger, amalgamated set of PPI was used in conjunction with a Markov clustering algorithm to organize the *S. cerevisiae* proteome into a large number of more reliable protein complexes *(18)*.

Having attempted to purify all the yeast proteins predicted to be soluble, we are now purifying the many yeast proteins predicted to be membrane-associated (*see* **Note 1**). Purification of membrane proteins often poses unique challenges *(19–22)* because they are often not solubilized by the extraction buffer that we normally use for the TAP method. With some modifications of our standard procedures and with the addition of various nonionic detergents to our buffers, we are able to solubilize and purify the majority of the yeast membrane proteins. Currently, we are attempting to purify the ~1,600 *S. cerevisiae* proteins that are predicted to contain signal peptides or with at least one transmembrane domain. Strains for the vast majority of these purifications are already present in the collection of yeast TAP-tagged strains *(3)*. As of March 2009, more than 1,200 yeast membrane proteins had been purified using at least two different detergents.

In this chapter, detailed procedures are described for the purification of TAP-tagged bait proteins and subsequent identification of interacting protein partners by mass spectrometry. This method has been optimized for a well-characterized yeast laboratory strain and can be used to purify the subunits of high- or low-abundance yeast protein complexes. This basic approach can potentially be adapted for use in other organisms.

2. Materials

2.1. Purification of TAP-Tagged Proteins from S. Cerevisiae Strains

2.1.1. Culturing TAP-Tagged S. Cerevisiae Strains and Cell Lysis

1. *S. cerevisiae* strains used for culturing in which a bait protein is TAP-tagged have been described *(3)*.
2. Yeast extract–peptone–dextrose (YEPD) medium: Dissolve 10 g of yeast extract (Bioshop), 20 g of peptone (Bioshop), and 20 g of glucose in 800 ml of distilled water. Autoclave the media on a liquid cycle for 30 min.
3. 2 M Tris–HCl (pH 7.9) stock solution: Dissolve 242.2 g of Tris base in 800 ml of double distilled water, adjust to pH 7.9 with concentrated HCl, and make up the volume to 1 L.

4. 5 M NaCl stock solution: Dissolve 292.2 g of NaCl in 1 L of double-distilled water.

5. 0.5 M ethylenediaminetetraacetic acid (EDTA) (pH 8.0) stock solution: Dissolve 186.1 g of EDTA in 800 ml of double-distilled water, adjust the pH to 8.0 using 1 M NaOH, and make up the volume to 1 L.

6. HEPES-KOH (pH 7.9) stock solution: Dissolve 238.3 g of HEPES in 800 ml of double-distilled water, adjust the pH to 7.9 using 5 M KOH, and make up the volume to 1 L.

7. 1 M $CaCl_2$: Dissolve 110.98 g of $CaCl_2$ in 800 ml of double-distilled water and make up the volume to 1 L.

8. 2 M KCl stock solution: Dissolve 149.1 g of KCl in 800 ml of double-distilled water and make up the volume to 1 L.

9. EDTA-free protease inhibitor tablets (Roche). Make sure to prepare YEB buffer with and without the EDTA-free protease inhibitor tablet.

10. YEB buffer with protease inhibitor: Dissolve one tablet of protease inhibitor in 50 ml YEB buffer.

11. Yeast extract buffer (YEB): Mix 36.6 ml of 2 M KCl, 3 ml of 0.5 M EDTA, 3 ml of 0.5 M ethyleneglycoltetraacetic acid (EGTA)-KOH (pH 7.9), and 30 ml of 1 M HEPES-KOH (pH 7.9) in 229 ml of double-distilled water. Store the solution at room temperature. Prior to use, add 750 μl of 1 M dithiothreitol (DTT; Bioshop) into the YEB buffer.

12. Dialysis buffer (DB): Mix 80 ml of 5 M NaCl, 20 ml of 2 M Tris–HCl (pH 7.9), 1.6 ml of 0.5 M EDTA, 800 ml of 100% glycerol, and 0.32 g of DTT with 3098.4 ml of double-distilled water. Store the solution at 4°C.

2.1.2. Purification of TAP-Tagged Proteins

1. Store 10% Triton X-100 solution (Sigma Aldrich) at 4°C.

2. IPP buffer: Add 50 μl of 2 M Tris–HCl (pH 7.9), 200 μl of 5 M NaCl, and 100 μl of 10% Triton X-100 to 9.65 ml of sterile distilled water.

3. TEV protease cleavage buffer: Add 250 μl of 2 M Tris–HCl (pH 7.9), 4 μl of 0.5 M EDTA, 200 μl of 5 M NaCl, 100 μl of 10% Triton X-100, and 10 μl of 1 M DTT to 9.7 ml of sterile distilled water.

4. Calmodulin binding buffer: Add 200 μl of 2 M Tris–HCl (pH 7.9), 800 μl of 5 M NaCl, 80 μl of 1 M $CaCl_2$, 400 μl of 10% Triton X-100, and 28 μl of β-mercaptoethanol solution to 38.9 ml of sterile distilled water.

5. Calmodulin wash buffer: Add 50 μl of 2 M Tris–HCl (pH 7.9), 200 μl of 5 M NaCl, 1 μl of 1 M $CaCl_2$, 100 μl of 10% Triton X-100, and 7 μl of β-mercaptoethanol solution to 9.6 ml of sterile distilled water.

6. Calmodulin elution buffer: Add 50 μl of 2 M Tris–HCl (pH 7.9), 200 μl of 5 M NaCl, 60 μl of 0.5 M EGTA, 7 μl of β-mercaptoethanol, and 100 μl of 10% Triton X-100 to 8.8 ml of sterile distilled water.

2.2. Protein Identification by MALDI-TOF Mass Spectrometry

2.2.1. SDS-Polyacrylamide Gel Electrophoresis (SDS-PAGE)

1. SDS stock solution (10% (w/v)): 10 g SDS is dissolved in 100 ml double-distilled water. The solution is stored at room temperature.
2. Electrode running buffer (5×): (125 mM Tris–HCl, 960 mM glycine, 0.5% (w/v) SDS, pH 8.3). 30 g of Tris–HCl, 144 g of glycine, and 10 g of SDS are dissolved in 2 L of double-distilled water without pH adjustment and stored at 4°C.
3. Acrylamide: 30% acrylamide monomer, 0.8% N,N′-methylenebis-acrylamide (Bio-Rad). The solution is filtered through Whatman No.1 filter paper (Fischer Scientific) and stored at 4°C in the dark.
4. Separating buffer (4×): 1.5 M Tris–HCl (pH 8.7), 0.4% SDS. Store at room temperature.
5. Stacking buffer (5×): 0.5 M Tris–HCl (pH 6.8), 0.4% SDS. Store at room temperature.
6. Ammonium persulfate (10% APS): Mix 1 g of APS (Bioshop) in 10 ml of double-distilled water. Aliquot the solution into several microcentrifuge tubes in a volume of 200 μl. Store the aliquots at −20°C.
7. Water saturated *n*-butanol: Equal volumes of *n*-butanol (Sigma Aldrich) and double-distilled water are mixed in a glass bottle and left at room temperature to separate. Use the topmost layer containing *n*-butanol saturated with water. Store the solution at room temperature.
8. Resolving gel (12.5% polyacrylamide): Mix 7.2 ml of 4× separating buffer with 12 ml acrylamide solution, 9.6 ml double-distilled water, 120 μl 10% APS solution, and 20 μl tetramethylethylenediamine (TEMED, Bioshop).
9. Stacking gel (4.5% polyacrylamide): Mix 4 ml of 5× stacking buffer with 2.4 ml acrylamide solution, 9.6 ml double distilled water, 150 μl 10% APS solution, and 15 μl TEMED.
10. Sample stock buffer (3×): Mix 1.25 ml of 0.5 M Tris–HCl (pH 6.8), 1 ml of 100% glycerol, 1 ml of 10% SDS (w/v), and 20 mg of bromophenol blue (Bio-Rad).
11. 2× sample buffer: Add 6 ml of 3× sample stock buffer, 0.6 ml of β-mercaptomethanol, and 3.4 ml of distilled water.

2.2.2. Silver-Staining the SDS-Polyacrylamide Gel

1. Fixer: 50% methanol and 10% acetic acid (AA) are added to 400 ml sterile distilled water.
2. Sensitizer: 20 mg of fresh sodium thiosulfate (Sigma) is dissolved in 1,000 ml distilled water.

3. Silver nitrate solution: 2 g of silver nitrate (Fischer Scientific) is dissolved in 1,000 ml distilled water.

4. Developing solution: 1.4 ml of 37% formaldehyde and 30 g of sodium carbonate are added to 1,000 ml of distilled water.

5. Stop solution: 5 ml AA is diluted in 500 ml of distilled water.

2.2.3. Gel-Based MALDI-TOF Mass Spectrometry

1. An ULTRAFlex II MALDI-TOF instrument for acquiring spectral data from in-gel trypsin digested yeast samples (Bruker Daltonics, Billerica, MA).

2. MALDI target plate (Bruker).

3. Bulk C18 reverse phase resin (Sigma).

4. Knexus automation, a Windows-based program from Genomics Solutions Bioinformatics for database searches.

5. α-Cyano-4-hydroxycinnamic acid matrix solution (Fluka Buchs SG, Switzerland).

6. 1 mM HCl: Add 1 μl of 10 N HCl into 10 ml HPLC-grade water. Make a fresh solution each day and store on ice until use.

7. Trypsin stock solution: Dissolve 100 μg Boehringer Mannheim unmodified sequencing-grade trypsin (Roche) in 1 ml of 1 mM HCl. Store the trypsin stock solution at −80°C.

8. Digestion buffer: Mix 9.12 ml of 100 mM NH_4HCO_3, 9.12 ml of HPLC-grade water, 960 μl of 1% $CaCl_2$ and 1 ml of trypsin stock solution. Prepared a fresh solution each day and store on ice until use.

9. 100 mM NH_4HCO_3: Dissolve 0.79 g NH_4HCO_3 in 100 ml HPLC-grade water.

10. 100 mM NH_4HCO_3 containing 10 mM DTT: Add 100 μl of 1 M DTT to 9.9 ml of 100 mM NH_4HCO_3. Prepare a fresh solution each day and store in an amber-colored bottle at room temperature in the dark until use.

11. 100 mM NH_4HCO_3 containing 55 mM iodoacetamide (Sigma): Add 0.103 g of iodoacetamide to 10 ml of 100 mM NH_4HCO_3. Prepare a fresh solution every day and store in an amber-colored bottle at room temperature in the dark until use.

12. 66% Acetonitrile (ACN), 1% AA: Mix 66 ml HPLC-grade ACN (Sigma), 33 ml of HPLC-grade water, and 1 ml AA. Store the solution in a glass bottle for up to 1 month at room temperature.

13. 75% ACN, 1% AA: Mix 75 ml HPLC-grade ACN, 24 ml HPLC-grade water, and 1 ml AA. Store the solution in a glass bottle for up to 1 month at room temperature.

14. 2% ACN, 1% AA: Mix 2 ml of HPLC-grade ACN, 97 ml HPLC-grade water, and 1 ml AA. Store the solution in a glass bottle for up to 1 month at room temperature.
15. 1% $CaCl_2$: Dissolve 1 g of $CaCl_2$ in HPLC-grade water to a final volume of 100 ml. Store the solution in a glass bottle for up to 1 month at room temperature.
16. 0.1% Trifluoroacetic acid (TFA) stock solution: Dilute 0.25 µl of TFA in 250 µl of HPLC-grade water.
17. Peptide calibration standard (Bruker) solution: Lyophilized peptide standard is dissolved with 125 µl of 0.1% HPLC-grade TFA.

2.3. Protein Identification by Liquid Chromatography Mass Spectrometry (LC-MS/MS)

1. LTQ tandem mass spectrometer (Finnigan Corp, San Jose, CA, USA) to run the samples, and XCalibur software to acquire tandem mass spectra and to control the instrument.
2. Digestion buffr: Mix 599 µl of 50 mM NH_4HCO_3 and 1 µl of 1 M $CaCl_2$. Store the solution at 4°C prior to use.
3. Immobilized trypsin solution: Mix 18.7 µl of digestion buffer, 1.8 µl PIERCE immobilized trypsin beads (PIERCE), 0.9 µl immobilized trypsin beads (Applied Biosciences), and 0.06 µl of 1 M $CaCl_2$. Make sure that the pH of the immobilized trypsin solution is ~8.0.
4. 150 µm fused silica (Polymicro Technologies, Pheoenix, AZ, USA).
5. C18 reverse phase packing material (Zorbax eclipse XDB-C18 resin; Agilent Technologies, Mississauga, ON, Canada).
6. Solvent A: 5% ACN, 0.5% AA, and 0.02% heptafluorobutyric acid (HFBA)
7. Solvent B: 100% ACN
8. Proxeon nano HPLC pump (Proxeon Biosciences).

3. Methods

3.1. Purification of S. Cerevisiae TAP-Tagged Fusion Proteins

3.1.1. Culturing TAP-Tagged S. Cerevisiae Strains and Cell Lysis

1. Inoculate a loop of a glycerol stock of a TAP-tagged *S. cerevisiae* strain *(3)* into 10 ml YEPD liquid medium in a 20-ml sterile culture tube.
2. Grow the culture overnight and the next day at 30°C with shaking at 250 rpm until the OD_{600} reaches 1.0–1.5.
3. Inoculate 10 ml of the overnight culture into 2 L fresh YEPD liquid medium in a 4-L flask (*see* **Note 2**). The culture is grown overnight at 30°C with shaking at 250 rpm until the OD_{600} reaches ~1.0–1.5.

4. Transfer 2 L *S. cerevisiae* culture from the shaker to clean 1-L centrifugation bottles.

5. Centrifuge the bottles containing the *S. cerevisiae* culture in a Beckman J-20XP Avanti centrifuge at $3,993g$ for 5 min at 4°C.

6. Discard the supernatant and remove excess liquid by inverting the bottles on paper towels. Keep the centrifugation bottles on ice.

7. Add 10 ml cold distilled water to the centrifugation bottles and resuspend the *S. cerevisiae* cell pellets using a clean 25-ml pipette.

8. Transfer the cell lysates into 50-ml polypropylene Falcon tubes.

9. Centrifuge the Falcon tubes containing the cell lysates in a Beckman J-20XP Avanti centrifuge at $3,993g$ for 5 min at 4°C.

10. Decant the cold water from the Falcon tubes, and resuspend the cell pellets with an equal volume of cold YEB buffer without protease inhibitor using a clean 25-ml pipette.

11. Centrifuge the Falcon tubes containing the cell lysates in an Eppendorf centrifuge 5810R using an A-4–62 rotor at $751 \times g$ (4,000 rpm) for 5 min at 4°C.

12. Add an equal volume of cold YEB buffer with protease inhibitor to the Falcon tubes containing the cell pellets and resuspend the pellets using a clean 25-ml pipette.

13. Repeat **step 11**.

14. Snap-freeze the Falcon tubes containing the cell pellets using liquid nitrogen. The frozen cell pellets are stored at –80°C for future use.

15. The Falcon tubes containing the frozen cell pellets are removed from the freezer and wrapped in several layers of paper towels.

16. Using a hammer, smash the Falcon tubes containing the frozen cell pellets into small pieces.

17. Transfer each yeast cell pellet (7–10 g) into a prechilled Krups coffee grinder (Krups, Model 203–70). Avoid transferring the broken pieces of plastic.

18. Add 25% dry ice into the coffee grinder containing 7–10 g yeast cell pellets.

19. Perform lysis by grinding the yeast cell pellets to a fine powder with dry ice using the Krups coffee grinder (*see* **Note 3**). This takes approximately 2–3 min.

20. Scrape the lysed powder into a 25-ml ultracentrifuge tube (25 × 89 mm, Beckman, Part No. 355642) placed on dry ice.

21. Resuspend the lysed powder by adding an equal volume of YEB buffer containing protease inhibitor into the 25-ml ultracentrifuge tubes (*see* **Note 4**).

22. Centrifuge the 25-ml ultracentrifuge tubes containing the cell lysates in a Beckman L8-M ultracentrifuge using a Type 70Ti rotor at 208,429 × g (45,000 rpm) for 1 h at 4°C.

23. The supernatants from the ultracentrifuge tubes are collected and transferred to dialysis tubes (Spectra/Por®, 29 mm in diameter). The supernatants are dialyzed against 4 L dialysis buffer containing 100 mM NaCl for 3 h at 4°C (*see* **Notes 5** and **6**).

24. After dialysis, the extracts are transferred from the dialysis tubes to sterile 25-ml ultracentrifuge tubes.

25. Centrifuge the 25-ml ultracentrifuge tubes containing the extracts in a Beckman L8-M ultracentrifuge using a Type 70Ti rotor at 208,429 × g (45,000 rpm) for 30 min at 4°C to remove any precipitated material.

26. Remove the extracts carefully from the Falcon tubes and transfer them to 15- or 50-ml (depending on the volume) sterile Falcon tubes.

27. Snap-freeze the Falcon tubes containing the samples in liquid nitrogen. The lysed frozen cell extracts are stored at –80°C for future use.

28. Thaw the frozen Falcon tubes containing the lysed cell extracts by placing the tubes in cold water. Centrifuge the tubes containing the lysed cell extracts in an Eppendorf centrifuge 5810R using a A-4–62 rotor at 751 × g (4,000 rpm) for 5 min at 4°C to remove any precipitated material.

3.1.2. Protein Purification

IgG-Sepharose Beads

1. All purification steps are performed at 4°C with precooled buffers and equipment.

2. The Falcon tubes containing the lysed frozen cell extracts are thawed by placing the tubes in cold water.

3. Prior to use, add 1 ml of IgG-Sepharose beads to a fresh 15-ml Falcon tube. The beads are washed with 3 × 5 ml of IPP buffer. Briefly mix the contents by tilting the Falcon tubes upside down.

4. Centrifuge the Falcon tubes containing the IgG-Sepharose beads with IPP buffer in an Eppendorf centrifuge 5810R using a A-4–62 rotor at 751 × g (4,000 rpm) for 5 min at 4°C. Discard the supernatants, taking care not to disturb the beads.

5. Add 100 µl of the washed IgG-Sepharose beads and 100 µl of IPP buffer to each Falcon tube containing cell extract. Briefly mix the contents by tilting the Falcon tubes upside down.

6. Rotate the Falcon tubes for 3 h at 4°C using a LabQuake shaker (Barnstead/Thermolyne).

7. Centrifuge the Falcon tubes containing the IgG beads and extracts in an Eppendorf centrifuge 5810R using a A-4–62 rotor at $751 \times g$ (4,000 rpm) for 5 min at 4°C. Remove the supernatants as much and as carefully as possible, taking care not to disturb the loose bead pellets.

8. Resuspend the bead pellets in the leftover supernatants and transfer the beads into 0.8 × 4 cm Bio-Rad polypropylene prep columns (Bio-Rad). Make use of P1000 pipette tips clipped at the ends to transfer the beads into the prep columns. Remove the bottom outlet plugs of the columns and allow the eluates to drain by gravity flow.

9. Wash the columns five times with 200 μl of IPP buffer and twice with 200 μl of TEV cleavage buffer.

10. Close the bottom outlet of the column and add 200 μl of 1× TEV cleavage buffer and 5 μl of TEV protease (2 mg/ml). Close the top of the column with a cap and rotate the column overnight at 4°C.

Calmodulin-Sepharose Beads

1. Remove the top and bottom outlet plugs of the columns after incubation with the TEV protease and recover the eluates by gravity flow into Eppendorf tubes. Wash the columns with 200 μl TEV cleavage buffer and collect the eluates into the same Eppendorf tubes containing the eluates recovered after TEV cleavage. Mix the contents in the Eppendorf tubes by gently pipetting up and down.

2. To the eluates add 400 μl of calmodulin binding buffer and 3 μl of 1 M $CaCl_2$.

3. Transfer the mixtures into 0.8 × 4 cm Bio-Rad polypropylene prep columns containing 200 μl of calmodulin-Sepharose beads (Amersham Biosciences) washed twice with 5 ml of calmodulin binding buffer.

4. Close the tops of the columns with caps and rotate for 2 h at 4°C using a LabQuake shaker.

5. Remove the top and bottom plugs of the columns and drain the eluates by gravity flow.

6. The beads are washed five times with 200 μl of calmodulin binding buffer followed by three washes with 200 μl of calmodulin wash buffer.

7. The bound proteins are eluted in six fractions of 100 μl into fresh Eppendorf tubes using calmodulin elution buffer (*see* **Note** 7).

8. The eluted fractions (600 μl) are distributed into two separate Eppendorf tubes in equal volumes of 300 μl.

9. Dry down 300 μl of the eluted fractions using a Speedvac (Eppendorf Vacufuge). Add 60 μl of 2× SDS sample buffer to the dried eluate. The eluate and the sample buffer mixture are boiled for 5 min. The proteins are separated by SDS-PAGE and visualized by silver staining. Protein bands excised from the gel are analyzed by MALDI-TOF mass spectrometry.

10. In parallel with **step 9**, the other 300-μl eluted fraction is dried down using a Speedvac. This dried sample is then subjected to LC-MS/MS for protein identification.

3.2. Protein Identification by MALDI-TOF Mass Spectrometry

The tagged and purified *S. cerevisiae* proteins are separated by SDS-PAGE and stained with silver. The protein bands are excised, reduced, alkylated, subjected to in-gel tryptic digestion, and analyzed by MALDI-TOF mass spectrometry. The mass spectra are searched using Knexus automation against the complete yeast protein database to generate probable identifications of the proteins. The identified proteins are associated back to the purified protein bands (**Fig. 1b**). The various steps, such as silver staining, in-gel trypsin digestion, extraction and purification of tryptic peptide fragments, spotting samples on MALDI target plates, acquisition of spectra, and protein identification, are described below:

3.2.1. SDS-Polyacrylamide Gel Electrophoresis

1. The steps described below are carried out using the Whatman Model V16 (Gibco BRL) gel system. The glass plates, spacers, and combs should be clean and free of dried gel fragments, grease, and dust.

2. For each gel, lay out one small and one large glass plate, separated by spacers.

3. Carefully slide the glass plates into the holder by making sure that the glass plates are pushed all the way to the bottom.

4. Slide in the combs and mark lines at 2–3 cm from the bottoms of the combs.

5. Remove the combs gently and pour 12.5% polyacrylamide resolving gels up to the marked lines. Overlay a thin layer of water-saturated *n*-butanol on the top of each gel. Allow the gels to polymerize for about 30 min.

6. Invert the gels to remove the *n*-butanol. Touch with filter paper to wick off residual liquid.

7. Pour the 4.5% polyacrylamide stacking gels to fill the remaining space between the glass plates. Insert the combs and allow the gels to polymerize for another 30 min.

8. Once the stacking gels are polymerized, carefully remove the combs and use a 3-ml syringe fitted with a 22G needle to wash the wells with running buffer.

9. Place the gels in the unit and add running buffer to the upper and lower chambers of the unit.

10. Load each well with 60-μl samples suspended in 2× SDS sample buffer. Include high and low range Precision Plus prestained protein molecular weight standards (Bio-Rad) in two of the wells for each gel.

11. Complete the assembly of the gel unit by attaching the power cords first to the apparatus, and then to the power supply.

12. Turn on the power supply and run the gels at 150 V through the stacking gels and 200 V through the resolving gels. The gels are run until the blue dye front reaches the bottom.

13. After the electrophoresis, disassemble the gel plate from the apparatus. Use a thin spatula to carefully pry the upper glass plates away from the gels.

14. Transfer the gels immediately to clean the staining solution containing fixer and proceed immediately with the silver-staining protocol described below.

3.2.2. Silver-Staining the SDS-Polyacrylamide Gel

1. Agitate the gel in clean fixer for 20 min on a rocking shaker.
2. Rinse the gel with 20% ethanol for 10 min.
3. The gel is then washed twice with 500 ml double-distilled water for 10 min. Note that thorough rinsing gives a uniform, low background.
4. Remove the double-distilled water and agitate the gel in 500 ml sensitizer solution for 1 min.
5. Remove the sensitizer solution and rinse the gel twice again with double-distilled water for 20 s.
6. Pour off the water and incubate the gel in 200 ml of 0.1% silver nitrate for 30 min.
7. Discard the silver nitrate and rinse the gel once with distilled water for 20 s to remove excess silver nitrate.
8. Wash the gel with 50–75 ml of freshly prepared developing solution for half a minute. Replace with fresh developing solution and agitate the gel slowly by hand constantly (*see* **Note 8**).
9. When the desired staining intensity is achieved, discard the developing solution and add 80 ml of stop solution to the gel.
10. Incubate the gel with the stop solution for a minimum of 20 min before proceeding to excise the protein bands from the silver-stained gels.

3.2.3. Protein Identification by Gel-Based MALDI-TOF Mass Spectrometry

Reduction and Alkylation of Protein Bands

1. Protein bands are excised from silver-stained gels with clean razor blades. The gel slices are excised as closely as possible to the boundaries of the protein bands and stored in −80°C freezers in 96-well polypropylene plates (Nunc).
2. Gel slices in 96-well polypropylene plates are thawed for approximately 20 min. The liquid that accumulates during thawing is carefully removed.

3. Gel slices are shrunk with 200 μl of 100% ACN for ~10 min on an IKA Schuttler MTS 4 orbital shaker (VWR Scientific) at 700 rpm (*see* **Note 9**). Remove all liquid.

4. The gel slices are reduced with 75 μl of 100 mM NH_4HCO_3 containing 10 mM DTT for 30 min in a 50°C heating block. Note that the gel pieces should be covered with liquid. When hydrated, there should still be some fluid left.

5. Remove all liquid by centrifuging the plate at 58 × *g* (500 rpm) in a Beckman Allegra X-12 centrifuge using a SX4750 μ plate carrier for 3–5 min, and repeat **step 3**.

6. The gel slices are alkylated in the 96-well microtiter plate with 75 μl of 100 mM NH_4HCO_3 containing 55 mM iodoacetamide for 20 min in the dark at room temperature.

7. Centrifuge the plate at low speed 58 × *g* (500 rpm) for 3–5 min. Remove all liquid. Repeat **step 3**.

In-Gel Tryptic Digestion

1. The gel slices are hydrated with 60 μl of digestion buffer containing trypsin for 30–45 min on ice. Note that the gel pieces should be fully hydrated with trypsin solution on ice. After 15–20 min, if needed, add more digestion buffer containing trypsin to allow complete hydration.

2. Add 20 μl (if needed) of digestion buffer without trypsin and incubate samples overnight at 37°C (*see* **Note 10**).

Extraction of Tryptic Peptides

1. The extracted peptides are transferred into clean 96-well polypropylene plates.

2. Add 100 μl of 100 mM NH_4HCO_3 to the gel slices and extract peptides by shaking at 700 rpm on an orbital shaker for 60 min at room temperature.

3. Briefly centrifuge, transfer the extracted peptides into the clean 96-well polypropylene plates, and add 2.5 μl of 100% AA to the extracted peptides in each well.

Purification of Tryptic Peptides

1. Purification is performed using bulk C18 reverse phase resin (Sigma). Add 1.5 g of dry resin to a reservoir.

2. Wash the dry resin two times with HPLC-grade methanol and two times with HPLC-grade 66% ACN, 1% AA prior to use. Add 75% ACN, 1% AA to prepare 5:1 resin slurry.

3. Add 2.5 μl of C18 reverse phase resin slurry to the extracted peptides in the wells of the 96-well plate. The resin should float on top of the liquid. Shake the plate on an orbital shaker at 500–700 rpm for 45 min at room temperature. Discard the liquid underneath the beads by using a 200-μl multichannel pipette.

4. Add 200 μl of 2% ACN, 1% AA. Shake the plate for 5–15 min on an orbital shaker at 500–700 rpm at room temperature.

5. Prepare in advance a 384-well Melt Blown Polypropylene (MBPP) Whatman filter plate (Whatman). Place the MBPP filter plate on top of a 384-well collection plate (Whatman). Wash the MBPP filter plate wells with 15 µl of 66% ACN, 1% AA. Centrifuge the MBPP filter plate for 1–2 min at 1,000g and discard the filtrates collected in the collection plate.

6. Centrifuge the 96-well plates from **step 4** for 5–15 min at 1,000g at room temperature. Remove the supernatants with a 200-µl multichannel pipette and discard them.

7. Elute the peptides by adding 30 µl of 66% ACN, 1% AA. Shake the 96-well plate briefly at high speed on an orbital shaker. Ensure that all of the resin has entered into the slurry. Incubate the plates for approximately 5–10 min at room temperature. The resin should make a slurry and slowly pellet to the bottom.

8. Transfer the liquid supernatant with 50-µl multichannel pipette from the 96-well plates to the 384-well MBPP filter plate. Place a 384-well collection plate (Whatman) under the MBPP filter plate and centrifuge for 3–5 min at 2,000g. The filtrates collected in the 384-well collection plate are either spotted immediately onto the Bruker MALDI target plate or sealed with sealing foil and stored at –70°C until spotting onto MALDI targets.

Spotting Samples onto the MALDI Target Plates

1. Separate the Bruker MALDI target plate from its base.

2. Remove the previously spotted matrix spots from the MALDI target plate by washing the plate with 100% methanol followed by rinsing the plates gently with HPLC-grade water. Wipe the plate with 100% methanol using Kimwipes (*see* **Note 11**).

3. Spot 1 µl of α-cyano-4-hydroxycinnamic acid matrix (Fluka Buchs SG, Switzerland) solution and 1 µl of a purified trypsin-digested sample onto each target spot of the MALDI target plate. Allow the samples to dry at room temperature.

4. In the bottom-most row of the MALDI target plate, spot 1 µl of α-cyano-4-hydroxycinnamic acid matrix solution and 1 µl of peptide calibration standard (Bruker).

5. After the spotting is completed, reseal the 384-well trypsin-digested peptide plate (s) and store at –80°C. Make sure that all the spots on the MALDI target plate are dry before acquiring the spectra.

Acquiring Spectra from MALDI Target Plates

1. Double click on the flex control icon of an ULTRAFlex II MALDI-TOF instrument. When the log on information appears, Click OKAY. Open the FLEX control method: RP_pepmix.par. Wait for the method to upload.

2. Insert the spotted MALDI target plate into the source chamber and click on the green load button on the instrument.

3. Wait until the system fully loads the MALDI plate. Once the plate is loaded, the system status light on the FLEX control should turn green and say "READY." The sample status light should turn green and say "IN."

4. The spectrum is acquired by clicking the start button and adjusting the laser intensity so that the peptide peaks can be visualized.

5. Click on "Add" to the sum buffer.

6. Click on the "calibration tab" and choose "peptidemix monoisotopic II" from the drop down menu, and a peptide reference list will appear.

7. Change the zooming factor to 0.5.

8. To manually calibrate the instrument, click on each peptide in the reference list. The red vertical line indicates the corresponding peptide. Click on the monoisotopic peak (left of the highest peak) to visualize each of the peptides.

9. Click on "accept fit result"; now the instrument is calibrated.

10. To automatically acquire the samples, click on the AutoXecute tab. Load the .txt file by clicking on the select button. Make sure that the method to be run is "autoXYMP." Check the box where it shows the autoX output.

11. Click on "START automatic run" and the system should automatically start acquiring the spectra.

MALDI-TOF Spectral Analysis and Protein Identification

1. MALDI spectral searches are performed using Knexus automation, a Windows-based program from Genomics Solutions Bioinformatics (Discovery Scientific, Inc., Vancouver, Canada) that selects the spectral peaks automatically and performs the searches.

2. After the installation of the Knexus program, the most recent Fasta sequences of the *S. cerevisiae* genome downloaded from the European Bioinformatics Institute database (http://www.ebi.ac.uk/) were uploaded into the machine using the Knexus Database Installation Wizard.

3. Protein identification is done using the ProFound search engine, which matches the observed peaks against the database of theoretical peaks.

4. The program runs ProFound on all the spectra using one set of conditions. In addition, a Java program was developed in-house, which automates the rerunning of Knexus. It uses 72 varying parameter sets and evaluates the aggregate results to produce a set of identified proteins in each case. Based on these results, it calculates an aggregate score for each protein.

5. Using graphical interface software developed in-house, the user is able to specify where the bands are located on the corresponding gel image. The entire area of the band is inputted so that its total intensity is measured (*see* **Note 12**). When all the data has been entered into the system for a given band, an annotated gel image and a plot are produced in JPEG format. Examples of affinity-purified protein complexes identified by in-gel trypsin digestion and MALDI mass spectral analysis are shown in **Fig.2**.

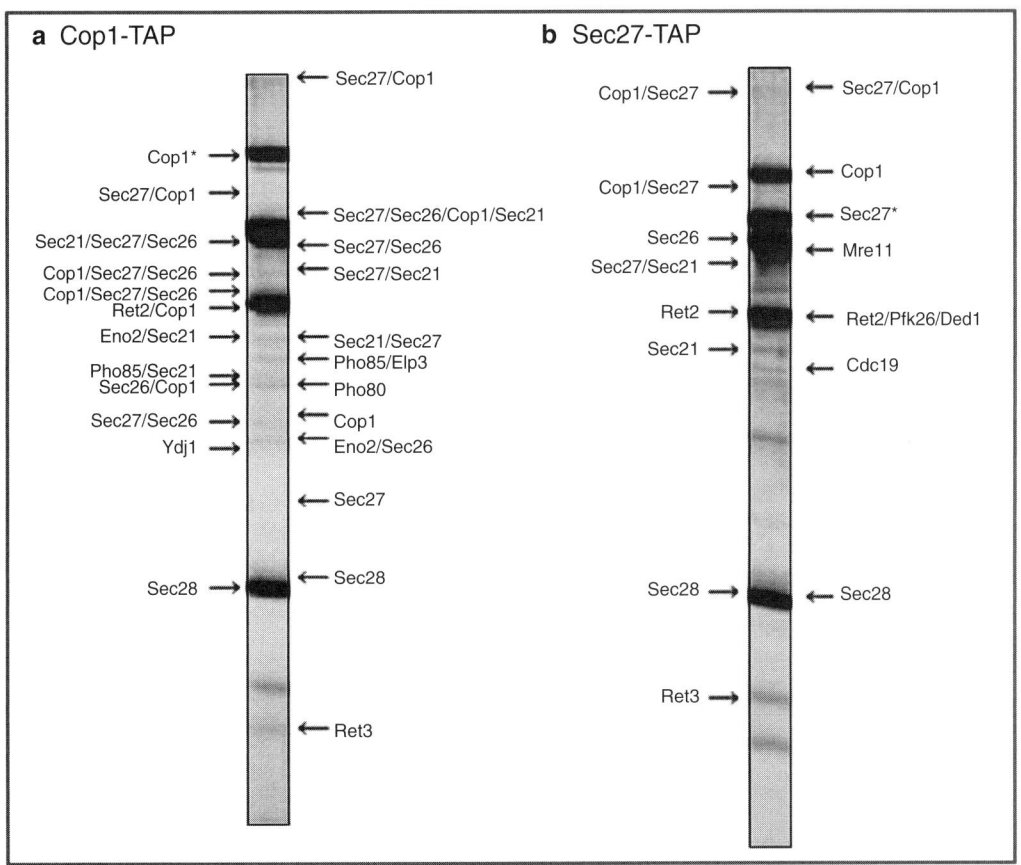

Fig. 2. Examples of silver-stained SDS polyacrylamide gel tracks containing COP1 coatomer protein complexes, which are involved in ER-to-Golgi and Golgi-to-ER transport, following affinity purification of *S. cerevisiae* TAP-tagged Cop1 (**a**) and Sec27 (**b**) proteins, respectively. Purifications of TAP-tagged Cop1 and Sec27 revealed other subunits of the COP1 coatomer protein complexes. The TAP-tagged bait proteins are specified at the top of each lane. Individual subunits of the purified complex and various co-purifying contaminants, all indicated by *arrowheads*, were identified by in-gel trypsin digestion followed by MALDI-TOF mass spectrometry. *Asterisks* designate the intact tagged bait protein recovered from each purification; proteolytic fragments of various coatomer subunits are also often identified. The known coatomer subunits include Cop1, Sec21, Sec26, Sec27, Sec28, Ret2, and Ret3. These purified preparations of coatomer complex also contained small amounts of the contaminating proteins Eno2, Pho80, Pho85, Elp3, Ded1, Mre11, Ydj1, Pfk26, and Cdc19.

3.3. Protein Identification by Gel-Free Liquid Chromatography-Tandem Mass Spectrometry

3.3.1. Proteolysis and Sample Preparation for LC-MS/MS

1. Three hundred μl eluted fractions (described in **step 10** in "Calmodulin-Sepharose Beads") in microcentrifuge tubes were dried down using a Speedvac.
2. The dried down samples are dissolved in 20 μl of the immobilized trypsin solution. Make sure to dissolve all the dried down sample in the immobilized trypsin solution. Using a P1000 pipette tip, pipette the samples up and down.
3. The samples are incubated overnight at 30°C with rotation or agitation.
4. Add 20 μl of the LC-MS solvent A to stop digestion.
5. The peptide mixtures are centrifuged for 5 min at maximum speed using an Eppendorf centrifuge.
6. Carefully transfer 10–20 μl of the supernatants containing the peptide mixtures into fresh microcentrifuge tubes and immediately analyze by LC-MS, or store the samples at –20°C prior to use.

3.3.2. LC-MS Spectral Analysis and Protein Identification

1. Protein samples are analyzed using single-dimension reverse phase chromatography coupled online to ion trap tandem mass spectrometry using standard conditions.
2. The microcolumns are packed with ~10 cm of 5-μm Zorbax eclipse XDB-C_{18} resin and are interfaced to a custom electrospray ion source.
3. The packed microcolumn is placed in line with the LC-MS instrument.
4. A Proxeon nano HPLC pump is used to deliver a stable tip flow rate of ~300 nl/min during the peptide separations.
5. Peptide elution is achieved using the following gradient which may be increased or decreased according to the complexity of the sample: 15% of solvent B from 0 to 30 min, 40% of solvent B from 31 to 49 min, 80% of solvent B from 50 to 55 min, 80% of solvent B from 56 to 60 min, 100% of solvent A from 61 to 65 min, 45% solvent B from 66 to 70 min, 80% solvent B from 71 to 75 min, 80% solvent B from 76 to 80 min, 100% solvent A from 81 to 85 min, 100% solvent A from 86 to 105 min. The flow rate at the tip of the needle is set to 300 nl/min for 105 min.
6. The mass spectrometer cycles run through successive series of 11 scans as the gradient progresses. The first scan in a series is a full mass scan and is followed by successive tandem mass scans of the most intense ions.
7. Proteins from the mixture are identified using the SEQUEST computer algorithm *(23)* and validated using the STATQUEST *(24)* probabilistic scoring program (*see* **Note 13**).

4. Notes

1. A similar TAP strategy is being used to purify the membrane-associated proteins except that all the purification steps are performed in the presence of a nonionic detergent.

2. Although low-abundance proteins can be detected using highly sensitive mass spectrometers, it is important to start with enough cells for the purification. This can be achieved by noting the expression level of the tagged protein *(3)* or by examining the literature to identify conditions that are suitable for the preparation of cell extracts specifically for the target protein. It is possible in some cases to successfully purify and identify by mass spectrometry a tagged protein that is not detectable by Western blotting.

3. If using the same coffee grinder to grind multiple protein samples, make sure not to clean the grinder with water because it will freeze. Use wipes and paper towels to clean the grinder.

4. Make sure to mix the YEB buffer with the lysed powder until it thaws. The tube is then inverted several times at room temperature to ensure proper mixing.

5. In some cases, however, it is necessary to use a higher salt concentration to reduce nonspecific binding of proteins.

6. It is not mandatory to perform the dialysis step in the preparation of extracts.

7. Make sure to elute the samples with calmodulin elution buffer, as it is a very critical step in releasing the complexes from the resin. Yields of the bound proteins can sometimes be improved either by increasing the salt concentration or by increasing the EGTA concentration in the calmodulin elution buffer.

8. When the developing solution becomes cloudy, replace it with new solution. Make sure to develop all the gels for about the same time.

9. The gel slices will become white and should feel gritty like grains of sand. Make sure to remove all the liquid when the gel slices are completely shrunk.

10. Make sure that in each well there is enough liquid for the gel slices to be completely submerged.

11. If the spots still remain on the MALDI target plate, sonicate the plate for 10 min with 100% methanol followed by 5 min sonication with HPLC water and 100% methanol, respectively.

12. The band location algorithm tries to identify the complete area of each band. Make sure to confirm that the location of the band that the computer identifies matches the band in the gel image.

13. The acquired spectra are used to search the *S. cerevisiae* protein database downloaded from the European Bioinformatics Institute (http://www.ebi.ac.uk/) using the SEQUEST search algorithm *(23)* to identify the proteins from which they originated. Confidence scores for all putative matches were evaluated and assigned using a probabilistic STATQUEST *(24)* scoring program. Proteins detected with two or more high-confidence peptide matches and with a minimum likelihood threshold cut-off of 90% or greater probability were considered as a positive identification.

Acknowledgments

The authors wish to thank Xinghua Guo and Gouqing Zhong for purifying the *S. cerevisiae* bait proteins and Shamanta Chandran, Peter Wong, and Constantine Christopoulos for assisting with mass spectrometry. This research was supported by grants from Genome Canada and the Ontario Genomics Institute to J.F.G. and A.E.

References

1. Puig, O., Caspary, F., Rigaut, G., Rutz, B., Bouveret, E., Bragado-Nilsson, E., Wilm, M., and Seraphin, B. (2001) The tandem affinity purification (TAP) method: A general procedure of protein complex purification. *Methods* 24, 218–229.
2. Rigaut, G., Shevchenko, A., Rutz, B., Wilm, M., Mann, M., and Seraphin, B. (1999) A generic protein purification method for protein complex characterization and proteome exploration. *Nat. Biotech.* 17, 1030–1032.
3. Ghaemmaghami, S., Huh, W.-K., Bower, K., Howson, R.W., Belle, A., Dephoure, N., O'Shea, E.K., and Weissman, J.S. (2003) Global analysis of protein expression in yeast. *Nature* 425, 737–741.
4. Gavin, A.C., Bosche, M., Krause, R., Grandi, P., Marzioch, M., Bauer, A., Schultz, J., Rick, J.M., Michon, A.M., Cruciat, C.M., Remor, M., Hofert, C., Schelder, M., Brajenovic, M., Ruffner, H., Merino, A., Hudak, M., Dickson, D., Rudi, T., Gnau, V., Bauch, A., Bastuck, S., Huhse, B., Leutwein, C., Heurtier, M.A., Copley, R.R., Edelmann, A., Querfurth, E., Rybin, V., Drewes, G., Raida, M., Bouwmeester, T., Bork, P., Seraphin, B., Kuster, B., Neubauer, G., and Superti-Furga, G. (2002) Functional organization of the yeast proteome by systematic analysis of protein complexes. *Nature* 415, 141–147.
5. Gavin, A.C., Aloy, P., Grandi, P., Krause, R., Boesche, M., Marzioch, M., Rau, C., Jensen, L.J., Bastuck, S., Dumpelfeld, B., Edelmann, A., Heurtier, M.A., Hoffman, V., Hoefert, C., Klein, K., Hudak, M., Michon, A.M., Schelder, M., Schirle, M., Remor, M., Rudi, T., Hooper, S., Bauer, A., Bouwmeester, T., Casari, G., Drewes, G., Neubauer, G., Rick, J.M., Kuster, B., Bork, P., Russell, R.B., and Superti-Furga, G. (2006) Proteome survey reveals modularity of the yeast cell machinery. *Nature* 440, 631–636.
6. Krogan, N.J., Cagney, G., Yu, H., Zhong, G., Guo, X., Ignatchenko, A., Li, J., Pu, S., Datta, N., Tikuisis, A.P., Punna, T., Peregrin-Alvarez,

J.M., Shales, M., Zhang, X., Davey, M., Robinson, M.D., Paccanaro, A., Bray, J.E., Sheung, A., Beattie, B., Richards, D.P., Canadien, V., Lalev, A., Mena, F., Wong, P., Starostine, A., Canete, M.M., Vlasblom, J., Wu, S., Orsi, C., Collins, S.R., Chandran, S., Haw, R., Rilstone, J.J., Gandi, K., Thompson, N.J., Musso, G., St Onge, P., Ghanny, S., Lam, M.H., Butland, G., Altaf-Ul, A.M., Kanaya, S., Shilatifard, A., O'Shea, E., Weissman, J.S., Ingles, C.J., Hughes, T.R., Parkinson, J., Gerstein, M., Wodak, S.J., Emili, A., and Greenblatt, J.F. (2006) Global landscape of protein complexes in the yeast Saccharomyces cerevisiae. *Nature* 440, 637–643.

7. Shoemaker, B.A., and Panchenko, A.R. (2007) Deciphering protein-protein interactions. Part I. Experimental Techniques and Databases. *PLoS Computational Biol.* 3, 0337–0344.

8. Uetz, P., Giot, L., Cagney, G., Mansfield, T.A., Judson, R.S., Knight, J.R., Lockshon, D., Narayan, V., Srinivasan, M., Pochart, P., Qureshi-Emili, A., Li, Y., Godwin, B., Conover, D., Kalbfleisch, T., Vijayadamodar, G., Yang, M., Johnston, M., Fields, S., and Rothberg, J.M. (2000) A comprehensive analysis of protein–protein interactions in Saccharomyces cerevisiae. *Nature* 403, 623–627.

9. Ito, T., Chiba, T., Ozawa, R., Yoshida, M., Hattori, M., and Sakaki, Y. (2001) A comprehensive two-hybrid analysis to explore the yeast protein interactome. *Proc. Natl. Acad. Sci. U S A* 98, 4569–4574.

10. Polanowska, J., Martin, J.S., Fisher, R., Scopa, T., Rae, I., and Boulton, S.J. (2004) Tandem immunoaffinity purification of protein complexes from Caenorhabditis elegans. *Biotechniques* 36, 778–780.

11. Veraksa, A., Bauer, A., and Tsakonas, S.A. (2005) Analyzing protein complexes in *Drosophila* with tandem affinity purification-mass spectrometry. *Dev. Dyn* 232, 827–834.

12. Butland, G., Peregrín-Alvarez, J.M., Li, J., Yang, W., Yang, X., Canadien, V., Starostine, A., Richards, D., Beattie, B., Krogan, N., Davey, M., Parkinson, J., Greenblatt, J., and Emili, A. (2005) Interaction network containing conserved and essential protein complexes in Escherichia coli. *Nature* 433, 531–537.

13. Tsai, A., and Carstens, R.P. (2006) An optimized protocol for protein purification in cultured mammalian cells using a tandem affinity purification approach. *Nat. Protoc.* 1, 2820–2827.

14. Knuesel, M., Wan, Y., Xiao, Z., Holinger, E., Lowe, N., Wang, W., and Liu, X. (2003) Identification of novel protein-protein interactions using a versatile mammalian tandem affinity purification expression system. *Mol. Cell Proteomics* 2, 1225–1233.

15. Brown, A.P., Affleck, V., Fawcett, T., and Slabas, A.R. (2006) Tandem affinity purification tagging of fatty acid biosynthetic enzymes in *Synechocystis* sp. PCC6803 and Arabidopsis thaliana. *J. Exp. Bot.* 57, 1563–1571.

16. Oliver, S.G. (2000) Guilt by association goes global. *Nature* 403, 601–603.

17. Collins, S.R., Kemmeren, P., Zhao, X.C., Greenblatt, J.F., Spencer, F., Holstege, F.C., Weissman, J.S., and Krogan, N.J. (2007) Towards a comprehensive atlas of the physical interactome of Saccharomyces cerevisiae. *Mol. Cell. Proteomics* 6, 439–450.

18. Pu, S., Vasblom, J., Emili, A., Greenblatt, J., and Wodak, S.J. (2007) Identifying functional modules in the physical interactome of Saccharomyces cerevisiae. *Proteomics* 7, 944–960.

19. Grisshammer, R., and Tate, C.G. (1995) Overexpression of integral membrane proteins for structural studies. *Q. Rev. Biophys.* 28, 315–422.

20. Loll, P.J. (2003) Membrane protein structural biology: the high throughput challenge. *J. Struct. Biol.* 142, 144–153.

21. Tate, C.G., and Grisshammer, R. (1996) Heterologous expression of G-protein-coupled receptors. *Trends Biotechnol.* 14, 426–430.

22. Dobrovetsky, E., Lu, M.L., Broza, R.A., Khutoreskaya, G., Bray, J.E., Savchenko, A., Arrowsmith, C.H., Edwards, A.M., and Koth, C.M. (2005) High-throughput production of prokaryotic membrane proteins. *J. Struct. Funct. Genomics* 6, 33–50.

23. Eng, J.K., McCormack, A.L., and Yates, J.R. III (1994) An approach to correlate tandem mass spectral data of peptides with amino acid sequences in a protein database. *J. Am. Soc. Mass Spectrom.* 5, 976–989.

24. Kislinger, T., Rahman, K., Radulovic, D., Cox, B., Rossant, J., and Emili, A. (2003) PRISM, a generic large scale proteomic investigation strategy for mammals. *Mol. Cell Proteomics* 2, 96–106.

Chapter 12

Protein Microarrays

Joseph Fasolo and Michael Snyder

Summary

Protein microarrays containing nearly the entire yeast proteome have been constructed. They are typically prepared by overexpression and high-throughput purification and printing onto microscope slides. The arrays can be used to screen nearly the entire proteome in an unbiased fashion and have enormous utility for a variety of applications. These include protein–protein interactions, identification of novel lipid- and nucleic acid-binding proteins, and finding targets of small molecules, protein kinases, and other modification enzymes. Protein microarrays are thus powerful tools for individual studies as well as systematic characterization of proteins and their biochemical activities and regulation.

Key words: Protein chips, Microarrays, High throughput, Protein interactions

1. Introduction

Protein microarrays contain a large number of proteins that have been spotted in an addressable format and at high density onto microscope slides, thereby allowing the individual analysis of large numbers of proteins simultaneously (1, 2). There are two types of protein microarrays: antibody microarrays that are used for protein profiling, and functional protein microarrays that contain sets of proteins or, for the case of yeast, nearly an entire yeast proteome. The functional arrays are used for a wide variety of applications and are presented here.

Protein arrays are typically prepared from large overexpression libraries in which proteins are overexpressed, purified in a 96-well format, and then spotted at high density onto microscope slides. The arrays are then probed using molecules containing fluorescent or other probes. To date, protein microarrays have been prepared for yeast (6,000), Arabidopsis (5,000), humans (8,000), and coronaviruses, as well as a number of bacteria *(3–7)*.

Thus far, protein microarrays have been used for a wide variety of applications. These include enzymatic assays and interactions with proteins, lipids, small molecules, and nucleic acids *(3, 8)*. One advantage of protein microarrays is that an entire proteome can be screened in an unbiased fashion, thereby leading to the discovery of novel activities that were not anticipated. For example, screening of a yeast protein chip using biotinylated liposomes revealed many new lipid-binding proteins including enzymes involved in glycolysis *(3)*, and screening of a yeast proteome chip with yeast DNA revealed a novel metabolic enzyme, Arg5,6, that is associated with DNA and appears to regulate mitochondrial gene expression *(9)*.

Protein microarrays have been used for many other applications as well. Recently, arrays that contain most yeast transcription factors have been prepared and probed using motifs that are conserved across yeast species but whose binding protein was not known *(10)*. This study assigned candidate binding proteins for specific sequences, one of which was confirmed. Protein microarrays have also been used to assess antibody specificity *(11)* and for identification of autoreactive antibody in patients with diseases such as autoimmune diseases and cancer and thereby find candidate biomarkers and insights into the disease state *(5)*. Lastly, protein microarrays can be used to identify substrates of modification enzymes such as kinases and ubiquitation enzymes *(12, 13)*.

2. Materials

2.1. Generation of C-Terminally Tagged Yeast Strains

1. pDONR221 vector containing the open reading frame (ORF) of interest.
2. pBG1805 vector: c-terminal His6X-3C-Protein A "zz" domain or pYES-DEST52 for His6X-V5 fusion proteins.
3. Flanking sequences for integration into Gateway pDONR221 plasmid
   (a) *att* B4: 5′ GGGGCAACTTTGTATAGAAAAGTTG 3′ (added to the 5′ PCR primer).
   (b) *att* B1: 5′ GGGGCTGCTTTTTTGTACAAACTTG 3′ (added to the 3′ PCR primer).
4. Sequencing primers for pBG1805 plasmid
   (a) F5: 5′ CATTTTCGGTTTGTATTACTTCTTATTC 3′.
   (b) R3: 5′ GGACCTTGAAAAAGAACTTC 3′.
5. *S. cerevisiae* genomic DNA (strain BY4700, *MATa ura3Δ0*).

6. *S. cerevisiae* Y258: *MAT*a, *pep*4, *his4-58*, *ura3-52*, *leu 2-3, 112*.
7. Pfx polymerase (Invitrogen) or Pfu Ultra polymerase (Stratagene).
8. BsrG1 endonuclease for screening positive clones.
9. Culture tubes.
10. Agarose.
11. LB plus Kanamycin, both liquid, and 2% bacto-agar.
    (a) 10 g tryptone.
    (b) 5 g yeast extract.
    (c) 5 g NaCl.
    (d) 20 g agar (for plates).
    (e) Make up to 1 L with distilled water (dH$_2$O).
    (f) Autoclave and allow to cool prior to addition of 50 μg/mL of ampicillin final concentration.
12. DNA ladder
13. *E. coli*: DH5α
14. TAE
    (a) 242 g Tris base.
    (b) 57.1 mL 100% glacial acetic acid.
    (c) 100 mL 0.5 M ethylenediaminetetraacetic acid (EDTA) (pH 8.0).
    (d) Make up with dH$_2$O up to 1 L.
15. LR clonase (Invitrogen).
16. BP clonase (Invitrogen).

2.2. Transformation Reagents

1. YPAD (Yeast extract-peptone-adenine-dextrose)
   (a) 10 g bacto yeast extract.
   (b) 20 g bacto peptone.
   (c) 50 mg adenine.
   (d) 20 g dextrose.
   (e) 20 g agar (for plates).
   (f) Make up with dH$_2$O up to 1 L.
2. 1 M lithium acetate (LiAc).
3. 50% polyethylene glycol (PEG).
4. 10 mg/mL ssDNA.
5. DNA.
6. dH$_2$O.
7. Water bath or heat block.
8. 96-well box.

9. 96-well polymerase chain reaction (PCR) plate.
10. Synthetic complete minus uracil media (Sc-ura)
    (a) 1.5 g yeast nitrogen base.
    (b) 5 g (NH4)2SO4.
    (c) 2 g Sc-ura drop out mix (commercially available).
    (d) 20 g raffinose (or dextrose for starter culture media and SD-ura plates).
    (e) 20 g agar (for plates).

2.3. High-Throughput Immunoblot Analysis

1. 3× YEP–GAL (yeast extract–peptone–galactose)
    (a) 30 g yeast extract.
    (b) 60 g peptone.
    (c) Make up to 700 mL with dH_2O.
    (d) Add 300 mL of sterile filtered 20% galactose to media after autoclaving.
2. 0.5 mm zirconia beads or acid-washed glass beads.
3. Lysis buffer 150
    (a) 50 mM Tris–HCL at pH 7.5.
    (b) 150 mM NaCl.
    (c) 1 mM ethyleneglycoltetraacetic acid (EGTA).
    (d) 10% glycerol.
    (e) 0.1% Triton X-100.
    (f) 0.5 mM dithiothreitol (DTT).
    (g) 1 mM phenylmethylsulfonyl fluoride (PMSF).
    (h) 1× complete protease inhibitor tablet (Roche).
4. Paint shaker (5G-HD, Harbil) or similar mechanism to break cells.
5. SDS-polyacrylamide gel electrophoresis (SDS-PAGE) gels.
6. Polyvinylidene fluoride (PVDF) or nitrocellulose membranes.
7. Multichannel pipette.
8. Semi-dry transfer apparatus.
9. HRP-IgG.
10. HA antibody (16B12, Covance).
11. Imaging Film (Kodak BioMax MR Film).

2.4. Preparation of Proteins for Protein Microarray

1. Wash buffer (150)
    (a) 50 mM Tris-HCl (pH 7.5).
    (b) 150 mM NaCl.
    (c) 10% glycerol.
    (d) 0.1% and Triton X-100.

2. Elution buffer
   (a) 50 mM Tris (pH 7.5).
   (b) 150 mM NaCl.
   (c) 25% glycerol.
   (d) 0.1% Triton X-100.
3. 96-well box.
4. PVDF 96-well filter plate (1.2-μm pore size).
5. IgG-Fast Flow 6 Sepharose.
6. 3C protease.
7. 384-well microplate (to array proteins).
8. Adhesive foil lids for sealing microplates.

2.5. Printing Arrays

1. Arrays suitable for assay being performed.
2. Bio-Rad ChipWriter Pro or comparable printer.

2.6. Purifying V5-Fusion Protein Probe

1. Phosphate buffered saline (PBS) lysis solution
   (a) PBS (0.01 M phosphate buffer, 0.0027 M potassium chloride, 0.137 M sodium chloride; pH 7.4).
   (b) 0.1% Triton X-100.
   (c) 0.5 mM DTT.
   (d) 2 mM MgCl2.
   (e) 500 mM NaCl.
   (f) 50 mM imidazole (pH 7.4).
   (g) 1 mM PMSF.
   (h) Complete protease inhibitor cocktail – EDTA-free (Roche).
   (i) Phosphatase inhibitor Cocktail 1 (Sigma).
2. Ni^{2+}, or Co^{2+} affinity resin.
3. G25 columns.
4. Microcentrifuge.
5. Wash buffer
   (a) PBS (0.01 M phosphate buffer, 0.0027 M potassium chloride, 0.137 M sodium chloride; pH 7.4).
   (b) 0.1% Triton X-100.
   (c) 500 mM NaCl.
   (d) 0.5 mM DTT.
   (e) 50 mM imidazole (pH 7.4).
   (f) 2 mM $MgCl_2$.
6. Elution buffer
   (a) PBS (0.01 M phosphate buffer, 0.0027 M potassium chloride, 0.137 M sodium chloride; pH = 7.4).

(b) 0.1% Triton X-100.

(c) 500 mM NaCl.

(d) 0.5 mM DTT.

(e) 500 mM imidazole (pH 7.4).

(f) 2 mM MgCl2.

(g) 20% glycerol (if freezing at −80°).

2.7. Probing Arrays

1. Blocking buffer
   (a) PBS (0.01 M phosphate buffer, 0.0027 M potassium chloride, 0.137 M sodium chloride; pH 7.4).
   (b) 1% bovine serum albumin (BSA).
   (c) 0.1% Tween-20.
2. Probe buffer
   (a) PBS (0.01 M phosphate buffer, 0.0027 M potassium chloride, 0.137 M sodium chloride; pH 7.4).
   (b) 2 mM $MgCl_2$.
   (c) 0.5 mM DTT.
   (d) 0.05% Triton X-100.
   (e) 50 mM NaCl.
   (f) 500 µM ATP (for kinases).
   (g) 1% BSA.

2.8. Identifying Kinase–Substrate Interactions

1. Lysis Buffer
   (a) 100 mM Tris–HCl pH 7.4.
   (b) 100 mM NaCl, 1 mM EGTA.
   (c) 0.1% 2-mercaptoethanol.
   (d) 0.1% Triton X-100, protease cocktail (Roche).
   (e) 1 mM EDTA.
   (f) 50 mM NaF.
   (g) 10 mM sodium glycerophosphate.
   (h) 1 mM Na_3VO_4.
2. Kinase Buffer
   (a) 100 mM Tris–HCl pH 8.0.
   (b) 100 mM NaCl.
   (c) 10 mM $MgCl_2$.
   (d) 20 mM glutathione.
   (e) 20% glycerol.
3. Superblock (Pierce)
4. [^{33}P]ATP

3. Methods

3.1. High-Throughput Preparation of Clones

Protein microarrays are typically constructed using overexpression libraries. To date, two comprehensive expression libraries have been prepared that are suitable for production of yeast protein microarrays. Initially, a library was prepared in which 5,800 yeast ORFs were fused at their amino terminal coding sequences to glutathione-S-transferase and HisX6 coding sequences *(3)*. Subsequently, a movable ORF, or MORF, library was prepared in which 5,700 yeast ORFs were fused at their carboxy terminal coding sequences to His6 and IgG binding domain of protein A *(14)*. The MORF library contains GATEWAY-compatible sequences flanking the inserts, so the yeast ORFs can be shuttled into any vector. In each library the fusion protein is expressed from the GAL1 galactose promoter. A protocol for library construction is as follows:

3.1.1. Constructing a Large Number of Clones for Protein Arrays

1. Design forward primers containing the att B4 sequence and reverse primers with the att B1 sequence (*see* **subheading 2.1** and **Note 4**) for the gene of interest.
2. Amplify the gene of interest by PCR using the previously mentioned genomic DNA and polymerase (*see* **subheading 2.1**).
3. Perform BP clonase reaction (5 µL) using 30–300 ng of PCR product added to 150 ng of pDONR221 plasmid and following manufacturer's instructions.
4. Incubate the reaction overnight at 25°C.
5. Combine half of the of the BP reaction (3 µL) with 150 ng of pBG1805 DNA or pYES-DEST52 (for V5 fusion) backbone, 0.6 µL of LR clonase, and 0.6 µL of LR clonase 10× buffer.
6. Incubate the reaction overnight at 25°C and use half of the reaction to transform DH5α cells.
7. Plate the cells on LB/ampicillin (*see* **Subheading 2.1**, **item 11**) plates overnight at 37°C.
8. Miniprep overnight cultures of several colonies of each transformant grown in LB/ampicillin liquid media.
9. Digest with 5 µL of DNA with BsrG1 overnight at 37°C to confirm the presence of insert and sequence clones that contain the insert using sequencing primers (*see* **Subheading 2.1**, **item 4**).

3.1.2. Transforming Yeast with C-Terminal Tagged ORF

1. Inoculate 20 mL YPAD (*see* **subheading 2.2**, **item 1**) with a starter culture of a fresh Y258 colony from YPAD agarose plate.
2. Inoculate 50 mL of YPAD/10 transformations with enough starter culture to attain an OD_{600} of 0.1.

3. Grow for several doubling times to an OD_{600} of ~0.6–0.8.
4. Transfer the cells to a 50-mL conical tube and spin in a tabletop centrifuge at 3,000 rpm/5 min/4°C.
5. Bring up to a volume of 50 mL with water to wash, and repeat centrifugation.
6. Decant water and add 0.5 mL 100 mM LiAc and transfer to a 1.5-mL snap-cap tube.
7. Spin at 13,000 rpm in a microfuge for 30 s and remove the solution with a pipette.
8. Add enough 100 mM LiAc to bring the volume up to 0.5 mL, and vortex to suspend the cells.
9. Aliquot 50 µL of the suspension into separate snap-cap tubes and centrifuge.
10. Remove liquid as before and prepare pellet for transformation.
11. Add 240 µL of 50% PEG solution, 36 µL of 1 M LiAC, 10 µL of 10 mg/mL ssDNA, 2 µL of miniprepped DNA fusion construct, and 72 µL of dH_2O.
12. Vortex the mixture and incubate at 30°C for 30 min.
13. Next, move the tubes to a water bath set to 42°C for 30 min.
14. Finally, spin the tubes at 10,000 × g for 15 s, and aspirate the solution with a pipette.
15. Add 500 mL of dH_2O to each tube and vortex briefly.
16. Pipette 50–100 µL of suspension onto Sc-ura/2% dextrose agar Petri dishes (*see* **subheading 2.2**, **item 10**).
17. Allow 48 h to recover and screen the transformants by immunoblot.

3.1.3. High-Throughput Assaying and Preparation of Protein Methods for Assay Protein Expression and High-Throughput Preparation of Proteins

High-Throughput Immunoblot Analysis

1. Inoculate 96-well starter cultures in 0.8 mL of SD-ura/2% dextrose.
2. On day 2, wash the cells with Sc-ura/2% raffinose (*see* **subheading 2.2**, **item 10**), and inoculate 5 µL of starter culture into a new 96-well box (2 mL/well) containing 0.8 mL of Sc-ura/2% raffinose with a 3.5-mm glass ball (PGC scientific) in each well for mixing and aeration.
3. Mix cells for 15 h at 30°C on a platform shaker and then induce by addition of 0.4 mL of 3× YEP-GAL (*see* **subheading 2.3**, **item 1**) for 6 h, and follow by centrifugation and washing with ice cold water and storage at −80°C.
4. Obtain the lysates by adding 200 µL of lysis buffer 150 (*see* **subheading 2.3**, **item 2**) and shaking for 6 min in a paint shaker (5G-HD, Harbil) at 4°C with 250 µL of acid-washed

glass beads (0.5 mm, Sigma), followed by centrifugation at 1,000 × g for 5 min.

5. Combine the crude lysates with 5× SDS loading buffer in a 96-well PCR plate, heat for 5 min at 95°C, and centrifuge again at 2,500 rpm 5 min.

6. Remove 12 μL from each well with a multichannel pipette and load onto SDS-PAGE gels.

7. Western transfer to either PVDF or nitrocellulose membranes using standard wet or dry transfer techniques.

8. Block the membrane with TBS/0.1% Tween–1% milk powder for 1 h.

9. Probe with anti-HA antibodies (16B12, 1:1,000, Covance) overnight in TBS/0.1% Tween–1% milk powder.

10. On day 2, wash the membrane three times with TBS/0.1% Tween, 10 min each wash.

11. Probe for 1 h at room temperature with HRP-conjugated sheep anti-mouse IgG antibody (Amersham).

12. Wash the membrane three times with TBS/0.1% Tween, 10 min each wash, and develop with Supersignal west chemiluminescent substrate (Pierce).

Purifying Proteins for Arrays

1. Inoculate 96-well starter cultures in 0.8 mL of Sc-ura medium.

2. On day 2, wash the cells with Sc-ura/2% raffinose, and inoculate 5 μL of starter culture into a new 96-well box (2 mL/well) containing 0.8 mL of Sc-ura/2% raffinose with a 3.5-mm glass ball (PGC scientific) in each well for mixing and aeration.

3. Mix cells for 15 h at 30°C on a platform shaker and then induce by addition of 0.4 mL of 3× YEP-GAL (*see* **Subheading 2.3**, **item 1**) for 6 h followed by centrifugation, washing with ice cold water, and storage at –80°C.

4. Obtain the lysates by adding 200 μL of lysis buffer 150 (*see* **Subheading 2.3**, **item 2**) and shaking for 6 min in a paint shaker (5G-HD, Harbil) at 4°C with 250 μL of acid-washed glass beads (0.5 mm, Sigma), followed by centrifugation at 2,500 rpm for 5 min.

5. Transfer the cleared lysate to a new 96-well box along with IgG Sepharose resin (6 Fast Flow; GE Biosciences) and incubate at 4°C for several hours with agitation.

6. Afterwards, wash the beads four times with wash buffer 150 (*see* **Subheading 2.4**, **item 1**)

7. Transfer the beads to a PVDF filter plate (1.2 μm pore size) and spin-through the excess wash. Suspend the resin is in 40 μL of elution buffer containing 1 μL of 3C protease for 18 h at 4°C with agitation.

8. The next day, add preconditioned glutathione beads to the reaction to remove the 3C protease and spin the eluted protein through the PVDF membrane into a clean 96-well freezer plate.

9. Array the proteins into 5 µL aliquots on multiple 384-well printing plates and cover with an adhesive aluminum seal for storage at −80°C until printing.

10. Include a control plate with assay-specific controls (*see* **Note 2**).

3.1.4. Printing Protein Microarrays

Once proteins are prepared they must be printed onto surfaces, typically microscope slides. A variety of different surfaces exist for preparing protein microarrays. They include nitrocellulose and aldehyde surface chemistries for chemical attachment through lysines and affinity attachment methods such as nickel-chelated slides for attaching His tagged proteins and glutathione for affinity attachment *(15, 3, 16)*. The different surfaces have advantages and disadvantages; nitrocellulose allows the attachment of large amounts of protein, and it along with chemical attachment methods results in random orientation of proteins away from the surface. It is likely that the activity of the attached molecule is lost for many molecules attached by these methods. Moreover, binding of probes to regions near the surface is likely to be sterically inhibited. Affinity attachment methods are more likely to retain activity and orient proteins away from the surface, although random presentation of protein surfaces from the slide is not attained. Different surfaces yield different backgrounds and it is prudent to test several surfaces whenever a new assay is employed.

1. Configure the contact arrayer and align the microscope slides (FAST, Path, or other types of slides *see* **Note 1**) to allow the maximum number of slides to be printed per run. This will vary depending on the arrayer used, but for the Bio-Rad ChipWriter Pro it is approximately 90 arrays/run.

2. Maintain a constant humidity of ~40% in the printing chamber, and a constant temperature of 4°C (this is achieved by placing the printer in a temperature-controlled cold room).

3. Remove two plates at a time from the −80°C storage, and allow them to thaw on ice for 5–10 min followed by centrifugation at $1{,}000 \times g$ at 4°C for 2 min.

4. Remove the foil seal from the top of the plate and place the plate into the chamber in the proper orientation for printing (see manufacturer's instructions).

5. After the print run, cover the plate with a new aluminum seal and return to the −80°C freezer.

6. Repeat for each plate.

7. After printing, store the arrays in a −20°C non-deicing freezer, which can remain stable for up to 1 year.

3.1.5. Probing for Protein–Protein Interactions with Protein Microarrays

Protein microarrays can be screened for a wide variety of activities. A common activity is interaction with other proteins. One of the best ways to perform this is to produce the protein of interest with an epitope tag (we prefer to use the V5 epitope for yeast) and probe a yeast proteome chip as follows:

3.1.6. Purifying the V5-Fusion Protein Probe

1. Grow 5–20 mL of yeast culture containing V5-fusion protein (pYES-DEST52 vector, Invitrogen, *see* **Note 4**) probe overnight in Sc-ura/2% dextrose.

2. Inoculate 40–400 mL of the Sc-ura/2% raffinose culture with sufficient starter culture to a final OD_{600} of 0.1.

3. Grow a large culture for a period of three doubling times and induce with 3×-YEP supplemented with 2% galactose, by adding enough to dilute the induction media by a factor of 3.

4. Induce cells at 30°C for 5 h.

5. Harvest cells using in a JA-10 (or comparable) rotor by spinning ≤400 mL volumes of cell suspension at 4,000 rpm/4°C/5 min.

6. Wash the cells once with 50 mL of cold dH_2O and transfer to a 50-mL conical tube. Wash again in cold buffer (without detergents or other additives) used for lysis (e.g., PBS, Tris, HEPES) and transfer to 2-mL snap-cap tubes for lysis.

7. Spin the cells at $20,000 \times g$/4°C/1 min to a pellet and pipette away the buffer.

8. Place tubes on ice and proceed with lysis step.

9. Lyse the cells with 0.5-mm zirconia beads in a 1:1:1 volumes of cell pellet, beads, and PBS lysis buffer (*see* **Subheading 2.6, item 1**). Vortex the mixture using a paint shaker several times at 2-min intervals at 4°C.

10. Centrifuge the lysate at $20,000 \times g$ in a tabletop microfuge for 10 min at 4°C, followed by ulracentrufugation for 30 min at $150,000g$ at 4°C.

11. Apply the clarified lysate to the previously mentioned affinity resin.

12. Wash the resin several times with wash buffer (*see* **Subheading 2.6, item 5**), followed by elution with elution buffer (*see* **Subheading 2.6, item 6**)

Probing the Arrays

1. Dilute the protein probe over a concentration range of 5–500 µg/mL, which must be optimized for each protein–protein interaction assay (*see* **Fig. 1**). The purified V5-fusion

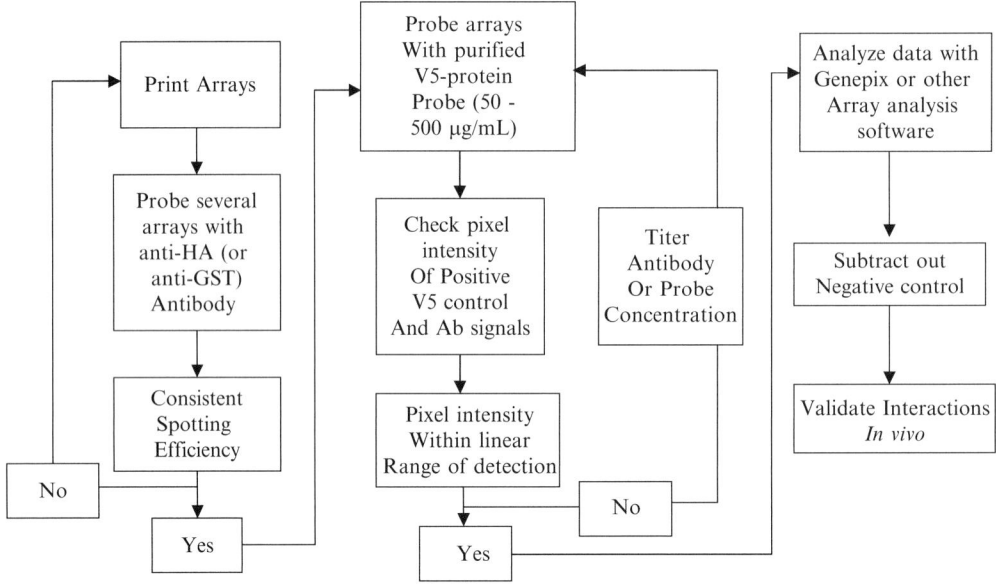

Fig. 1. Flowchart for protein–protein interaction assay optimization.

protein is diluted into the probe buffer (*see* **Subheading 2.7, item 2**).

2. Remove the arrays from the freezer (−20°C) and bring to 4°C in the refrigerator, just prior to use, for about 15–20 min. Block the arrays in blocking buffer (*see* **Subheading 2.7, item 1**) for 1 h by shaking at 50 rpm on a stage at 4°C.

3. After blocking, transfer the arrays to a humidity chamber, and add 90 µL of diluted probe directly to the array surface. Overlay the arrays with a raised lifter slip and incubate static (no shaking) in the humidor for 1.5 h.

4. Wash the arrays 3 times for 1 min each in probe buffer in three 50-mL conical tubes (*see* **Note 4**).

5. To detect interactions, dilute the V5-AlexaFluor 647 antibody to 260 ng/mL in probe buffer and mix thoroughly by shaking.

6. After washing the arrays several times as indicated, add antibody solution directly to array and overlay with a raised lifter slip as before. Incubate the arrays for 30 min/static/4°C.

7. Finally, wash the arrays for 1 min (3×) in probe buffer, and spin in a 50-mL conical tube at 800*g* in a tabletop centrifuge for 5 min at room temperature. Air-dry the arrays in a slide holder in the dark for 30 min prior to scanning the array at 647 nm.

3.1.7. Identifying Protein Kinase Substrate with Protein Arrays

Another useful application of protein microarrays is to identify targets of protein modification enzymes. Recently, a large study has been performed to identify the targets of 87 distinct protein kinases of yeast *(12)*. A protocol for screening for targets of protein kinases is as follows:

1. Grow cells in 50–500-mL cultures, harvest, and lyse, with glass beads in the lysis buffer (*see* **Subheading 2.7**, **item 1**) as in "Purifying Proteins for Arrays."
2. Kinases-GST fusions are bound to glutathione beads and eluted into the kinase buffer (*see* **Subheading 2.7**, **item 2**).
3. Block the proteome arrays in a Superblock (Pierce) with 0.1% Triton X-100 for 1 h at 4°C and probe in duplicate for every kinase.
4. Optimize conditions: Dilute the kinase into kinase buffer plus 0.5 mg/mL BSA, 0.1% Triton X-100, and 2 L of [^{33}P]ATP (33.3 nM final concentration).
5. Overlay each kinase in buffer on two arrays, cover with a coverslip, and place in a humidified chamber at 30°C for 1 h.
6. Wash the slides twice with 10 mM Tris–HCl pH 7.4, 0.5% SDS and once with double-distilled water before being spun dry and exposed to X-ray film (Kodak).
7. For each experiment, incubate two additional arrays with kinase buffer in the absence of kinase, which will serve as autophosphorylation reference slides.

4. Notes

1. Care needs to be taken when printing samples with high glycerol content on surfaces other than nitrocellulose (nickel, ultrgap, etc.) in high humidity >40% because the spot diameter will differ and this could result in samples bleeding together.
2. Probe-specific control spots must be added to each array. These include known concentrations of 3C protease (if used) as well as Alexa Fluor V5 antibody and V5-fusion constructs. These will be used to ascertain optimal probing conditions and antibody titer for protein probing experiments.
3. Primers for sequencing of V5-fusions in pYES-DEST52 can be obtained from Invitrogen's Website. ORFs can be shuttles from any of the destination (DEST) vectors using the pDONR221 as an intermediary.
4. When probing for protein–protein interactions, it is advisable to wash the arrays using three separate 50-mL conical tubes

per slide. Allow the slides to remain submerged in the first chamber until the lifter slip falls off (do not pull it off with force). Afterward, proceed with 1 min washes, gently removing slides with forceps from each bath by gripping the non-membraneous portion of the slide.

References

1. Zhu H, et al. (2003) Proteomics. *Annu. Rev. Biochem.* 72:783–812.
2. Zhu H, Snyder M. (2003) Protein chip technology. *Curr. Opin. Chem. Biol.* 7(1): 55–63.
3. Zhu H, et al. (2001) Global analysis of protein activities using proteome chips. *Science* 293(5537):2101–5.
4. Popescu SC, et al. (2007) Differential binding of calmodulin-related proteins to their targets revealed through high-density Arabidopsis protein microarrays. *Proc. Natl. Acad. Sci. U S A* 104(11):4730–5.
5. Hudson ME, et al. (2007) Identification of differentially expressed proteins in ovarian cancer using high-density protein microarrays. *Proc. Natl. Acad. Sci. U S A* 104(44):17494–9.
6. Zhu H, et al. (2006) Severe acute respiratory syndrome diagnostics using a coronavirus protein microarray. *Proc. Natl. Acad. Sci. U S A* 103(11):4011–6.
7. Rolfs A, et al. (2008) Production and sequence validation of a complete full length ORF collection for the pathogenic bacterium Vibrio cholerae. *Proc. Natl. Acad. Sci. U S A* 105(11):4364–9.
8. Huang J, et al. (2004) Finding new components of the target of rapamycin (TOR) signaling network through chemical genetics and proteome chips. *Proc. Natl. Acad. Sci. U S A* 101(47):16594–9.
9. Hall DA, et al. (2004) Regulation of gene expression by a metabolic enzyme. *Science* 306(5695):482–4.
10. Ho SW, et al. (2006) Linking DNA-binding proteins to their recognition sequences by using protein microarrays. *Proc. Natl. Acad. Sci. U S A* 103(26):9940–5.
11. Michaud GA, et al. (2003) Analyzing antibody specificity with whole proteome microarrays. *Nat. Biotechnol.* 21(12):1509–12.
12. Ptacek J, et al. (2005) Global analysis of protein phosphorylation in yeast. *Nature* 438(7068):679–84.
13. Gupta R, et al. (2007) Ubiquitination screen using protein microarrays for comprehensive identification of Rsp5 substrates in yeast. *Mol. Syst. Biol.* 3:116.
14. Gelperin DM, et al. (2005) Biochemical and genetic analysis of the yeast proteome with a movable ORF collection. *Genes Dev.* 19(23):2816–26.
15. MacBeath G, Schreiber SL. (2000) Printing proteins as microarrays for high-throughput function determination. *Science* 289(5485):1760–3.
16. Ramachandran N, et al. (2004) Self-assembling protein microarrays. *Science* 305(5680):86–90.

Chapter 13

Analysis of Protein–Protein Interactions Using Array-Based Yeast Two-Hybrid Screens

Seesandra V. Rajagopala and Peter Uetz

Summary

The yeast two-hybrid system (Y2H) is a powerful tool to identify protein–protein interactions. Here we describe array-based two-hybrid methods that use defined libraries of open reading frames (ORFs) as opposed to random genomic or cDNA libraries. The array-based Y2H system is well suited for interactome studies of existing ORFeomes or subsets thereof, preferentially in a recombination-based cloning system. Array-based Y2H screens efficiently reduce false positives by using built-in controls, retesting, and evaluation of background activation. Hands-on time and the amount of used resources grow exponentially with the number of tested proteins; this is a disadvantage for large genome sizes. For large genomes, random library screen may be more efficient in terms of time and resources, but not as comprehensive as array screens, and they require an efficient sequencing facility. However, large array screens require some extent of automation although they can be carried out manually on smaller scales. Future-generation Y2H plasmid constructs including tightly regulated expression systems and features that facilitate biochemical characterization will provide more efficient and powerful tools to identify interacting proteins.

Key words: Yeast two-hybrid system, Protein–protein interactions, Two-hybrid array

1. Introduction

Specific interactions between proteins form the basis of many essential biological processes. Comprehensive analysis of protein–protein interactions is a challenging task of proteomics and has been best explored in budding yeast. Protein interactome analysis at a genome scale was first achieved by using yeast two-hybrid (Y2H) screens *(1)* and next by large-scale mass spectrometric analysis of affinity-purified protein complexes *(2, 3)*. The Y2H system is a genetic method that detects binary protein–protein interactions in vivo. Classical two-hybrid screens used random libraries (genomic or cDNA) to identify novel interactions for a protein of interest. However, more recently, an array-based variation of this original principle has been increasingly used (**Fig. 1**). This approach can be applied to a few proteins but also to whole genomes. Advantages of arrays are their built-in controls and their systematic nature. However, random library screens may be more comprehensive and may yield fewer false negatives. Protocols for random library screens are not included in this chapter but may be found in *(4)*.

1.1. The Principle of the Yeast Two-Hybrid System

The Y2H system is a genetic method extensively used to detect binary protein–protein interactions in vivo (in yeast cells). This system was developed by Stanley Fields *(5)* based on the observation that protein domains can be separated and recombined and still can retain their properties. In particular, transcription factors can frequently be split into the DNA-binding domain (DBD) and activation domains (ADs). In the two-hybrid system, a DNA-binding domain (here: from the yeast Gal4 protein) is fused to a protein "B" (for bait) for which one wants to find interacting partners (**Fig. 1**, step 3). A transcriptional activation domain is then fused to some or all the predicted open reading frames (ORFs or "preys") of an organism. Bait and prey fusion proteins are then co-expressed in the same yeast cell. Usually both protein fusions are expressed from plasmids that can be manipulated easily and then transformed into yeast cells. If the two proteins "**B**" and ORF interact, a transcription factor is reconstituted, which in turn activates one or more appropriate reporter genes. The expression of the reporter allows the cell to grow only under certain conditions. For example, the HIS3 reporter encodes imidazoleglycerolphosphate (IGP) dehydratase, a critical enzyme in histidine biosynthesis. In the Y2H screening strain lacking an endogenous copy of HIS3, expression of a HIS3 reporter gene is driven by a promoter that contains a Gal4p-binding site, so the bait protein fusion can bind to it. However, since the bait fusion does not contain a transcriptional activation domain it remains inactive. If a protein ORF with an attached activation domain binds to the bait, this activation domain can recruit the basal transcription machinery, and expression of the reporter gene ensues. These cells can now grow in the absence of histidine in the media because they can synthesize their own.

1.2. Applications

Originally, the two-hybrid system was invented to demonstrate the association of two proteins *(5)*. Later, it was demonstrated that completely new protein interactions can be identified with this system, even when there are no candidates for an interaction with a given bait. Over time, it has become clear that the ability to

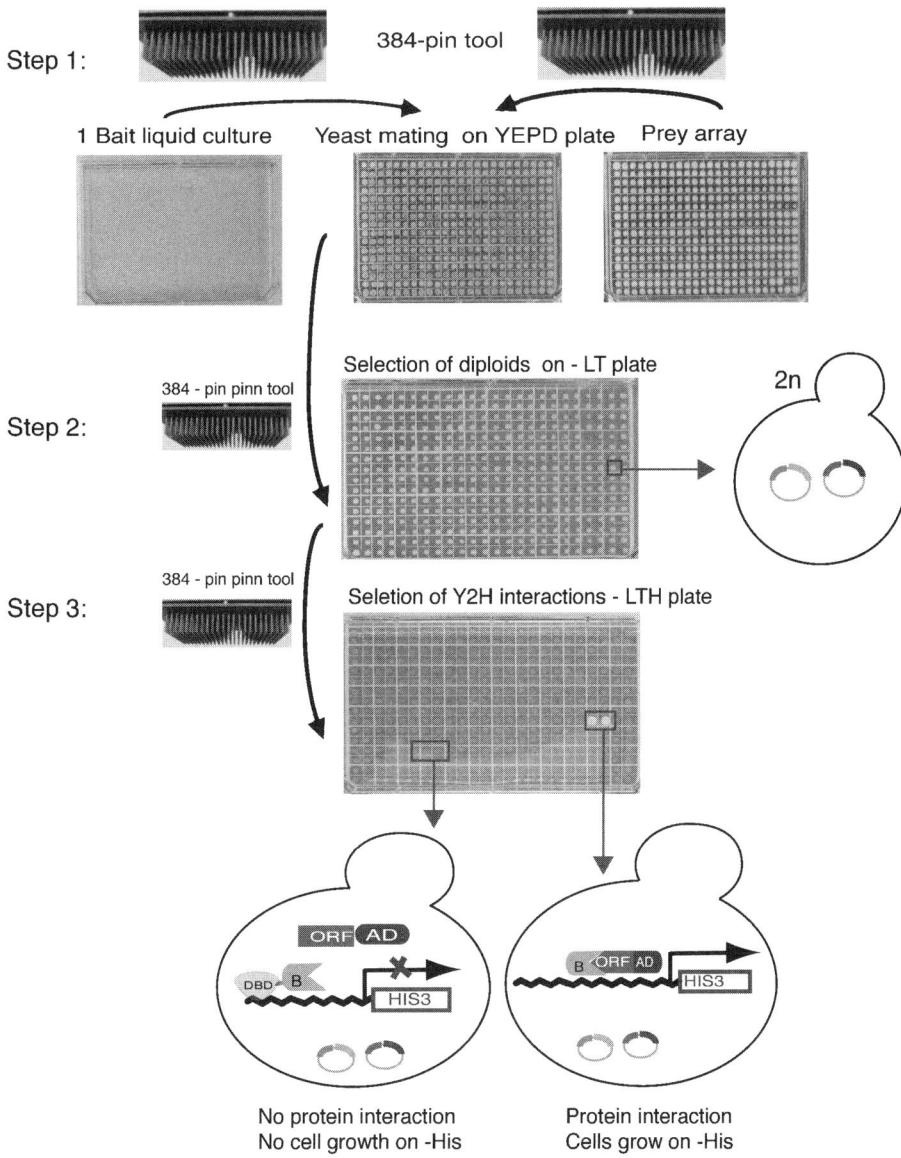

Fig. 1. Scheme of an array-based two-hybrid screen. **Step 1:** Yeast mating is the first step in the Y2H screening, which combines the bait and prey plasmids. First the bait (DNA-binding domain (DBD) fusion) liquid culture is pinned onto YEPD agar plates using a 384-pin pinning tool, and then the prey array (activation-domain (AD) fusion) is pinned on top of the baits using the sterile pinning tool. Then the mating plates are incubated at 30°C for 16 h. **Step 2:** The yeast mating plates are pinned onto −Leu −Trp (−LT) medium plates using a sterile 384-pin pinning tool. On −LT plates only diploid cells will grow. Selection on −LT media ensures that both the prey and bait plasmids are combined in the diploid yeast cells. **Step 3:** The diploid cells are pinned onto −Leu −Trp −His medium plates for protein interaction detection. Only if bait and prey proteins interact, and an active transcription factor is reconstituted transcription of a reporter gene is activated (*lower right panel*). In the array-based Y2H system, a bait protein is tested against the whole systematic prey library. This library consists of individual yeast colonies at specific positions of an array (e.g., in 384 format); each colony carries a specific prey construct of a specific ORF. Systematic testing is done by robotic transfer of yeast cells starting with the mating of bait and prey strains. Diploid cells are selected on specific plates and, finally, transferred to plates selecting for the activation of a reporter gene. Here the activation of the His3p gene is detected by scoring the growth on His-deficient plates. The *rectangles* on the selective plate mark negative and positive interactions between the bait and the prey at these specific positions of the array (test is done in duplicates). After Uetz et al. *(1)*.

conveniently perform unbiased library screens is the most powerful application of the system. With whole-genome arrays, such unbiased screens can be expanded to defined, nonredundant sets of proteins. Arrays, like traditional two-hybrid screens, can also be adapted to a variety of related questions, such as the identification of mutants that prevent or allow interactions *(6)*, the screening for drugs that affect protein interactions *(7, 8)*, the identification of RNA-binding proteins *(9)*, or the semiquantitative determination of binding affinities *(10)*. The system can also be exploited to map binding domains *(11)*, to study protein folding *(12)*, or to map interactions within a protein complex (proteosome *(13)*, flagellum *(14)*). Finally, recent large-scale projects have been successful in systematically mapping interactions within whole proteomes or subsets thereof (yeast *(1, 15)*; worm *(16)*; fly *(17)*; human *(18, 19)*). These studies have shown for the first time that most proteins in a cell are actually connected to each other *(20)*.

In combination with structural genomics, gene expression data, and metabolic profiling, the enormous amount of data in these interaction networks should allow us eventually to model complex biological phenomena in molecular detail. An ultimate goal of this work is to understand the interplay of DNA, RNA, and proteins, together with small molecules, in a dynamic and realistic way.

1.3. Limitation of Yeast Two-Hybrid Screening

False negatives: Two-hybrid screens are not perfect. It is quite unlikely that you will detect all physiologically relevant interactors of your bait protein. False negatives may arise from steric hindrance of the two fusion proteins, so that physical interaction or subsequent transcriptional activation is prevented. Other explanations for false negatives include instability of proteins or failure of nuclear localization; the absence of a prey protein from a library; and inappropriate post-translational modification of a bait or prey, prohibiting an interaction. We estimate that the false negative rate in array-based two-hybrid screens is on the order of 75%; i.e., up to 75% of all "true" interactions may be missed *(14)*. This large number can be reduced by several strategies. For example, we investigated the interactome of bacterial flagella by using ORFs from both *Treponema pallidum* and *Campylobacter jejuni* which had estimated false negative rates of 76% and 77%, respectively. However, a combination of both datasets recovered 33% of all known flagellar interactions and thus had a false negative rate of 67%. When protein domains and fragments are used, this number can be further reduced, although it may be difficult to recover more than 50% of all interactions using the Y2H.

False positives: Like many assay systems, the two-hybrid system has the potential to produce false positives. "False positives" may be of technical or biological nature. The technical false positive is an apparent two-hybrid interaction that is not based on the assembly of two hybrid proteins (that is, the reporter gene(s) gets activated without a protein–protein interaction between bait and prey being involved). Frequently, such false positives are associated with bait proteins that act as transcriptional activators. Some bait or prey proteins may affect general colony viability and hence enhance the ability of a cell to grow under selective conditions and activate the reporter gene. Mutations or other random events of unknown nature may be invoked as potential explanations as well. A number of procedures have been developed to identify or avoid false positives, including the utilization of multiple reporters, independent methods of specificity testing, or simply repeating assays to make sure a result is reproducible *(21–23)* (described below).

The biological false positive means a bona fide two-hybrid interaction with no physiological relevance. Those include the partners that can physically interact but that are never in close proximity to one another in the cell because of distinct subcellular localization or expression at different times during the life cycle of a cell or organism. Examples may include paralogs that are expressed in different tissues or at different developmental stages. The problem is that the "false positive" nature can rarely be proven, as there may be unknown conditions under which these proteins do interact with a biological purpose. Overall, hardly any false positives can be explained mechanistically (although many may simply do interact nonphysiologically!).

While it often remains difficult to prove the biological significance of an interaction, many studies have attempted to validate them by independent methods. Finding an interaction by several methods certainly increases the probability that it is biologically significant. Recently, Uetz et al. *(24)* evaluated the all the Y2H interactions of Kaposi Sarcoma Herpes virus (KSHV) by co-immunoprecipitation (coIP) and found that about 50% of them can be confirmed. Similarly, when subsets of the large-scale human Y2H interactome were evaluated, 78 and 65% of them could be verified by independent methods *(18, 19)*.

1.4. Array-Based Screens

In an array, a number of defined prey proteins are tested for interactions with a bait protein (**Fig. 1**). Usually, the bait protein is expressed in one yeast strain and the prey is expressed in another yeast strain of different mating type. The two strains are then mated so that the two proteins are expressed in the resulting diploid cell (**Fig. 1**). The assays are done side-by-side under identical conditions, so they can be well controlled, i.e., compared. As the identity of the preys is usually known, no sequencing is required after positives have been identified. However, the prey clones need to be obtained or made upfront. This can be done for a few genes or for a whole genome, e.g., an ORFeome (i.e., all ORFs of a genome).

In an array, each element has a known identity, and therefore it is immediately clear which two proteins are interacting when positives are selected. In addition, it is often immediately clear if an interaction is stronger than another one (but see below). Most importantly, since all these assays are done in an ordered array, background signals can be easily distinguished from true signals (**Fig. 1**, step 3). Until recently, it was much easier to construct a random library and screen it rather than to construct many individual clones and screen them individually. However, now whole genomes become increasingly available as ordered clone sets in a variety of vectors. Modern cloning systems also allow direct transfer of entry clones into many specialized vectors (see below). For most model organisms such genome-scale clone collections are already available or will be soon. One of the first applications of such clone collections is often a protein interaction screen, so it is likely that a prey library is already available for your favorite organism!

In fact, in some cases only an array screen may do the job. For example, if you have a bait protein that activates transcription on its own, a carefully controlled array may be the only way to distinguish between signal and background (*see* **Fig.1**, step 3). Similarly, weak interactors may be detectable only when compared with a uniformly weak or no background.

1.5. Pooled Array Screening

A completely different screening strategy, the pooling strategy, has the potential to accelerate screening a lot but might also have the disadvantage of increasing the number of false negatives. This may have been a reason why pooled screens in *Campylobacter jejuni* resulted in more false negatives than in the one-by-one screens of *Treponema pallidum* *(14, 25)*.

In the first step, sets of proteins (pools, rather than single proteins) are tested for interactions against each other (for example, pools of 10 known preys in each colony of the prey array). In the second step, positive sets are taken and the proteins defining this set are individually tested for their binary interactions (as in the classical array-based Y2H strategy). Depending on the pool size, the first level of screening is therefore usually fast and, as only a few interactions are expected for each protein, only a few pools need to be tested for binary interactions in the second step. Such a pooling strategy was established by Zhong et al. *(26)*. Pools of prey proteins were tested against single bait proteins, and it was shown that the pools could contain 96 or even more proteins. The authors calculated that the Y2H array screening of the whole yeast genome (~6,000 proteins) would require only 1/24 of the time and effort when using the proposed pooling strategy as opposed to a one-by-one strategy. Prey-based pools are advantageous over bait-based pools because 10–20% of bait–BD fusions can activate the two-hybrid reporter gene (for example, the HIS3 gene) without the presence of any prey–AD fusion. Addition of 3-amino-triazole, an enzymatic inhibitor of His3, can inhibit autoactivation. Thus, it is possible to optimize two-hybrid selection conditions based on the autoactivation level of the individual bait (by varying 3-amino-triazole concentration in the medium).

Recently Jin et al. developed a smart pool-array (SPA) system *(27)* where they show that the pooling-deconvolution principle can be applied to pool prey (AD) strains, permitting efficient screening of individual baits with high accuracy and coverage. In an SPA scheme, 16 (=2^4) strains are mixed into four pairs of pools (pairs 0–3), with eight (=2^3) strains per pool. This pooling scheme has two important properties of deconvolution and redundancy. First, deconvolution is possible because every strain is pooled into four different pools (one from each pair), so if one of the 16 strains is two-hybrid positive (for a given bait), then four of the eight pools will yield a positive colony. Thus we can deconvolute the identity of the two-hybrid positive strain owing to its presence only in a specific combination of four pools and absence in the other pools.

The second important property of SPA is the built-in redundancy; each AD strain is tested four times against the bait, resulting in a situation that is equivalent to four separate individual screens. This inherent replication can facilitate removal of false positives because false positives are unlikely to be observed reproducibly. Likewise, replicated screens will cover more true positives, because losing the same true positive repeatedly as a result of experimental variation is also less likely. In general, this scheme generates $2n$ pools from 2^n strains, with a built-in "screen redundancy" of n. Thus, when the number of strains is increased exponentially, for example, from 16 (=2^4) to 32 (=2^5), the number of pools only needs to be increased linearly, that is, from 8 (= 2 × 4) to 10 (= 2 × 5).

1.6. General Requirements for a Screen and Alternatives

Although the protocols in this chapter are based on the DNA-binding and activation domain of the yeast Gal4 protein, other DNA-binding domains and activation domains can be used.

In the LexA two-hybrid system, the DNA-binding domain is provided by the entire prokaryotic LexA protein, which normally functions as a repressor in *E. coli* when it binds to LexA operators. In the Y2H system, the LexA protein does not act as a repressor, as the promoter with its binding sites is not constitutively active. An activation domain often used in the LexA two-hybrid system is the heterologous 88-residue acidic peptide B42 that strongly activates transcription in yeast. An interaction between the target protein (fused to the DNA-BD) and a library-encoded protein (fused to an AD) creates a novel transcriptional activator with binding affinity for LexA operators.

In general, every component of the "classic" two-hybrid system can be replaced by different components: For example, the reporter gene does not need to be *HIS3*. Alternatively, *LEU2*, an enzyme involved in leucine biosynthesis, can be used. The reporter does not have to be a biosynthetic enzyme at all; green fluorescent protein (GFP) has been successfully used as a reporter gene *(28)*, beta-galactosidase (lacZ) is common *(29)*, and many others are under investigation. Finally, the two-hybrid system does not need to be based on transcription. Johnsson and Varshavsky (1994) developed a related system that is based on reconstituting artificially split ubiquitin, a protein that tags other proteins for degradation. As long as the function of a protein can be used as a selective marker, it is theoretically possible to divide it into fragments, and drive the reassociation of the two fragments by exogenous "Bait and Prey" proteins which are attached to each half. Several other variations have been developed and are described elsewhere *(30, 31)*.

1.7. Genome-Wide Yeast Two-Hybrid Screening (See Table 1)

The construction of an entire proteome array of an organism that can be screened in vivo under uniform conditions is a challenge. When proteins are screened at a genome scale, automated robotic procedures are necessary (*see* below). The procedure can be modified for manual use or for use with alternative screenings strategies such as synthetic lethal screens. With minor modifications, the array can be used to screen for protein interactions with DNA, RNA, or even small-molecule inhibitors of the protein–protein interactions.

The protocols described here were established for yeast proteins, but they can be applied to any other genome or subset thereof; for example, viral and bacterial genomes have been screened for interactions in our lab. Different high-throughput

Table 1
What you need for a yeast two-hybrid screen (examples)

Bait plasmid(s)	pGBKT7 (Clontech), pOBD2, pDEST22 (Invitrogen), pGBKT7g *(24)*
Prey plasmid(s)	pGADT7 (Clontech), PAS1, pDEST22 (Invitrogen), pGADT7g *(24)*
Bait yeast strain	AH109
Prey yeast strain	Y187
Yeast media	YEPD, selective liquid media and agar plates
Pin tool	Optional but necessary when large numbers are tested

cloning methods used to generate two-hybrid clones, i.e., proteins with AD fusions (preys) and the DBD fusions (baits) are therefore included below. The process involves the construction of the prey and bait array (described in **Subheadings 3.2–3.4**) and screening of the array by either manual or robotic manipulation (described in **Subheading 3.5.1–3.5.4**), including the selection of positives and scoring of results.

High-throughput screening projects deal with a large number of proteins; therefore hands-on time and amount of resources become an important issue. Options to reduce the screening effort are discussed. A prerequisite for array-based genome-wide screen is the existence of a cloned ORFeome; we will briefly mention strategies to create such ORFeomes. Many ORFeome projects are currently being done. We expect readily available complete ORFeomes for all major model organisms in the near future.

2. Materials

2.1. Yeast Media

1. YEPD liquid medium: 10 g yeast extract, 20 g peptone, 20 g glucose. Make up to 1 L with sterile water, and autoclave.

2. YEPD solid medium: 10 g yeast extract, 20 g peptone, 20 g glucose, 16 g agar. Make up to 1 L with sterile water and autoclave. After autoclaving, cool the medium to 60–70°C, then add 4 ml of 1% adenine solution (1% in 0.1 M NaOH), pour 40 ml into each sterile Omnitray plate (Nunc) under sterile hood, and let them solidify.

3. Medium concentrate: 8.5 g yeast nitrogen base, 25 g ammonium sulfate, 100 g glucose, 7 g dropout mix (see below). Make up to 1 L with sterile water, and filter-sterilize (Millipore).

2.2. Yeast Minimal Media (Selective) Plates

1. For 1 L of selective medium, autoclave 16 g agar in 800 ml water, cool the medium to 60–70° C, and then add 200 ml medium concentrate. Depending on the required selective plates, you have to add the missing amino acids or 3AT (3-amino-1,2,4-triazole).

2. –Trp plates (media lacking tryptophan): Add 8.3 ml leucine and 8.3 ml histidine from the stock solution (*see* below).

3. –Leu plates (media lacking leucine): Add 8.3 ml tryptophan and 8.3 ml histidine solution from the stock solution.

4. –Leu -Trp plates (media lacking tryptophan and leucine): Add 8.3 ml histidine from the stock solution.

5. −Leu −Trp −His plates (media lacking tryptophan, leucine, and histidine): Nothing needs to be added.

6. −Leu −Trp -His + 3 mM 3AT plates: Add 6 ml of 3AT (3-amino-1,2,4-triazole, 0.5 M) to a final concentration of 3 mM.

7. Dropout mix (-His, -Leu, -Trp): Mix 1 g methionine, 1 g arginine, 2.5 g phenylalanine, 3 g lysine, 3 g tyrosine, 4 g isoleucine, 5 g glutamic acid, 5 g aspartic acid, 7.5 g valine, 10 g threonine, 20 g serine, 1 g adenine, and 1 g uracil and store under dry, sterile conditions.

8. Amino acid stock solutions: Histidine (His): dissolve 4 g of histidine in 1 L sterile water and filter-sterilize. Leucine (Leu): dissolve 7.2 g of leucine in 1 L sterile water and sterile filter. Tryptophan (Trp): dissolve 4.8 g of tryptophan in 1 L sterile water and filter-sterilize.

2.3. Yeast Transformation

1. Salmon sperm DNA (Carrier DNA): Dissolve 7.75 mg/ml salmon sperm DNA (Sigma) in sterile water, autoclave for 15 min at 121°C, and store at −20°C.

2. Dimethylsufoxide (DMSO, Sigma).

3. Competent host yeast strains, e.g., AH109 (for baits), and Y187 (for preys).

4. Lithium acetate (LiOAc) (0.1 M).

5. Selective plates (depending on the selective markers, described in **Subheading 2.2**).

6. 96PEG solution: Mix 45.6 g PEG (Sigma), 6.1 ml of 2 M LiOAc, 1.14 ml of 1 M Tris-HCl, pH 7.5, and 232 μl 0.5 M EDTA. Make up to 100 ml with sterile water and autoclave.

7. Plasmid clones or linearized vector DNA and PCR product (for homologous recombination).

2.4. Bait Self Activation Test

1. YEPD liquid medium and selective media agar in single-well microtiter plates (Omnitray plates, Nunc).

2. −Trp −Leu plates (*see* **Subheading 2.2**).

3. Selective plates without Trp, Leu, and His, but with different concentrations of 3-AT, e.g., 0 mM, 1 mM, 3 mM, 10 mM, 50 mM and 100 mM (−LTH/3-AT plates).

4. Bait strains and the prey strain carrying the empty prey plasmid, e.g., Y187 strain with pDEST22 plasmid (Invitrogen).

2.5. Two-Hybrid Screening Protocol

1. 20% (v/v) bleach (1% sodium hypochlorite).

2. 95% (v/v) ethanol.

3. Single-well microtiter plates (e.g., OmniTray; Nalge Nunc) containing solid YEPD + adenine medium (*see* **Subheading 2.1**),

–Leu –Trp (–LT), –His –Leu –Trp (–LTH), and –His –Leu –Trp + different concentrations of 3AT.

4. 384-Pin replicator for manual screening or robot (Biomek FX).
5. Bait liquid culture (DBD fusion-expression yeast strain).
6. Yeast prey array on solid YEPD plates.

2.6. Retest of Protein Interactions

1. 96-well microtiter plates (U- or V-shaped).
2. YEPD medium and YEPD agar in Omnitrays (Nunc).
3. Selective agar plates (–LT, –LTH with 3-AT).
4. Prey yeast strain carrying empty prey plasmid, e.g., pDEST22 in Y187 strain.
5. Bait and prey strains to be retested.

2.7. Beta-Galactosidase Filter Lift Assay

1. Selective plate (–LT) with diploid yeast colonies (from **Subheading 3.5.4**). The diploid cells carry the bait and prey combinations to be tested for activation of the beta-galactosidase reporter.
2. Omnitray plate.
3. Nitrocellulose membrane and Whatman paper.
4. Z-buffer: 60 mM Na_2HPO_4 (anhyd.), 60 mM NaH_2PO_4, 10 mM KCl, 1 mM $MgSO_4$.
5. X-GAL solution: 40 mg/ml X-Gal (5-bromo-4-chloro-3-indolyl-b-D-galactopyranoside) in dimethylformamide (DMF).

3. Methods

3.1. Strategic Planning

Before starting an array-based screen, the size and character of the array must be designed and the ultimate aims of the experiment need to be considered. Factors that may be varied include the form of protein array (e.g., full-length protein or single domain, choice of epitope tags etc.). Similarly, the arrayed proteins may be related (e.g., a family or pathway of related proteins, orthologs of a protein from different species, the entire protein compliment of a model organism). In our experience, certain protein families work extremely well (e.g., splicing proteins), while others do not appear to work at all (e.g., many metabolic enzymes). We recommend to carry out a small-scale pilot study, incorporating positive and negative controls, before committing to a full-scale project.

Although high-throughput screening projects can be performed manually, automation is strongly recommended. Highly repetitive tasks are not only boring and straining but also error-prone when done manually. If you do not have local access to robotics, you may have to collaborate with a laboratory that has.

3.2. Generation of a Protein Array Suitable for High-Throughput Screening

Once the set of proteins to be included in the array is defined, the coding genes need to be PCR-amplified and cloned into Y2H bait and prey vectors. In order to facilitate the cloning of a large number to proteins, site-specific recombination-based systems are commonly used (e.g., Gateway *(32)*, *see* **Fig. 2b**). Gateway cloning requires expensive enzymes and vectors, although both may be produced in the lab.

3.2.1. Cloning by Homologous Recombination in Yeast

An alternative to site-directed systems is the cloning by homologous recombination directly in yeast *(33)*. A two-step PCR protocol is used to make DNA with sufficient homology to vector DNA at the terminal ends to allow homologous recombination in the yeast cell (**Fig. 2a**). In the "first round," PCR reaction the ORF is amplified with primers that contain ~20-nucleotide tails which are homologous to sequences in the two-hybrid vectors. In the second-round PCR, ~50-nucleotide tails are attached to the first-round PCR products that are homologous to the destination vector cloning site (**Fig. 2**). The PCR product is then transformed into the yeast cells together with the linearized vector, and the recombination event between them takes place inside the yeast cell. The advantage of this strategy is its much reduced cost.

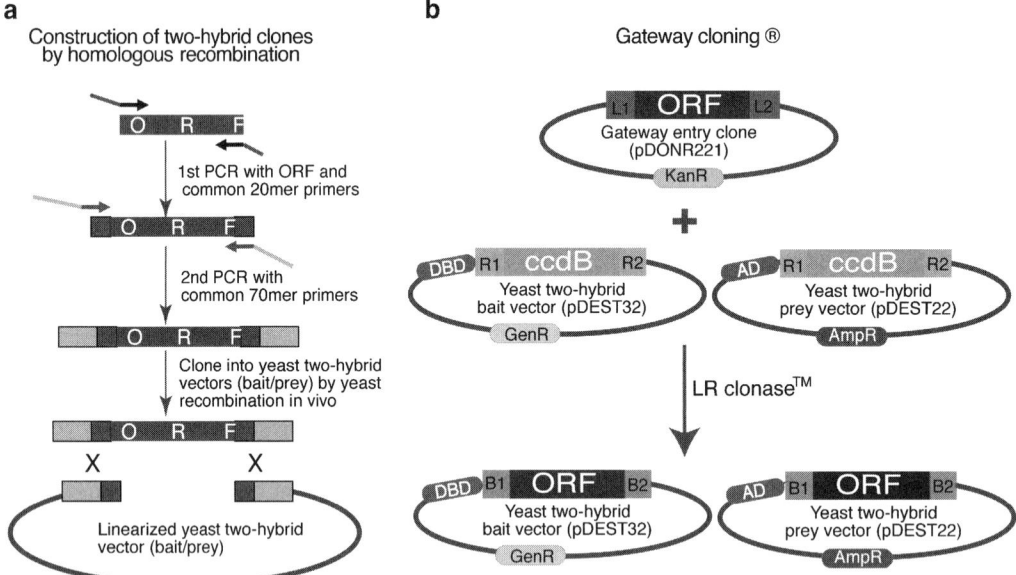

Fig. 2. Cloning strategies for creating baits and preys. (**a**) Homologous recombination. ORFs are amplified (first PCR) with specific primers that generate a product with common 5' and 3' 20-nucleotide tails. A second PCR generates a product with common 5' and 3' 70-nucleotide tails. The common 70-nucleotide ends allow cloning into linearized two-hybrid expression vectors by co-transformation into yeast. The endogenous yeast recombination machinery performs the recombination reaction and results in a circular plasmid. (**b**) The Gateway-based Y2H expression clones are made by combining the ORFs of interest from a Gateway entry vector (such as pDONR201 or pDONR207) and the Y2H expression vectors (such as pDEST22 and pDEST32) in Gateway LR Clonase (Invitrogen) enzyme mix, which transfers the ORF from an entry clone to Y2H expression vectors.

The disadvantage is that the resulting plasmids are not available as purified DNA but have to be recovered from yeast, which can be time consuming and inefficient.

3.2.2. Gateway Cloning

Gateway (Invitrogen) cloning provides another fast and efficient way of cloning the ORFs *(32)*. It is based on the site-specific recombination properties of bacteriophage lambda *(34)*; recombination is mediated between so-called attachment sites (att) of DNA molecules: between attB and attP sites or between attL and attR sites. The first step to Gateway cloning is inserting the gene of interest into a specific entry vector. One way of obtaining the initial entry clones is by recombining a PCR product of the ORF flanked by attB sites with the attP sites of a pDONR vector (Invitrogen). The resulting entry clone plasmid contains the gene of interest flanked by attL recombination sites. These attL sites can be recombined with attR sites on a destination vector, resulting in a plasmid for functional protein expression in a specific host (for example, pDEST22 and pDEST32, **Fig. 2b**).

3.2.3. The ORFeome

The starting point of an systematic array-based Y2H screening is the construction of an ORFeome. An ORFeome represents all ORFs of a genome – in our case, the selected gene set individually cloned into entry vectors. More and more ORFeomes are available and can be directly used for generating the Y2H bait and prey constructs. These ORFeome range from small viral genomes, e.g., KSHV and VZV *(24)*, to several bacterial genomes such as *Bacillus anthracis* or *Yersinia pestis*. These bacterial ORFeomes are available from the Pathogen Functional Genomics Resource Center (http://pfgrc.jcvi.org/). Clone sets of multicellular eukaryotes, e.g., *C. elegans(35)*, human *(36)*, or plant *(37)*, have also been described. However, not all genes of interest are already available in entry vectors. Both entry vector construction and the subsequent destination vector cloning can be done for multiple ORFs in parallel. The whole procedure can be automated using 96-well plates so that whole ORFeomes can be processed in parallel.

3.3. The Prey Array

The Y2H array is set up from an ordered set of AD-containing strains (preys) rather than BD-containing strains (baits), because the former do not generally result in self-activation of transcription. The prey constructs are assembled by transfer of the ORFs from entry vectors into specific prey vectors by recombination. Several prey vectors for the Gateway system are available. In our lab we use the Gateway-compatible pGADT7g vector, a derivative of pGADT7 (Clontech), or pDEST22 (Invitrogen) (**Fig. 2b**). We have also used pAS1 and other vectors for homologous recombination (**Fig. 2a**). These prey constructs are transformed into haploid yeast cells (described in **Subheading 3.4**); we use the Y187 strain (mating type alpha) (**Table 2**). Finally, individual yeast colonies, each carrying one specific prey construct, are arrayed on agar plates in a 96- or 384-format in duplicates or quadruplicates.

3.3.1. Bait Construction

Baits are also constructed by recombination-based transfer of the ORFs into specific bait vectors or, alternatively, directly by homologous recombination in yeast. Bait vectors used in our lab are the Gateway-adapted pGBKT7g (Clontech) or pDEST32 (Invitrogen) for Gateway cloning and pOBD2 vector for homologous recombination cloning (**Fig. 2a**). The bait constructs are also transformed into haploid yeast cells (described in **Subheading 3.4**); we use the AH109 strain (mating type a) (**Table 2**). After self-activation testing, the baits can be tested for interactions against the Y2H prey array (*see* **Note 1**).

3.4. High-Throughput Yeast Transformation

This method is recommended for the high-throughput transformation of the bait or prey plasmid clones into respective yeast strains, and is optimized on the method of Cagney et al. *(33)*. This protocol is suitable for 1,000 transformation, it can be scaled up and down as required, and most of the steps can be automated. Selection of the transformed yeast cells requires leucine- or tryptophan-free media (–Leu or –Trp depending on the selective marker on the plasmid). Moreover, at least one of the haploid strains must contain a two-hybrid reporter gene (here: HIS3 under GAL4 control).

1. Prepare competent yeast cells: Inoculate 250 ml YEPD liquid medium with yeast strains freshly grown on YEPD agar medium in a 2-L flask and grow in a shaker (shaking at 200 rpm) at 30°C. Remove the yeast culture from the shaker when the cell density reaches OD 1.0–1.3. This usually takes 12–16 h.

Table 2
Yeast strains and their genotypes

Yeast strains	Genotypes
Y187	MATα, ura3- 52, his3- 200, ade2- 101, trp1- 901, leu2- 3, 112, gal4Δ, met–, gal80Δ, URA3::GAL1UAS -GAL1TATA -lacZ (after Harper et al. *(39)*)
AH109	MATa, trp1-901, leu2-3, 112, ura3-52, his3-200, gal4Δ, gal80Δ, LYS2::GAL1UAS-GAL1TATA-HIS3, GAL2UAS-GAL2TATA-ADE2, URA3::MEL1UAS-MEL1 TATA-lacZ (after James et al. *(40)*)

2. Spin out the cells at 2,000 × *g* for 5 min at room temperature; pour off the supernatant.
3. Dissolve the cell pellet in 30 ml of LiOAc (0.1 M); make sure pellet is completely dissolved and there are no cell clumps.
4. Spin the cells in a 50-ml Falcon tube at 2,000 × *g* for 5 min at room temperature, pour off the supernatant, and dissolve the cell pellet in a total volume of 10 ml LiOAc (0.1 M).
5. Prepare the yeast transformation mix without yeast cells by mixing the components listed below in a 200-ml sterile bottle.

Component	For 1,000 reactions
96PEG	100 ml
Salmon sperm DNA	3.2 ml
DMSO	3.4 ml

6. Add the competent yeast cells prepared above (**steps 1–4**) to the yeast transformation mix; shake the bottle vigorously by hand, or vortex for 1 min.
7. Pipette 100 µl of the yeast transformation mix into a 96-well transformation plate (we generally use Costar 3596 plates) by using a robotic liquid handler (e.g., Biomek FX) or a multistep pipette.
8. Now add 25–50 ng of plasmid; keep one negative control (i.e., only yeast transformation mix).
9. Seal the 96-well plates with plastic or aluminum tape and vortex for 2–3 min. Care should be taken to seal the plates properly; vigorous vortexing might cause cross-contamination.
10. Incubate the plates at 42°C for 30 min.
11. Spin the 96-well plate for 5 min at 2,000 × *g*; aspirate the supernatant and dry by tapping on a cotton napkin a couple of times.
12. Add 150 µl of selective liquid media to each well (depending on the selective marker on the plasmid construct (for example, trytophan- or leucine-free liquid media). Seal the plates with AirPore tape (Qiagen) to protect from evaporation.
13. Incubate at 30°C for 36–48 h.
14. Pellet the cells by spinning at 2,000 × *g* for 5 min, discard the supernatant, and add 10 µl sterile H_2O to each well.
15. Transfer the cells to selective agar plate to select yeast with transformed plasmid (single-well Omnitrays 128 × 86 mm from Nunc are well suited for robotic automation). Typically,

we use a 96-pin tool (see reagent setup for the sterilization of the pin tool). As an alternative to the pin tool, one can use a multichannel pipette to transfer the cells. Allow the yeast spots to dry on the plates.

16. Incubate at 30°C for 2 days. Colonies start appearing after 24 h.

3.5. Screening and Retesting

3.5.1. Self-Activation Test

Prior to the two-hybrid analyses, the bait yeast strains should be examined for self-activation. Self-activation is defined as detectable bait-dependent reporter gene activation in the absence of any prey interaction partner. Weak to intermediate strength self-activator baits can be used in two-hybrid array screens because the corresponding bait–prey interactions confer stronger signals than the self-activation background. In case of the *HIS3* reporter gene, the self-activation background can be titrated by adding different concentrations of 3-AT, a competitive inhibitor of *HIS3*. Self-activation of all the baits is examined on plates containing different concentrations of 3-AT. The lowest concentration of 3-AT that suppresses growth in this test is used for the interaction screen (see below), because it avoids background growth while still detecting true interactions.

The aim of this test is to measure the background reporter activity (here: *HIS3*) of bait proteins in absence of an interacting prey protein. This measurement is used for choosing the selection conditions used for Y2H screening described in **Subheading 3.5.2**.

1. Bait strains are arrayed onto a single-well Omnitray agar plate; either the standard 96-spot format or the 384-spot format is used (*see* **Note 2**).

2. The arrayed bait strains are mated with a prey strain carrying the empty prey plasmid, e.g., Y187 strain with pDEST22 (Invitrogen). Mating is conducted according to the standard screening protocol as described in **Subheading 3.5.2**. Note that here an array of baits is tested whereas in a "real" screen (**Subheading 3.5.2**) an array of preys is tested.

3. After selecting for diploid yeast cells (on –LT agar), the cells are transferred to media selecting for the HIS3p reporter gene activity as described in **Subheading 3.5.2**. The -LTH transfer may be done to multiple plates with increasing concentrations of 3-AT. Suggested 3-AT concentrations are 0, 1, 3, 10, 25, 50, and 100 mM.

4. These –LTH + 3-AT plates are incubated for 1 week at 30°C. The self-activation level of each bait is assessed: the lowest 3-AT concentration that completely prevents colony growth is noted. As this concentration of 3-AT

suppresses reporter activation in the absence of an interacting prey, this 3-AT concentration is added to –LTH plates in the actual interaction screens as described in **Subheading 3.5.2**.

3.5.2. Screening for Protein Interactions Using a Yeast Protein Array

The Y2H prey array can be screened for protein interactions by a mating procedure that can be carried out manually or using robotics. A yeast strain expressing a single candidate protein as a DBD fusion is mated to all the colonies in the prey array (**Fig. 1**, step 1: shown for one prey plate). After mating, the colonies are transferred to a diploid-specific medium, and then to the two-hybrid interaction selective medium. To manually screen with more than one bait, replicate copies of the array are used. For large numbers of baits, robotic screening is recommended.

In many cases, a hand-held 384-pin replicating tool can be used for routine transfer of colonies for screening. For large projects, however, a robotic workstation (e.g., Biomek 2000, or Biomek FX, Beckman Coulter) may be used to speed up the screening procedures and to maximize reproducibility. A 384-pin steel replicating tool (e.g., High-Density Replication Tool; V&P Scientific) can be used to transfer the colonies from one plate to another. Between the transfer steps, the pinning tool must be sterilized, *see* below).

Note that not all plasticware is compatible with robotic devices, although most modern robots can be reprogrammed to accept different consumables. In the procedure described here, the prey array is gridded on 86 × 128 mm single-well microtiter plates (e.g., OmniTray, Nalge Nunc International) in a 384-colony format (*see* **Fig. 1**).

1. *Sterilization*: Sterilize a 384-pin replicator by dipping the pins into 20% bleach for 20 s, sterile water for 1 s, 95% ethanol for 20 s, and sterile water again for 1 s. Repeat this sterilization after each transfer. Note: Immersion of the pins into these solutions must be sufficient to ensure complete sterilization. When automatic pinning devices are used, the solutions need to be checked and refilled occasionally (especially ethanol which evaporates faster than the others).

Day 1:

2. *Preparing prey array for screening*: Use the sterile replicator to transfer the yeast prey array from selective plates to single-well plates containing solid YEPD medium and grow the array overnight in a 30°C incubator (*see* **Note 3**).

3. *Preparing bait liquid culture (DBD fusion-expressing yeast strain):* Inoculate 20 ml of liquid YEPD medium in a 250-ml conical flask with a bait strain and grow overnight in a 30°C shaker (*see* **Note 4**).

Day 2:

4. *Mating procedure*: Pour the overnight liquid bait culture into a sterile Omnitray plate. Dip the sterilized pins of the pin replicator (thick pins of ~1.5 mm diameter should be used to pin baits) into the bait liquid culture and place directly onto a fresh single-well plate containing YEPD agar media. Repeat with the required number of plates and allow the yeast spots to dry onto the plates for 10–20 min.

5. Pick up the fresh prey array (i.e., AD) yeast colonies with sterilized pins (thin pins of ~1 mm diameter should be used to pin the preys) and transfer them directly onto the baits on the YEPD plate, so that each of the 384 bait spots per plate receives different prey yeast cells (i.e., a different AD fusion protein). Incubate overnight at 30°C to allow mating (**Fig. 1** step 1, *see* **Note 5**).

6. *Seletion of Diploids*: For the selection of diploids, transfer the colonies from YEPD mating plates to plates containing −Leu −Trp medium using the sterilized pinning tool (thin pins should be used in this step). Grow for 2–3 days at 30°C until the colonies are >1 mm in diameter (**Fig. 1** step 2, *see* **Note 6**).

7. *Interaction selection:* Transfer the colonies from −Leu −Trp plates to a single-well microtiter plate containing solid −His −Leu −Trp agar, using the sterilized pinning tool. If the baits are self-activating, they have to be transferred to −His −Leu −Trp + a specific concentration of 3AT (**Subheading 3.5.1**). Incubate at 30°C for 6–8 days.

8. Score the interactions by looking for growing colonies that are significantly above background by size and are present as duplicate colonies.

9. The plates should be examined every day. Most two-hybrid-positive colonies appear within 3–5 days, but occasionally positive interactions can be observed later. Very small colonies are usually designated as background; however, there is no absolute measure to distinguish between the background and real positives. When there are many (e.g., >20) large colonies per array of 1,000 positions, we consider these baits as "random" activators. In this case, the screening should be repeated, or the interactions should be retested.

10. Scoring can be done manually or using automated image analysis procedures. When using image analysis, care must be taken not to score contaminated colonies as positives.

3.5.3. Protein Interaction Retesting

A major consideration when using the Y2H system is the number of false positives. The major sources for false positives are nonreproducible signals that arise through little-understood mechanisms. In Y2H screens, more than 90% of all interactions can be nonreproducible background *(38)*. Thus, simple retesting by repeated mating can identify most false positives. We routinely use at least duplicate tests, although quadruplicates should be used if possible (*see* **Fig. 1**). Retesting is done by manually mating the interaction pair to be tested and by comparing the activation strength of this pair with the activation strength of a control, usually the bait mated with the strain that contains the empty prey vector.

Testing for reproducibility of interactions greatly increases the reliability of the Y2H interaction data. This method is used for specifically retesting interaction pairs detected in an array screen.

1. Re-array bait and prey strains of each interaction pair to be tested into 96-well microtiter plates. Use separate 96-well plates for baits and preys. For each retested interaction, fill one well of the bait plate and one corresponding well of the prey plate with 150 µl YEPD liquid medium.

2. For each retested interaction, inoculate the bait strain into a well of the 96-well bait.

3. Plate bait and prey strain at the corresponding position of the 96-well prey plate, for example, bait at position B2 of the bait plate and prey at position B2 of the prey plate. In addition, inoculate the prey strain with the empty prey vector (e.g., strain Y187 with plasmid pDEST22) into 20 ml YEPD liquid medium.

4. Incubate the plates overnight at 30°C.

5. Mate the baits grown in the bait plate with their corresponding preys in the prey plate. In addition, mate each bait with the prey strain carrying an empty prey vector as a background activation control. The mating is done according to **Subheading 3.5.2**, using the bait and prey 96-well plates directly as the source plates (*see* **Note 7**).

6. The transfers to selective plates and incubations are done as described in **Subheading 3.5.2**. As before, test different baits with different activation strengths on a single plate and pin the diploid cells onto –LTH plates with different concentrations of 3-AT. For choosing the 3-AT range, the activation strengths (**Subheading 3.5.1**) serve as a guideline.

7. After incubating for ~1 week at 30°C on –LTH/3-AT plates, the interactions are scored; positive interactions show a clear colony growth at a certain level of 3-AT, whereas no growth should be seen in the control (bait mated with empty vector strain).

3.5.4. Beta-Galactosidase Filter Lift Assay (Alternative Reporter Genes)

Y2H interactions can be reproduced using other reporter genes in addition to the one used in the actual screen depending on the different reporter genes present in the yeast strains used. Examples include beta-galactosidase or *ADE2* (for selection on adenine-deficient medium). Because of the use of different promoters, these reporter genes have different activation requirements, and Y2H interactions reproduced with different reporter genes are assumed to be more reliable. However, the use of multiple reporters may result in the loss of weaker Y2H positives. The beta-galactosidase reporter has the advantage of giving a semiquantitative output of the activation strength. Other reporters might be advantageous and can be transformed into yeast as additional plasmids, or by using alternative strains which contain the reporter as integrated construct. For example, the strain AH109 carries an alpha-galactosidase reporter gene which produces an enzyme that is secreted into the medium. Therefore, these cells do not require cell lysis for detection. The following method was adapted from the Breeden lab (http://labs.fhcrc.org/breeden/Methods/index.html).

1. Use the same diploid plate as in **Subheading 3.4.2**. As a control, the bait strains are mated with a prey strain containing an empty vector (following mating steps of **Subheading 3.4.2**).
2. Cut a nitrocellulose membrane to the dimensions of an Omnitray plate (Nunc). Place the nitrocellulose membrane on top of diploid yeast colonies and leave for 10 s.
3. Use tweezers to lift the filter and slowly submerge in liquid nitrogen for 1 min.
4. Place the membrane on an empty Omnitray plate (Nunc) to thaw.
5. Cut a Whatman paper to same size as nitrocellulose membrane. Soak the Whatman paper with 2 ml Z-buffer, to which 35 μl X-solution had been added.
6. Overlay the nitrocellulose filter with the Whatman paper and remove air bubbles.
7. Incubate at 30°C for 10–60 min.
8. Evaluate: A blue stain indicates the activation of the beta-galactosidase reporter and, therefore, a positive interaction.

4. Notes

1. Bait and prey must be transformed into yeast strains of opposite mating types to combine bait and prey plasmids by mating and to co-express the fusion proteins in diploids. Bait and prey plasmids can go into either mating type. However, this decision also depends on existing bait or prey libraries to which the new library may be mated later.
2. Baits are first grown at the different positions of a 96-well plate as liquid culture, and then cells are transferred (manually or with the use of a robot) to solid agar single-well plates (Omnitray plates). In this step, the 96-well format can also be converted into the 384-well format. This will position each bait in quadruplicates on the 384-well formatted plate. Full media agar (YEPD

agar) can be used; however, for long-term storage of the array, selective agar is suggested to prevent loss of plasmids.

3. In a systematic array-based Y2H screening, duplicate or quadruplicate prey arrays are usually used. In a random genomic library screening, the entire experiment should be done in two copies to ensure reproducibility. Ideally, the template prey array should be kept on selective plates. The template of the prey array should be used to make "working" copies on YEPD agar plates for mating. The template can be used for 1–2 weeks; after 2 weeks it is recommended to copy the array onto fresh selective plates. Preys or bait clones tend to lose the plasmid if stored on YEPD for longer periods, which may reduce the mating and screening efficiency.

4. If the bait strains are frozen, they are streaked or pinned on selective solid medium plates and grown for 1–2 days at 30°C. Baits from this plate are then used to inoculate the liquid YEPD medium. It is important to make a fresh bait culture for Y2H mating, as keeping the bait culture on rich medium (YEPD) for a long time may cause loss of plasmids. Usually we grow baits overnight for mating.

5. Mating will usually take place in <15 h, but a longer period is recommended because some bait strains show poor mating efficiency. Adding adenine into the bait culture before mating increases the mating efficiency of some baits.

6. This step is an essential control step to ensure successful mating because only diploid cells containing the Leu2 and Trp1 markers on the prey and bait vectors, respectively, will grow in this medium. This step also helps the recovery of the colonies and increases the efficiency of the next interaction selection step.

7. First the baits are transferred from their 96-well plate to one or more YEPD plates (interaction test and control plate) using a 96-well replication tool. Let the plate dry for 10–20 min. Then transfer the preys from their 96-well plate onto the bait plate.

References

1. Uetz, P., Giot, L., Cagney, G., Mansfield, T.A., Judson, R.S., Knight, J.R., et al. (2000). A comprehensive analysis of protein-protein interactions in *Saccharomyces cerevisiae*. *Nature* 403, 623–627.
2. Ho, Y., Gruhler, A., Heilbut, A., Bader, G.D., Moore, L., Adams, S.L., et al. (2002). Systematic identification of protein complexes in *Saccharomyces cerevisiae* by mass spectrometry. *Nature* 415, 180–183.
3. Gavin, A.C., Bosche, M., Krause, R., Grandi, P., Marzioch, M., Bauer, A., et al. (2002). Functional organization of the yeast proteome by systematic analysis of protein complexes. *Nature* 415, 141–147.
4. Fu, H. (2004). Protein-Protein Interactions. Methods and Applications., Volume 261 (Totowa, NJ: Humana Press).
5. Fields, S., and Song, O. (1989). A novel genetic system to detect protein-protein interactions. *Nature* 340, 245–246.
6. Schwartz, H., Alvares, C.P., White, M.B., and Fields, S. (1998). Mutation detection by a two-hybrid assay. *Hum Mol Genet* 7, 1029–1032.

7. Vidal, M., and Endoh, H. (1999). Prospects for drug screening using the reverse two-hybrid system. *Trends Biotechnol* 17, 374–381.
8. Vidal, M., and Legrain, P. (1999). Yeast forward and reverse 'n'-hybrid systems. *Nucleic Acids Res* 27, 919–929.
9. SenGupta, D.J., Zhang, B., Kraemer, B., Pochart, P., Fields, S., and Wickens, M. (1996). A three-hybrid system to detect RNA-protein interactions in vivo. *Proc Natl Acad Sci U S A* 93, 8496–8501.
10. Estojak, J., Brent, R., and Golemis, E.A. (1995). Correlation of two-hybrid affinity data with in vitro measurements. *Mol Cell Biol* 15, 5820–5829.
11. Rain, J.C., Selig, L., De Reuse, H., Battaglia, V., Reverdy, C., Simon, S., et al. (2001). The protein-protein interaction map of *Helicobacter pylori*. *Nature* 409, 211–215.
12. Raquet, X., Eckert, J.H., Muller, S., and Johnsson, N. (2001). Detection of altered protein conformations in living cells. *J Mol Biol* 305, 927–938.
13. Cagney, G., Uetz, P., and Fields, S. (2001). Two-hybrid analysis of the *Saccharomyces cerevisiae* 26S proteasome. *Physiol Genomics* 7, 27–34.
14. Rajagopala, S.V., Titz, B., Goll, J., Parrish, J.R., Wohlbold, K., McKevitt, M.T., et al. (2007). The protein network of bacterial motility. *Mol Syst Biol* 3, 128.
15. Ito, T., Chiba, T., Ozawa, R., Yoshida, M., Hattori, M., and Sakaki, Y. (2001). A comprehensive two-hybrid analysis to explore the yeast protein interactome. *Proc Natl Acad Sci U S A* 98, 4569–4574.
16. Li, S., Armstrong, C.M., Bertin, N., Ge, H., Milstein, S., Boxem, M., et al. (2004). A map of the interactome network of the metazoan *C. elegans*. *Science* 303, 540–543.
17. Giot, L., Bader, J.S., Brouwer, C., Chaudhuri, A., Kuang, B., Li, Y., et al. (2003). A protein interaction map of *Drosophila melanogaster*. *Science* 302, 1727–1736.
18. Rual, J.F., Venkatesan, K., Hao, T., Hirozane-Kishikawa, T., Dricot, A., Li, N., et al. (2005). Towards a proteome-scale map of the human protein-protein interaction network. *Nature* 437, 1173–1178.
19. Stelzl, U., Worm, U., Lalowski, M., Haenig, C., Brembeck, F.H., Goehler, H., et al. (2005). A human protein-protein interaction network: a resource for annotating the proteome. *Cell* 122, 957–968.
20. Schwikowski, B., Uetz, P., and Fields, S. (2000). A network of protein-protein interactions in yeast. *Nat Biotechnol* 18, 1257–1261.
21. Serebriiskii, I., Estojak, J., Berman, M., and Golemis, E.A. (2000). Approaches to detecting false positives in yeast two-hybrid systems. *Biotechniques* 28, 328–336.
22. Serebriiskii, I.G., and Golemis, E.A. (2001). Two-hybrid system and false positives. Approaches to detection and elimination. *Methods Mol Biol* 177, 123–134.
23. Koegl, M., and Uetz, P. (2007). Improving yeast two-hybrid screening systems. *Brief Funct Genomic Proteomic* 6, 302–312.
24. Uetz, P., Dong, Y.A., Zeretzke, C., Atzler, C., Baiker, A., Berger, B., et al. (2006). Herpesviral protein networks and their interaction with the human proteome. *Science* 311, 239–242.
25. Parrish, J.R., Yu, J., Liu, G., Hines, J.A., Chan, J.E., Mangiola, B.A., et al. (2007). A proteome-wide protein interaction map for Campylobacter jejuni. *Genome Biol* 8, R130.
26. Zhong, J., Zhang, H., Stanyon, C.A., Tromp, G., and Finley, R.L., Jr. (2003). A strategy for constructing large protein interaction maps using the yeast two-hybrid system: regulated expression arrays and two-phase mating. *Genome Res* 13, 2691–2699.
27. Jin, F., Avramova, L., Huang, J., and Hazbun, T. (2007). A yeast two-hybrid smart-pool-array system for protein-interaction mapping. *Nat Methods* 4, 405–407.
28. Cormack, R.S., Hahlbrock, K., and Somssich, I.E. (1998). Isolation of putative plant transcriptional coactivators using a modified two-hybrid system incorporating a GFP reporter gene. *Plant J* 14, 685–692.
29. Rossi, F., Charlton, C.A., and Blau, H.M. (1997). Monitoring protein-protein interactions in intact eukaryotic cells by beta-galactosidase complementation. *Proc Natl Acad Sci U S A* 94, 8405–8410.
30. Drees, B.L. (1999). Progress and variations in two-hybrid and three-hybrid technologies. *Curr Opin Chem Biol* 3, 64–70.
31. Frederickson, R.M. (1998). Macromolecular matchmaking: advances in two-hybrid and related technologies. *Curr Opin Biotechnol* 9, 90–96.
32. Walhout, A.J., Temple, G.F., Brasch, M.A., Hartley, J.L., Lorson, M.A., van den Heuvel, S., et al. (2000). GATEWAY recombinational cloning: application to the cloning of large numbers of open reading frames or ORFeomes. *Methods Enzymol* 328, 575–592.
33. Cagney, G., Uetz, P., and Fields, S. (2000). High-throughput screening for protein-protein interactions using two-hybrid assay. *Methods Enzymol* 328, 3–14.
34. Landy, A. (1989). Dynamic, structural, and regulatory aspects of lambda site-specific recombination. *Annu Rev Biochem* 58, 913–949.
35. Lamesch, P., Milstein, S., Hao, T., Rosenberg, J., Li, N., Sequerra, R., et al. (2004). *C. elegans* ORFeome version 3.1: increasing the coverage

of ORFeome resources with improved gene predictions. *Genome Res* 14, 2064–2069.
36. Rual, J.F., Hirozane-Kishikawa, T., Hao, T., Bertin, N., Li, S., Dricot, A., et al. (2004). Human ORFeome version 1.1: a platform for reverse proteomics. *Genome Res* 14, 2128–2135.
37. Gong, W., Shen, Y.P., Ma, L.G., Pan, Y., Du, Y.L., Wang, D.H., et al. (2004). Genome-wide ORFeome cloning and analysis of Arabidopsis transcription factor genes. *Plant Physiol* 135, 773–782.
38. Uetz, P. (2002). Two-hybrid arrays. *Curr Opin Chem Biol* 6, 57–62.
39. Harper, J.W., Adami, G.R., Wei, N., Keyomarsi, K., and Elledge, S.J. (1993). The p21 Cdk-interacting protein Cip1 is a potent inhibitor of G1 cyclin-dependent kinases. *Cell* 75, 805–816.
40. James, P., Halladay, J., and Craig, E.A. (1996). Genomic libraries and a host strain designed for highly efficient two-hybrid selection in yeast. *Genetics* 144, 1425–1436.

Chapter 14

Analysis of Membrane Protein Complexes Using the Split-Ubiquitin Membrane Yeast Two-Hybrid (MYTH) System

Saranya Kittanakom*, Matthew Chuk*, Victoria Wong, Jamie Snyder, Dawn Edmonds, Apostolos Lydakis, Zhaolei Zhang, Daniel Auerbach, and Igor Stagljar

Summary

Recent research has begun to elucidate the global network of cytosolic and membrane protein interactions. The resulting interactome map facilitates numerous biological studies, including those for cell signalling, protein trafficking and protein regulation. Due to the hydrophobic nature of membrane proteins such as tyrosine kinases, G-protein coupled receptors, membrane bound phosphatases and transporters it is notoriously difficult to study their relationship to signaling molecules, the cytoskeleton, or any other interacting partners. Although conventional yeast-two hybrid is a simple and robust technique that is effective in the identification of specific protein-protein interactions, it is limited in its use for membrane proteins. However, the split-ubiquitin membrane based yeast two-hybrid assay (MYTH) has been described as a tool that allows for the identification of membrane protein interactions. In the MYTH system, ubiquitin has been split into two halves, each of which is fused to a protein, at least one of which is membrane bound. Upon interaction of these two proteins, the two halves of ubiquitin are reconstituted and a transcription factor that is fused to the membrane protein is released. The transcription factor then enters the nucleus and activates transcription of reporter genes. Currently, large-scale MYTH screens using cDNA or gDNA libraries are performed to identify and map the binding partners of various membrane proteins. Thus, the MYTH system is proving to be a powerful tool for the elucidation of specific protein-protein interactions, contributing greatly to the mapping of the membrane protein interactome.

1. Introduction

Membrane proteins constitute approximately 30% of the entire proteome and play pivotal roles in many cellular processes that are linked to human disease. They perform essential processes

---

*Saranya Kittanakom and Matthew Chuk equally contributed to this Chapter.

in the cell such as regulating the exchange of solutes and signals between different compartments, and mediating activities such as hormone action. Notably, the majority of drugs on the market today either directly target membrane proteins or are taken into the cell via specific membrane proteins. Therefore, the mapping of membrane protein interactions is invaluable in drug discovery and development.

Conventional yeast-two hybrid (Y2H) *(1, 2)* has been used extensively to detect protein-protein interactions. It can be used to identify novel protein interactions or to map domains or amino acid residues that are essential for the particular interaction of interest *(1)*. The Y2H assay utilizes the observation that the DNA-binding domain and the activation domain of a transcription factor can associate and activate transcription despite their fusion to different proteins as long as they are in proximity *(1)*. This transcriptional activation is observed through the use of an appropriate reporter gene(s). Although this system is both powerful and robust, the interaction is confined to the nucleus of the cell thereby excluding the study of membrane bound proteins.

Recently, yeast biochemical techniques have begun to facilitate the characterization of interactions among membrane proteins. Among these is the split-ubiquitin membrane yeast two-hybrid (MYTH) *(3, 4)*. This system is based on ubiquitin, an evolutionarily conserved 76 amino acid protein that serves as a tag for proteins targeted for degradation by the 26S proteasome. The presence of ubiquitin is recognized by ubiquitin specific proteases (UBPs) located in the nucleus and cytoplasm of all eukaryotic cells. Ubiquitin can be split and expressed as two halves, the amino-terminal (aa 1-34) and the carboxyl terminal (aa 35-76). These two halves have a high affinity for each other in the cell and can reconstitute to form pseudo-ubiquitin that is recognizable by UBPs (**Fig. 1**) *(3)*.

In MYTH the C-terminal moiety of ubiquitin (Cub) (AA 35-76) along with an artificial transcription factor (TF) consisting of the bacterial LexA-DNA binding domain and the *Herpes simplex* VP16 transactivator protein is fused to the protein of interest (the Bait). The Prey protein(s) are fused to the N-terminal moiety of ubiquitin (Nub, aa 1-34) at either the N or C terminus (**Fig. 1A**). Wild type Nub has an isoleucine at position 13 (NubI). NubI and Cub have high affinity for each other and reassemble spontaneously in the cell to be recognized by the UBPs. By replacing Ile-13 of wild-type NubI with glycine (NubG), the affinity between NubG and Cub is decreased and the two halves only reconstitute as a pseudo-ubiquitin protein if they are brought into proximity through an interaction between the Bait and Prey proteins. Pseudo-ubiquitin is recognized by UBPs that cleave after the carboxy terminus of ubiquitin. This releases the transcription

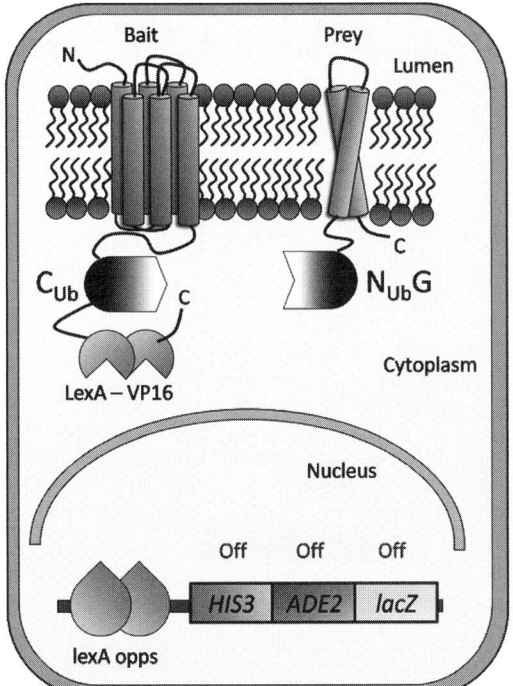 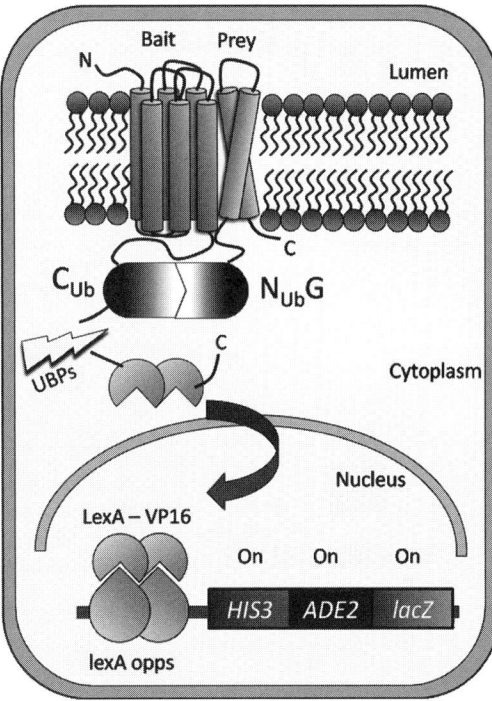

Fig. 1. Schematic presentation of split-ubiquitin membrane yeast-two hybrid (MYTH) system **(a)** A Bait membrane protein of interest is tagged with Cub followed by an artificial transcription factor LexA-VP16, while the membrane or cytoplasmic Prey protein is fused to the NubG domain. The reporter genes in MYTH/iMYTH systems are *HIS3/ADE2* and *LacZ*. They are not activated if there is no interaction between Bait and Prey. **(b)** The interaction of the Bait and Prey proteins allow for the reconstitution of pseudo-ubiquitin, leading to recognition and proteolytic cleavage by ubiquitin specific proteases (UBPs). The resultant release of the transcription factor (TF) leads to its subsequent entry into the nucleus and activation reporter genes.

factor, which then enters the nucleus and activates the reporter genes (**Fig. 1B**).

Human proteins can be studied in this system by expressing them from a plasmid containing the Bait-Cub-TF fusion construct (traditional MYTH). Yeast proteins can either be studied using a similar plasmid expression approach or by integrating the Cub-TF cassette directly at the genomic locus (integrated MYTH or iMYTH), thus placing the fusion protein under the control of its natural promoter and avoiding the possible complications of over-expression [5]. Large-scale screens are performed by transforming a Bait-containing yeast strain with a library of Prey plasmids and plating transformants on media that selects for cells that have activated the reporter genes of the system (typically *His3*, *Ade2* and *LacZ*) (**Fig. 2**) *(3, 6–9)*. Interactions are confirmed through a series of assays including Bait-dependency tests and co-immunoprecipitation. In addition, bioinformatics tools can also be used to assemble, analyze and validate the data sets.

This technique can be used to identify not only interactions among membrane proteins but also to identify and establish roles

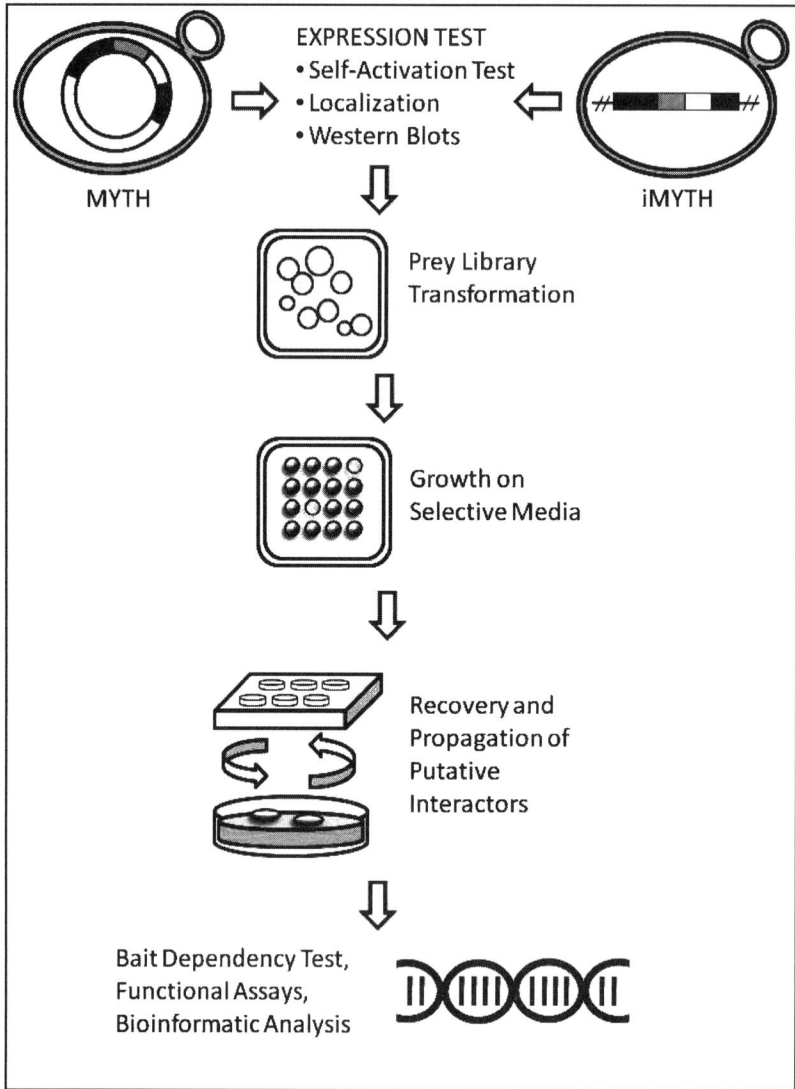

Fig. 2. *Schematic of MYTH/iMYTH screen.* The Bait undergoes various tests to ensure that the Cub-TF cassette has been correctly expressed before large-scale screening with Prey libraries. Putative interactors that successfully grow on selective media are recovered and propagated in bacteria. Interactors are identified through sequencing and undergo Bait dependency test and other functional assays. Chimeric oligonucleotide primers that have homologous regions to membrane protein of interest and cassette are used to amplify the cassette. The newly-amplifed cassette is then introduced into the yeast strain, which integrates into the genome via homologous recombination. The "Bait" undergoes various expression tests to ensure the cassette has been correctly integrated before being used in library screens.

for small molecules, chaperones or signalling molecules involved in the interaction and trafficking of membrane proteins. In the future this technique could also be applied to screens for potential drugs or inhibitors that act on specific membrane protein targets, as well as to study ligand dependent effects on membrane proteins of interest. Thus, MYTH and iMYTH represent robust and powerful technologies that should prove to be invaluable tools in the elucidation of complete membrane-protein interactomes, and in the development of drugs for the treatment of numerous membrane-linked diseases.

2. Materials

2.1. Generation of Bait Plasmid by gap repair for MYTH

1. Vector backbone for either Type I or Type II (see **Note 1**)
   a) Type I : pCCW-STE, pBT3-STE
   b) Type II : pTLB1, pBT3-N
2. Oligonucleotides for PCR
   a) Type I MYTH (**Fig. 3A**)

   These primers are for use with pCCW-STE and pBT3-STE.
   Forward: 5'GCC AGG CCT TTA ATT AAG GCC GCC TCG GCC ATC-21 gene-specific-nt-3'
   Reverse: 5'CCC CGA CAT GGT CGA CGG TAT CGA TAA GCT-21 gene-specific-nt-3'

   b) Type II MYTH (**Fig. 3B**)

   These primers are for use with pTLB1.
   Forward: 5'CTA AGA GGT GGT ATG CAC AGA TCA GCT TTG-21 gene-specific-nt-3'
   Reverse: 5'GCT CCG CGG AAG GCC TCC ATG GGT ATA TCT GCA-21 gene-specific-nt-3'
   These primers are for use with pBT3-N.
   Forward: 5'GCC AGG CCT TTA ATT AAG GCC GCC TCG GCC CCA-21 gene-specific-nt-3'
   Reverse: 5'GAC CTA TTA AGA TCT GAC GTC AGC GCT CCG CG- 21gene-specific-nt-3'
3. Restriction enzymes specific for the backbone of interest
4. PCR reaction Mix: 100 ng/μl template, 2 μM primers, 20 mM Tris-HCl (pH 8.8), 10 mM $(NH_4)_2SO_4$, 10 mM KCl, 0.1% Triton X-100, 0.1 mg/ml BSA, 2 mM $MgSO_4$, 0.1 U/ml Taq (Fermentas), 0.025 U/ml Pfu (Fermentas)
5. *Saccharomyces cerevisiae*:

   THY.AP4 [*MATa leu2, ura3, trp1*:: (lexAop-*lacZ*) (lexAop)-*HIS3* (lexAop)-*ADE2*]
6. Culture tubes
7. Yeast SD media lacking leucine, both liquid and 2% bacto-agar plates (see **Note 2**)
8. Agarose
9. LB supplemented with Kanamycin (see **Subheading 2.11.1**), both liquid and 2% bacto-agar plates (see **Subheading 2.10.5**)
10. DNA ladder

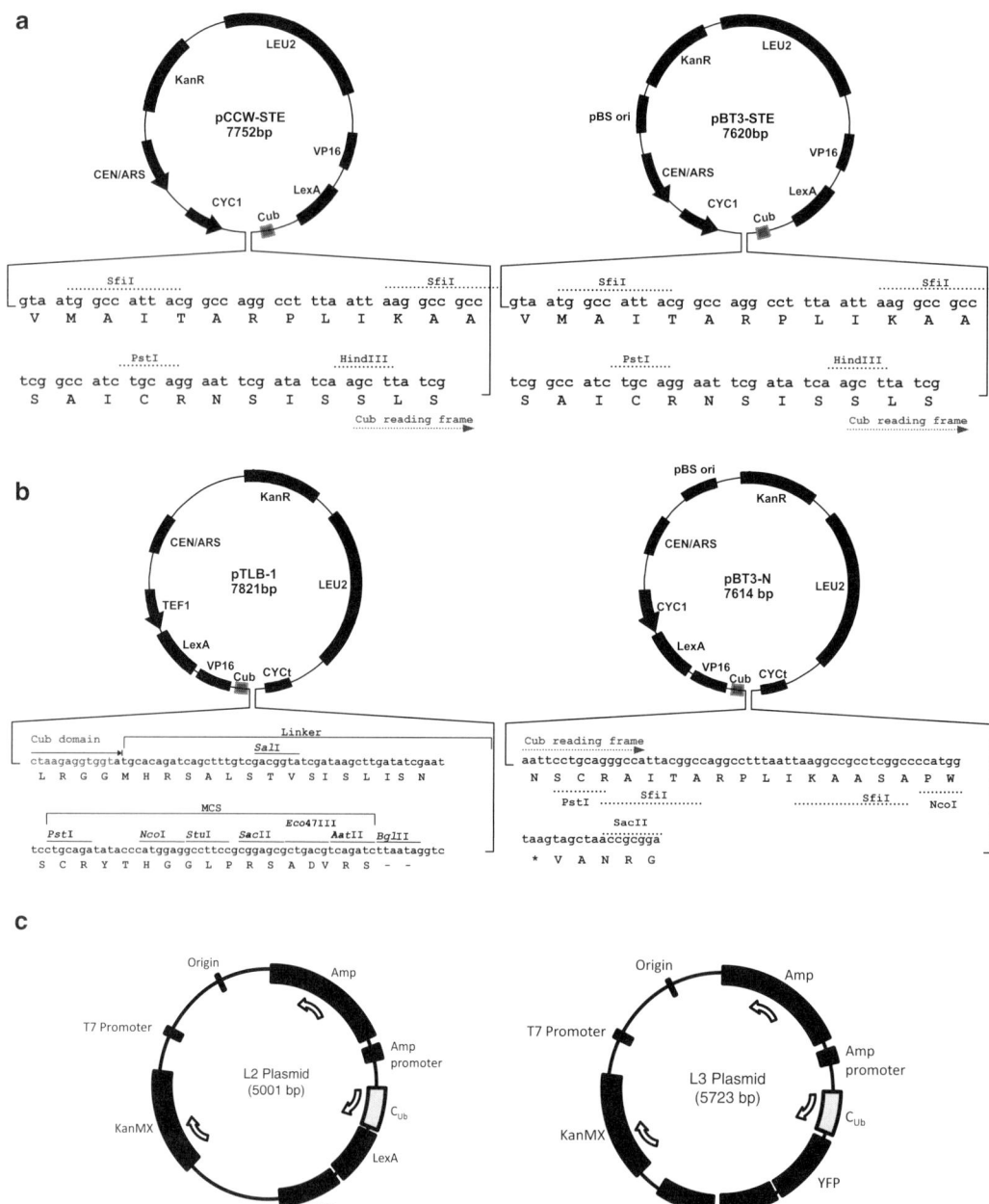

Fig. 3. *Bait vectors for MYTH/iMYTH systems.* Three sets of Bait vectors used to generate Cub-TF constructs. (a) Type I membrane protein with its amino terminus in the lumen and its carboxy terminus in the cytosol is fused C-terminally to the Cub-TF reporter cassette (Bait-Cub-LexA-VP16) in the vector pCCW-STE and pBT3-STE. The expression of this type I Bait is driven by the weak promoter, CYC1. (b) cDNA of Type II membrane protein with its amino terminus in the cytosol and its carboxy terminus in the lumen are expressed as a fusion C-terminally to the LexA-VP16-Cub reporter cassette (LexA-VP16-Cub-Bait). The strong promoter TEF1 drives a high level of expression of a heterologous gene in pTLB1 and the weak promoter CYC1 in pBT3-N. (c) The L2 cassette contains C-terminal half of Ubiquitin (Cub) (aa 35-76 respectively), followed by an artificial transcription factor (LexA-VP16). The L3 cassette is identical to the L2 cassette with the addition of a Yellow Fluorescent Protein (YFP) in-between the Cub and the artificial transcription factor. These cassettes are used to integrate the Cub-(YFP)-TF into the yeast genome for iMYTH and are expressed using the membrane protein's own endogenous promoter.

11. *E. coli*:

    DH5α [F-80dlacZ M15, endA1, recA1, hsdR17(r_k^-,m_k^+), supE44, thi1, gyrA96,relA1, (lacZYAargF)U169, λ⁻]

12. TAE (*see* **Subheading 2.11.6**)

2.2. Generation of Bait Strain for iMYTH

1) Plasmid for amplification of Cub-TF cassette (**Fig. 3C**)

    a) **L2**: contains the Cub - TF cassette followed by a *KanMX* marker flanked by two loxP sites and an ampicillin-resistance gene.

    b) **L3**: contains the Cub-YFP-TF cassette followed by a *KanMX* marker flanked by two loxP sites and an ampicillin-resistance gene.

2) Oligonucleotides for PCR

    a) L2/L3 Forward: 5'-gene specific-nt-ATG TCG GGG GGG ATC CCT CCA-3'

    b) L2/L3 Reverse: 5'-gene specific-nt-ACT ATA GGG AGA CCG GCA GA-3'

    c) KanMX Reverse: 5'-GAG CGT TTC CCT GCT CGC AGG TCT GCA G-3'

    Verification Forward: A 20 nucleotide primer corresponding to sequence approximately 150 to 200 bp upstream of the stop codon of the gene of interest.

3) PCR reaction mix: 0.1 ng/μL L2 or L3 template, 2 μM primers, 20 mM Tris-HCl (pH 8.8), 10 mM $(NH_4)_2SO_4$, 10 mM KCl, 0.1% Triton X-100, 0.1 mg/mL BSA, 0.2 mM dNTPs, 1.5 mM $MgSO_4$, 0.1 units/μL Taq (Fermentas), 0.025 unit/μL Pfu (Fermentas)

4) *Saccharomyces cerevisiae*:

    a) THY.AP4 (*see* **Subheading 2.1.5**)

    b) NMY32 [*MATa HIS3Δ trp1-901 leu2-3,112 ade2 LYS2::(lexAop)4-HIS3 URA3::(lexAop)8-lacZ (lexAop)8-ADE2 GAL4*)]

    c) L40 [*MATa HIS3 200 trp1-901 leu2-3, 112 ade2 LYS2:(lexAop)4-HIS3 URA3::(lexAop)8-lacZ GAL4*]

5) YPAD (*see* **Subheading 2.10.1**)

6) YPAD supplemented with G418 (*see* **Subheading 2.11.3**)

2.3. Verification of Bait Protein Function/Expression by "NubG/NubI" Test

1) Bait strain of interest (either carrying plasmid or integrated Bait)

2) Control Prey plasmids

    a) pOst1-NubG (negative control)

    b) pOst1-NubI (positive control)

    c) pFur4-NubG (negative control)

    d) pFur4-NubI (positive control)

3) Appropriate selective media (*see* **Note 2**)

4) 3-amino-1,2,4triazole (3-AT)

5) Parafilm

6) Reagents for "Basic Yeast Transformation" (*see* **Subheading 2.6**)

2.4. Library Transformation and Identification of Interactors (High Throughput Screen)

1) Large-scale library Transformation
   a) Plasmid libraries : NubG-X cDNA, X-NubG cDNA, NubG-X random genomic DNA, X-NubG random genomic DNA A variety of libraries are commercially available from Dualsystems Biotech Inc. (http://www.dualsystems.com)
   b) Liquid media, selective or rich (*see* **Note 2**)
   c) Bait strain
   d) Disposable cuvettes
   e) 0.9% NaCl (*see* **Subheading 2.11.4**)
   f) Single-stranded carrier DNA (*see* **Subheading 2.11.9**)
   g) Lithium acetate (LiOAc)/Tris EDTA buffer (pH 7.5) (TE) master mix (*see* **Subheading 2.11.5, 2.11.7 and 2.11.10**)
   h) 15 ml disposable screw-cap tubes
   i) Dimethy sulfoxide (DMSO)
   j) 1.5 ml microfuge tubes
   k) 150 mm petri dishes
   l) 100 mm petri dishes
   m) Parafilm

2) X-gal Test
   a) X-gal plates (*see* **Subheading 2.10.6**)
   b) Sterile ddH$_2$O

3) Recovery of Prey plasmids
   a) Positive colonies from X-gal test
   b) Reagents for "Yeast Mini-Prep" (*see* **Subheading 2.7**)

4) Amplification of putative interactors
   a) Plasmids from positive colonies from the X-gal test
   b) Reagents for "Basic *E. coli* Transformation" (*see* **Subheading 2.9**)

5) Sequencing and Bioinformatics
   a) Plasmids isolated from putative interactors
   b) Sequencing primers (*see* **Note 3**)

2.5. Bait Dependency Test

1) Original Bait plasmid/strain
2) Plasmids encoding putative interactors
3) Unrelated Bait plasmid (*see* **Note 4**)

4) Reporter *S. cerevisiae* strain (ie. THY.AP4, NMY32 or L40)

5) Appropriate selective media (*see* **Note 2**)

6) X-gal plates (*see* **Subheading 2.10.6**)

7) Sterile ddH$_2$O

2.6. Basic Yeast Transformation

1) Appropriate yeast strain
2) Appropriate selective liquid media (*see* **Note 2**)
3) Disposable cuvettes
4) Sterile ddH$_2$O
5) 0.1M Lithium Acetate (LiOAc) (*see* **Subheading 2.11.7**)
6) 1.5 ml microfuge tubes
7) Basic yeast transformation mix, (*see* **Subheading 2.11.12**)

2.7. Yeast Mini-Prep

1) Commercial Plasmid Mini-Prep kit
2) Appropriate selective liquid media (*see* **Note 2**)
3) 0.5 mm glass beads
4) 1.5 ml microfuge tubes
5) Sterile ddH$_2$O

2.8. E. coli Mini-Prep

1) *E. coli* containing plasmid to be prepped
2) Any commercially available mini-prep kit
3) LB broth with appropriate antibiotic (*see* **Note 2**)

2.9. Basic E. coli Transformation

1) Competent *E. coli* (*see* **Note 5**)
2) Plasmid of interest for transformation
3) LB broth (*see* **Subheading 2.10.5**)
4) LB plates with appropriate antibiotic (*see* **Note 2**)

2.10. Media

1. **YPAD (1 L liquid and plates)** *(11)*: 10 g bacto yeast extract, 20 g bacto peptone, 20 g D-glucose monohydrate, 40 mg adenine sulphate, 20 g agar (omit if preparing liquid medium). Dissolve all ingredients to a final volume of 1L. Autoclave at 121°C, 15 psi for 30 minutes. For plates allow agar to cool to 55°C and pour into sterile petri dishes. Store at 4°C.

2. **2x YPAD (1L liquid)** *(11)*: 20 g bacto yeast extract, 40 g bacto peptone, 40 g D-glucose monohydrate, 40 mg adenine sulphate. Dissolve all ingredients to a final volume of 1L. Autoclave at 121°C, 15 psi for 30 minutes.

3. **10x Amino Acid Drop-out Mix (1 L)**: 400 mg Adenine, 200 mg Arginine, 200 mg Histidine, 300 mg Isoleucine, 1000 mg Leucine, 300 mg Lysine, 1500 mg Methionine, 500 mg Phenylalanine, 2000 mg Threonine, 400 mg Tryptophan, 300 mg Tyrosine, 200 mg Uracil, 1500 mg Valine. Leave out

the desired amino acid(s). Bring the final volume to 1L with ddH$_2$O and dissolve with stirring. Autoclave at 121°C, 15 psi for 30 minutes and store at 4°C.

4. **Synthetic Dropout (SD) Media (1 L liquid and plates)** *(11)*: 6.7 g Yeast Nitrogen Base without amino acids with ammonium sulphate, 100 ml of 10x Amino Acid Drop-out Mix (*see* **Subheading 2.10.3**), 20 g D-glucose, 20 g agar (omit if preparing liquid medium). Bring to final volume of 1 L and dissolve with stirring. Autoclave at 121°C, 15 psi for 30 minutes. For plates, allow agar to cool to 55°C and pour into sterile petri dishes. Store at 4°C.

5. **LB (1 L liquid and plates)** *(11)*: 10 g bacto tryptone, 5 g yeast extract, 5 g NaCl, 20 g agar (omit if preparing liquid media) and adjust to a final volume of 1 L. Autoclave at 121°C, 15 psi for 30 minutes. For plates, cool to 55°C supplement with appropriate antibiotic (*see* **Note 2**) stir and pour into sterile Petri dishes. Store at 4°C.

6. **X-gal Plates** *(12)*

   a) Agar Base (1 L): 6.7 g Yeast Nitrogen Base without amino acids with ammonium sulphate, 20 g D-glucose monohydrate, 100 ml appropriate 10x Amino Acid Drop Out Mix (*see* **Subheading 2.10.3**), 20 g bacto agar. Bring final volume to 900 ml. Autoclave at 121°C, 15 psi for 30 minutes.

   b) Sodium Phosphate Solution: 7 g sodium phosphate dibasic, 3 g sodium phosphate monobasic (*see* **Note 6**). Bring final volume to 100 ml and dissolve with stirring. Autoclave at 121°C, 15 psi for 30 minutes.

   c) X-gal solution: 100 mg X-gal, 1 ml N,N-dimethyl formamide. Allow Agar Base and Sodium Phosphate solution to cool to 65°C. Add 100 ml Sodium Phosphate solution and 0.8 ml X-gal solution to Agar Base, mix and pour into sterile Petri dishes. Wrap in aluminium foil and store at 4°C.

2.11. Solutions

1. **50 mg/ml stock (1000x stock) Kanamycin** *(13)*: Dissolve 50 mg kanamycin monosulphate in 1 ml ddH$_2$O. Filter sterilize through a 0.22 μm pore filter. Aliquot as required and store at -20°C until use. Use at a final concentration of 50 mg/L.

2. **100 mg/ml stock (1000x stock) Ampicillin** *(13)*: Dissolve 100 mg ampicillin sodium salt in 1 ml ddH$_2$O. Filter sterilize through a 0.22 μm pore filter. Aliquot as required and store at -20°C until use. Use at a final concentration of 100 mg/L.

3. **200 mg/ml stock (1000x stock) Geneticin (G418)**: Dissolve 200 mg G418 sulphate in 1 ml ddH$_2$O. Filter sterilize through a 0.22 μm pore filter. Aliquot as required and store at

4°C until use. Use at a final concentration of 200 mg/L. G418 is temperature sensitive and stable for at least two months in solution at 4°C.

4. **0.9% NaCl:** Dissolve 0.9 g NaCl in 100 ml ddH2O with stirring. Autoclave at 121°C, 15 psi for 30 minutes and store at room temperature.

5. **Tris EDTA buffer pH 7.5 (TE) (1 L 10x stock)** *(14)*: 100 ml of 1M Tris-HCl pH 7.5, 20 ml of 0.5 M EDTA pH 8.0. Add water to 1L. Filter sterilize through a 0.22 μm pore filter and store at room temperature.

6. **Tris-acteate-EDTA electrophoresis Buffer (TAE) (1 L 50x stock)** *(13)*: 242 g Tris Base, 57.1 ml glacial acetic acid, 100 ml of 0.5 M EDTA (pH 8.0). Add water to 1 L. The working solution is at 1x (40 mM Tris-acetate, 1 mM EDTA).

7. **1M Lithium Acetate** *(11):* Dissolve 10.2 g lithium acetate in ddH$_2$O up to 100 ml. Autoclave at 121°C, 15 psi for 30 minutes and store at room temperature.

8. **50% PEG-4000 (w/v)** *(11)*: Dissolve 50 g PEG-4000 in 35 ml ddH$_2$O with stirring (*see* **Note 7**). Bring final volume to 100 ml. Autoclave at 121°C, 15 psi for 30 minutes. Add sterile water after autoclaving to bring volume back to 100 ml. Parafilm bottle neck and store at 4°C.

9. **Single-stranded carrier DNA** *(14):* Sterilize a 250 ml bottle and magnetic stir bar by autoclaving. Dissolve 200 mg Salmon Sperm DNA, Type III Sodium Salt in 100 ml sterile ddH$_2$O at room temperature. Disperse large chunks of DNA by drawing the solution up and down several times with a sterile 25-ml pipette. Stir solution overnight at 4°C if it remains undissolved after 2-3 hours at room temperature. Aliquot solution into sterile 1.5-ml microfuge tubes. Boil at 100°C for 5 min and chill immediately in an ice-water bath. Store at -20°C. Boil for an additional 5 min at 100°C before use.

10. **LiOAc/TE Master Mix** *(14)*: 1.1 ml 1M LiOAc (*see* **Subheading 2.11.7**), 1.1 ml 10x TE pH 7.5 (*see* **Subheading 2.11.5**), 7.8 ml sterile ddH$_2$O. Vortex thoroughly to mix.

11. **LiOAc/PEG Master Mix** *(14)*: 1.5 ml 1M LiOAc (*see* **Subheading 2.11.7**), 1.5 ml 10x TE pH 7.5 (*see* **Subheading 2.11.5**), 12 ml 50% PEG-4000 (w/v) (*see* **Subheading 2.11.8**). Vortex thoroughly to mix.

12. **Basic Yeast Transformation Mix (TRAFO Mix)**. Per reaction add: 240μl 50% PEG-4000 (w/v) (*see* **Subheading 2.11.8**), 36 μl 1M LiOAc (*see* **Subheading 2.11.7**), 50 μl ss DNA (*see* **Subheading 2.11.9**), and 34 μl sterile ddH$_2$O. Add PEG first as it shields the cells from the detrimental effects caused by the high concentrations of LiOAc used.

13. 1M 3-AT: Dissolve 8.4 g of 3-Amino-1,2,4,triazole (3-AT) in 80 ml ddH$_2$O, bring final volume to 100 ml. Filter sterilize, aliquot as required and store at -20°C.

3. Methods

3.1. Bait Plasmid Construction by gap repair in MYTH (see Note 1) (modified from (11))

1. Forward primers contain 35-40 nucleotides identical to the sequence upstream of the chosen restriction site followed by 18-20 nucleotides specific to the gene of interest (*see* **Subheading 2.1**)
2. Amplify the gene of interest by PCR (*see* **Subheading 2.1** for reaction condition).
3. Transform PCR product along with the appropriate linearized Bait vector into the appropriate target yeast strain. Follow the basic yeast transformation protocol described in **Subheading 3.6** using the following DNA mixtures: A 250-500 fmol PCR product, 50 fmol linearized Bait vector, B 50 fmol linearized Bait vector, C 50 fmol undigested empty Bait vector (to control for transformation efficiency).
4. Plate transformants on appropriate selective medium (*see* **Note 2**) and incubate at 30°C for 2-3 days until colonies appear
5. Extract plasmids with the yeast mini-prep protocol (*See* **Subheading 3.8**).
6. Verify Bait construct by sequencing (*See* **Note 3**).

3.2. Generation of Bait Strain for iMYTH

When studying the interactions of *S. cerevisiae* proteins, the Integrated Split-Ubiquitin Membrane Yeast two-hybrid (iMYTH) approach is the method of choice. In iMYTH, the Bait protein of interest is endogenously tagged with the Cub-(YFP)-TF reporter sequence. The use of an endogenously tagged Bait presents a significant advantage over traditional MYTH, since Bait protein expression is under the control of its natural promoter, and can be studied under more physiologically relevant, 'wild-type' conditions. In addition, the method helps reduce the rate of false-positives, which are observed with higher frequency when using over-expressed Baits *(5)*.

1. Design chimeric forward and reverse primers homologous to both the yeast gene of interest and the L2/L3 plasmids. The forward primer contains 45 nucleotides corresponding to the 3'-end of the gene of interest (excluding the stop codon) followed by the L2/L3 forward priming sequence (*see* **Subheading 2.2**). The reverse primer contains 45 nucleotides corresponding to the reverse complement of the sequence 150-200 nucleotides downstream of your gene of

interest, fused to the L2/L3 reverse priming sequence (*see* **Subheading 2.2**).

2. Amplify the tagging cassette by PCR (*see* **Subheading 2.2** for reaction conditions).

3. Transform the amplified cassette into the appropriate target yeast strain (THY.AP4, NMY32 or L40). Follow the Basic Yeast Transformation protocol described in subheading 3.7, however after heat shock resuspend the cells in 4 mL YPAD media and allow to incubate at 30°C overnight prior to plating on YPAD+G418 media.

4. Select colonies which grow on YPAD+G418 and verify successful integration via colony PCR and sequencing (use the KanMX Reverse and Verification Forward primers described in section 2.2).

3.3. Verification of Bait Protein Function/Expression by "NubG/NubI" Test (Figure 4A)

1) Proceed with the Basic Yeast Transformation (*see* **Subheading 3.7**) and transform with the following control plasmids:
   a) pOst1-NubG (negative control)
   b) pOst1-NubI (positive control)
   c) pFur4-NubG (negative control)
   d) pFur4-NubI (positive control)

2) Plate on appropriate selective media (*see* **Note 2**) and incubate at 30°C for 2-3 days.

3) Pick a single colony from each transformation and resuspend in 150 µl sterile ddH$_2$O (undiluted sample).

4) Prepare 4 serial 10-fold dilutions from the undiluted sample, up to 1:10,000.

5) Spot 5 µl of each dilution onto the following plates (*see* **Note 2**).
   a) Selective media to indicate transformation has occurred (eg. SD-Trp-Leu or SD-Trp)
   b) Selective media containing optimized concentration(s) of 3-AT (*see* **Note 8**) to indicate reporter gene activation has occurred.

6) Allow plates to dry, wrap in parafilm and incubate at 30°C for 2-3 days.

3.4. Large-scale Library Transformation (High Throughput Screen) (see Note 9) (modified from (14))

Library Transformation (*see* **Note 9**)

1) Inoculate 10 ml of selective media with a single colony (*see* **Note 2**) and grow overnight at 30°C with shaking.

2) The following morning, dilute into 200 ml of selective media to an OD$_{600}$ of 0.15 and continue growing at 30°C with shaking until an OD$_{600}$ of 0.6-0.8 (at least two cell divisions) is reached.

3) Prepare the LiOAc/TE master mix (*see* **Subheading 2.11.10**).

4) Prepare the LiOAc/PEG master mix (*see* **Subheading 2.11.11**).

5) Once the culture reaches the desired OD, divide into four 50 ml screw-cap centrifuge tubes and centrifuge at 2500g for 5 minutes.

6) Remove the supernatant and resuspend each pellet in 25 ml sterile ddH$_2$O. Centrifuge at 2500g for 5 min. Remove the supernatant.

7) Resuspend each pellet in 1 ml of LiOAc/TE master mix (*see* **Subheading 2.11.10**) and transfer to a 1.5 ml microfuge tube. Cells are now competent.

8) Add the following to four 15 ml cap centrifuge tubes and vortex thoroughly to mix:
   a) Library plasmid 7 µg
   b) ssDNA (*see* **Subheading 2.11.9**) 100 µl
   c) Competent cells (from step 11) 600 µl
   d) LiOAc/PEG mix (*see* **Subheading 2.11.11**) 2.5 ml

9) Incubate at 30°C for 45 minutes. Mix briefly by inversion every 15 minutes.

10) Add 160 µl DMSO to each tube and mix immediately by inversion.

11) Incubate at 42°C for 20 minutes.

12) Pellet cells at 2500g for 5 minutes.

13) Remove supernatant and resuspend each pellet in 3 ml 2x YPAD (*see* **Subheading 2.10.2**).

14) Pool the cell suspensions and allow to recover at 30°C for 90 minutes with shaking

15) Centrifuge at 2500g for 5 min. Remove supernatant and resuspend in 4.8 ml of 0.9% NaCl (*see* **Subheading 2.11.4**)

16) Plate 300 µl aliquots of cell suspension onto 16-17 of 150 mm selective plates (*see* **Note 2**).

17) Use 10 µl of the resuspended cells to prepare 1:100, 1:1000 and 1:10000 dilutions in 0.9% NaCl (*see* **Note 9**).

18) Plate 100 ml of each dilution onto 100 mm plates containing selective media appropriate for the determination of transformation efficiency (*see* **Note 2**).

19) Allow plates to dry, seal with parafilm and incubate at 30°C for 3-4 days.

X-gal Test (modified from *(11)*) (**Fig. 4B**)

1) Pick positive colonies from the selective plates and replica plate on the same selective medium used in the library screen as well as on X-gal plates (*see* **Subheading 2.10.6**).

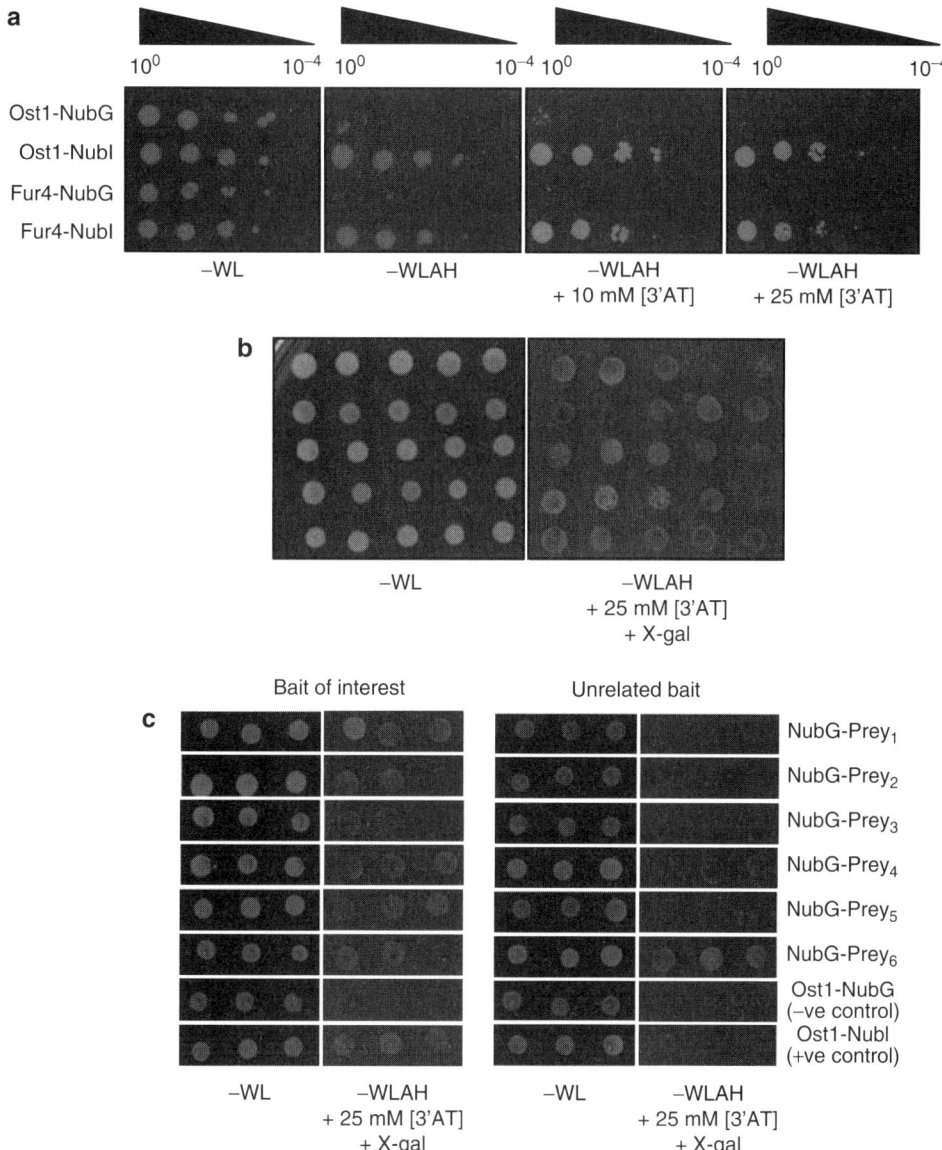

Fig. 4. *Spot assays MYTH system to assess Bait expression and system functionality.* (a) An example of a NubG/NubI test. THY.AP4 cells were co-transformed with TF-Cub Bait and Prey plasmids expressing either ER localized (Ost1p) or membrane localized (Fur4p) NubG/NubI-tagged proteins. Serial dilutions were prepared as indicated and cells were spotted on to media to select for transformation (SD-WL) and reporter gene activation (SD-WLAH). (b) X-gal Test. After large-scale screening, transformants were plated onto both SD-WL and selective media containing X-gal. Dark blue colonies were selected for further processing. (c) Bait-dependency test. Plasmids expressing putative interactors identified in the screen were retransformed into reporter strains along with plasmid expressing either the Bait of interest or an unrelated Bait. Preys that exclusively interacted with the Bait of interest were considered to be specific and subjected to further study.

2) Seal with parafilm and incubate at 30°C for 2-3 days to allow the blue colour to develop.

Recovery of Prey plasmids (putative interactors)

1) Perform yeast mini-prep (*see* **Subheading 2.7, 3.8**) on colonies that were positive in the X-gal test.

Amplification of putative interactors in E. coli

1) Proceed with the Basic *E. coli* Transformation protocol (*see* **Subheading 3.10**) using plasmids recovered in 3.4.3

   Proceed onto the Basic *E. coli* Mini-prep protocol (*see* **Subheading 3.9**).

Sequencing and Bioinformatics

1) Sequence putative interactors using an appropriate sequencing primer (*see* **Note 3**).

2) Sequences are typically aligned to sequences available on the National Center for Biotechnology Information (NCBI) database using the Basic Local Alignment Search Tool (BLAST) alignment tool available at the NCBI website, http://www.ncbi.nlm.nih.gov/blast.

3.5. Bioinformatics validation of MYTH experimental data

There are various perspectives that could be examined to computationally evaluate the credibility of MYTH's experimentally identified interactions. To begin with, one could investigate whether the Bait and Preys' genotypes follow common expression patterns, and whether their Gene Ontology (GO) terms justify the relevant interactions in terms of biological context. Consideration of gene expression profiles to validate protein-protein interactions is based on the rational that if two proteins follow positively correlated expression patterns they are more likely to interact than if they are negatively correlated *(15, 16)*. A GO terms *(17)* enrichment study can examine whether a combination of the relevant assigned functions, processes and especially localization, biologically makes sense. For instance, a reported physical interaction between a plasma membrane and a nuclear protein would most probably signify a false positive case. In addition, analyzing the proteins' structural information can provide more evidence. The overall proteins' topology and the termini orientation could serve as one of the validation criteria. For most membrane proteins, we lack detailed structure knowledge, due to the challenges related to the hydrophobic membrane environment. However, relevant Hidden Markov Models (HMMs) give rise to very high quality topology predictions for trans-membrane proteins especially when coupled with experimental identifications, and can be employed to our validation study *(18, 19)*. Finally, an integration of all those approaches using artificial intelligence techniques combines their strongest characteristics and results in a more mature estimation. Machine learning algorithms that can be used for this purpose range from Bayes classifiers, and logistic regression, to Support Vector Machines (SVMs). The latter is becoming increasingly popular for its capacity to integrate diverse sources of information, and has been used already in relevant studies *(20)*. It is noteworthy that each of the discussed approaches should be weighted and considered accordingly. For example, it can be plausible for two proteins with different expression profiles to interact as in cases of transient interactions *(21)*, and because membrane proteins generally have lower expression levels than soluble proteins *(22)*. Appropriately combining multiple computational approaches can prove useful in the validation of interactions identified through MYTH.

3.6. Bait Dependency Test (modified from (11)) (Figure 4C)

1) Proceed with the Basic Yeast Transformation protocol (*see* **Subheading 3.7**).
2) Transform Bait strains (i.e. original Bait in addition to an unrelated Bait) with plasmids encoding putative interactors.
3) Plate transformants on appropriate selective media.
4) Pick a single colony after 2-3 days and resuspend in 150 μl sterile ddH$_2$O.
5) Spot 5 μl of the resusupended cells onto appropriate selective and X-gal plates (*see* **Note 2**).
6) Incubate at 30°C until colony growth/development of blue colour (2-3 days) is observed.

3.7. Basic Yeast Transformation (23)

1) Inoculate a single colony of the appropriate yeast strain (THY. AP4, NMY32, or L40) into 5 ml of YPAD and incubate overnight at 30°C with shaking (200 rpm).
2) The next morning, dilute overnight culture to OD$_{600}$ of 0.15 in YPAD. Incubate at 30°C with shaking (200 rpm) until an OD$_{600}$ of 0.6-0.8 is reached (3-4 h).
3) Harvest the cells by centrifugation at 2500g for 5 minutes and remove the supernatant.
4) Resuspend the pellet in 25 ml of sterile ddH$_2$O and centrifuge again at 2500g for 5 minutes. Remove supernatant.
5) Resuspend the pellet in 1 ml of 0.1 M LiOAc and transfer to a 1.5 ml microfuge tube.
6) Pellet cells at 13200 g for 15 sec and remove supernatant.
7) Resuspend the pellet in 500 μl of 0.1 M LiOAc.
8) Use 50 μl of competent yeast cells.
9) Add 350 μl of the TRAFO mix (*see* **Subheading 2.11.12**) to each cell containing microfuge tube.

10) Add 1-2 μl of desired DNA and vortex tube for 1 minute to thoroughly mix all components.

11) Incubate at 30°C for 30 minutes.

12) Heat shock at 42°C for 40 minutes.

13) Centrifuge at 13200 g for 15 sec and remove the TRAFO mix.

14) Add 40 μl of sterile dH$_2$O and plate on selective plates (*see* **Note 2**).

15) Allow plates to dry, wrap in parafilm and incubate plates at 30°C for 2-3 days to recover transformants.

3.8. Basic Yeast Mini-Prep (modified commercial kit)

1) Inoculate 5 ml of SD containing the appropriate dropout (*see* **Note 2**) with a single colony.

2) Grow overnight with shaking at 30°C.

3) Extract the plasmid DNA using any commercial Plasmid Extraction kit with the following modifications:

   a) Add about 100 μl 0.5 mm glass beads to a microfuge tube.

   b) After resuspending the yeast pellet in the resuspension buffer, transfer the cell suspension to the tubes from step a.

   c) Vortex at maximum speed for 10 minutes then continue with the commercial protocol.

   d) Elute the plasmid DNA in the smallest recommended volume as per the kit's protocol.

3.9. Basic E. coli Mini-Prep

1) Inoculate 5 ml of LB broth containing the appropriate antibiotic with a single colony.

2) Grow overnight with shaking at 37°C.

3) Extract the plasmid DNA using any commercial plasmid mini-prep kit.

3.10. Basic E. coli Transformation

1) Add 15 μl of competent *E.coli* to chilled microfuge tubes

2) Add 1-2 μl DNA to the mixture and pipette gently to mix (15 μl from a Yeast Mini-Prep plasmid extraction).

3) Chill on ice for 30 minutes.

4) Heat shock at 42°C for 45 seconds.

5) Chill immediately on ice for 2 minutes.

6) Add 100 μl of LB and incubate at 37°C for 1 hour.

7) Centrifuge at 13200 g for 15 seconds

8) Remove all but 40 μl of the LB, use the remaining LB to resuspend the cells and plate on LB containing the appropriate antibiotic (*see* **Note 2**)

3.11. Further Validation

Interactions of interest identified from a screen could be further characterized using various techniques such as co-immunoprecipitation, fluorescence microscopy or *in vitro* transport of radiolabeled substrates, which could be used to study transporter proteins. An example of validation approaches is taken from a publication in Molecular Cell 2007, 26(1), 15-25. "Mapping protein-protein interactions for the yeast ABC transporter Ycf1p by integrated split-ubiquitin membrane yeast two-hybrid analysis" *(5)*. The ABC transporter Ycf1p was shown to interact with Tus1p, a guanine nucleotide exchange factor identified in the random yeast genomic X NubG library screening *(5)*.

3.11.1. Co-immunoprecipitation

A practical way to validate the interaction between Bait and putative interactors is co-immunoprecipitation using specific antibody and/or tagged-protein copurification. In the Ycf1p studies, co-immunoprecipitation was performed to test whether Ycf1p could form a complex with Tus1p. Yeast strain THY.AP4 containing either chromosomally integrated *YCF1*-CYT or *YBT1*-CYT as a negative control, was transformed with a plasmid containing Tus1-hemagglutinin (HA) tagged. Protein extracts were immunoprecipitated with either anti-HA (to pull down Tus1p and associated protein) or anti-LexA (to pull down Ycf1p/Ybtp-CYT and associated proteins) antibodies. Immunoprecipitate was resolved by 10% SDS-PAGE and analysed by western blot using either anti-LexA or anti-HA antibodies, respectively (**Fig. 5A**).

3.11.2. Fluorescence Microscopy

Protein localization was determined by fluorescence microscopy. To address whether the interactor Tus1p had an effect on the localization of Ycf1p, the authors examined Ycf1p-GFP in strains deleted for either *Ycf1* or *Tus1*. Fluorescence images taking by fluorescence microscopy show that deletion of either *Ycf1* or *Tus1* does not alter the localization of Ycf1p-GFP as compared to wild-type (**Fig. 5b**).

3.11.3. In Vitro Transport Assay of Radiolabeled Substrate

To assay the transport activity of Ycf1p *in vitro*, vacuoles were purified from WT or *ycf1Δ* strains and the transport activity of Ycf1p's substrate, [^{3}H]-estradiol-17-glucaronide[^{3}H] E217G was measured in the presence of GTP. Transport was measured as the amount of the radioactive substrate ([^{3}H] estradiol-glucaronide) sequestered within the vacuoles over 10 min. To test whether Tus1p might be the component in the cytosol responsible for stimulation of transport, cytosol derived from a *tus1Δ* strain was tested. Interestingly, this cytosol failed to stimulate transport, suggesting that Tus1p is indeed necessary for stimulatory activity. Together, these results show that Tus1p is essential for the cytosol-dependent stimulation of Ycf1p activity *in vitro* (**Fig. 5C**).

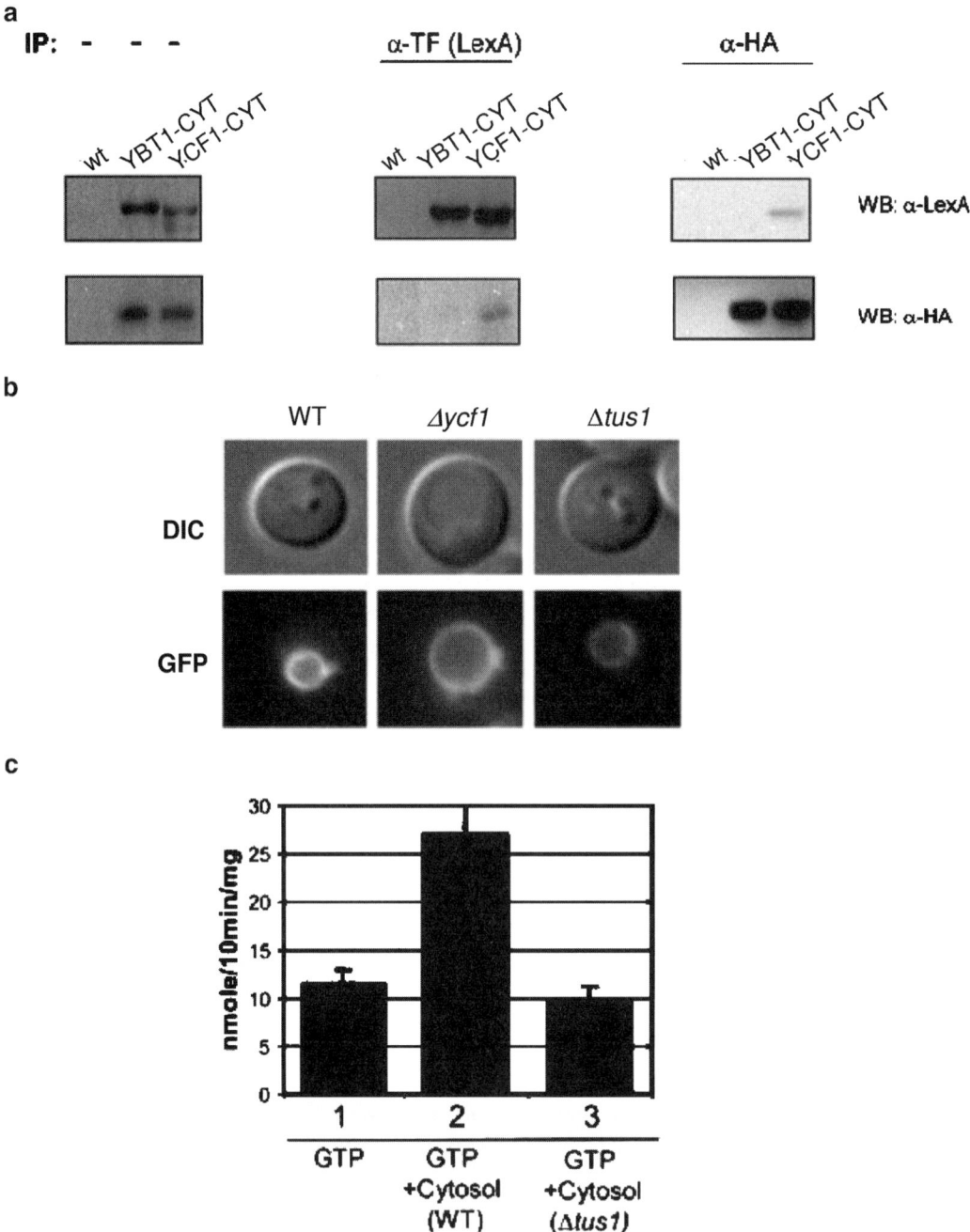

Fig. 5. *Validation techniques* (a) Co-immunoprecipitation. HA-tagged Tus1p was transformed into THY.AP4 expressing either endogenously tagged YCF1-CYT or YBT1-CYT. Protein extracts prepared from these and a wild-type control strain were immunoprecipitated (IP) with anti-HA antibodies (recognizing Tus1p) or anti-LexA antibodies (recognizing YCF1/YBT1-CYT) as indicated. Immunoprecipitated proteins were resolved by SDS-PAGE (10% gel) and analyzed by western blot using anti-HA or anti-LexA antibodies. (b) Fluorescence microscopy. The GFP fluorescence pattern of Ycf1p-GFP expressed exogenously from the plasmid was monitored in wild-type and deletion strains as shown. (c) Transport assay. *In vitro* Ycf1p-dependent transport of [^{3}H] E2b17G into vesiculated vacuoles prepared from a wild-type strain was measured in the presence of GTP +/- cytosolic extract derived form wild-type or mutant strains.

4. Notes

1. Membrane proteins are classified based on the orientation of their N and C termini. Type I membrane proteins present their C terminus in the cytosol. Type II membrane proteins present their N terminus in the cytosol. Hence, Type I and Type II MYTH respectively. In this system, the cDNA encoding the Bait protein of interest should be cloned into the vector such that it is in frame with the Cub-TF cassette located either downstream (Type I proteins) or upstream (Type II proteins). Forward primers should contain 35-40 nt identical to the sequence upstream of the chosen restriction site, followed by 18-20 gene-specific nt. For the reverse primer, it is essential to omit the native stop codon from the gene of interest for type I transmembrane proteins; however, it should be kept for type II transmembrane proteins. It is also essential to design the fusion such that the 40 nt homology region of the cDNA is in frame with the Cub-TF sequence (*see* **Subheading 2.1**). All of the type I and II Bait vectors contain the weak *CYC1* promoter, which drives low levels of heterologous protein expression except pTLB1, which contains the strong *TEF1* promoter. All of these vectors also contain the yeast Ste2 leader sequence that improves targeting of the heterologous Bait protein to the yeast plasma membrane. In addition, these vectors are also centromeric plasmids that contain an autonomously replicating sequence (ARS) origin of replication and one centromeric locus (CEN), which results in one to two copies of the plasmid per cell. These low copy number plasmids autonomously replicate in both *Escherichia coli* and *Saccharomyces cerevisiae*. The vectors are selected for by the *KanR* and *LEU2* genes, allowing growth on media containing kanamycin (bacteria) or lacking leucine (yeast), respectively *(11)*.

2. Selective media versus rich media (*see* **Subheading 2.10, 2.11**). If selection is not required, Yeast Extract-Peptone-Adenine-Dextrose (YPAD) media is the medium of choice. However, selective media is used when selection is required (*see* **Table** 1 for media requirements). Bait vectors for the MYTH system contain the kan^R and *LEU2* genes which allow growth in media containing kanamycin (bacteria) or lacking leucine (yeast), respectively. Prey vectors that are commercially available from Dualsystems Biotech for the MYTH system typically contain the *ampR* and *TRP1* genes, which allow growth in media containing ampicillin (bacteria) or lacking tryptophan (yeast), respectively *(11)*. Bait strains in the iMYTH system contain the KanMX cassette, which confers resistance to geneticin, G418 (*see* **sections 3.1, 3.2**).

Table 1
Media requirements

System	Organism	To Select	Media (liquid or 2% agar)
MYTH	Yeast	Bait plasmids	SD-Leu
		Prey plasmids	SD-Trp
		Bait and Prey plasmids	SD-Trp-Leu
		Interacting Baits and Preys	SD-Trp-Leu-Ade-His
	Bacteria	Bait plasmids	LB supplemented with kanamycin
		Prey plasmids	LB supplemented with ampicillin
iMYTH	Yeast	Integrated Baits	YPAD supplemented with G418
		Interacting Baits and Preys	SD-Trp-Ade-His
	Bacteria	Prey plasmids	LB supplemented with ampicillin

3. The following primers are routinely used to sequence Preys from libraries available from Dualsystems Biotech Inc.

   NubG-X cDNA: 5' CCG ATA CCA TCG ACA ACG TTA AGT CG 3'
   X-NubG cDNA: 5' CGA CTT AAC GTT GTC GAT GGT ATC GG 3'

4. As in other genetic selection systems, false positives will also be isolated from the MYTH system. To eliminate these isolates, the Prey plasmids isolated are retransformed with the original Bait plasmid. These isolates are also retransformed with an unrelated Bait protein. Only Preys that yield a $His^+/Ade^+/LacZ^+$ phenotype (if the NMY32 or THY.AP4 yeast strains are used) upon transformation with the original Bait plasmid (but not the unrelated Bait) on selective media should be considered for further analysis (*see* **Table 2** for a list of common false interactors, Dualsystem Biotech website) *(14)*.

5. Competent *E.coli* at a transformation efficiency of 10^7 transformants/μg DNA or higher is required for the recovery of plasmids extracted from yeast. These efficiencies are readily obtainable using standard protocols for the generation of competent cell grown under low temperatures. Alternatively, competent cells can be obtained from a commercial source.

6. In addition to auxotrophic markers, the strains THY.AP4, NMY32 and L40 also contain the color reporter *LacZ*. This encodes the bacterial enzyme β-galactosidase that is able to convert the substrate 5-Bromo-4-Chloro-3-Indolyl-β-D-Galactopyranoside (X-gal) into a blue compound. Hence, this test can be used to assess the strength of interaction of the individual transformants isolated from the library screen (**Fig. 4B**) *(14)*. Occasionally it takes 4-5 days for the colonies to develop the blue colour.

Table 2
False-positive interactors

Interactor	Frequency	Comment
H$^+$-ATPase	Frequent	Mostly vacuolar ATPases, may be connected with sorting of particular Bait proteins to the vacuoles of yeast.
PLP1	Frequent	
Ubiquitin	Frequent	Frequently isolated from x-NubG libraries, less frequently from NubG-x libraries. Confirmed false positive, interacts with the Cub portion of the Bait via the wild type ubiquitin. Isolated sequences often encode partially truncated ubiquitin, thereby creating a wild type Nub (N-terminal part of ubiquitin) fused to the NubG portion.
HDAC	Rare	May bind to hydrophobic patches located on the surface of the Cub portion of the Bait.
Translocon components	Rare	May interact with the Bait upon translocation through the membrane due to spatial proximity. This is not a true false positive, as it reflects a biologically relevant interaction.
Signal peptidases	Rare	May reflect an interaction of the signal peptidase with type I Baits upon cleavage of the signal sequence peptide.

7. It is crucial that the 50% PEG-4000 (*see* **Subheading 2.11.7**) solution be at a 50% concentration weight/volume. It is recommended that the bottle's neck be wrapped in parafilm and the solution is stored at 4°C to prevent evaporation. Transformation efficiencies decrease substantially if the solution is greater than 50%.

8. The "NubG/NubI" test is used to verify the expression of the Bait in yeast after its sequence has been confirmed. The following control Prey plasmids: pOst1-NubG, pOst1-NubI, pFur4-NubG, and pFur4-NubI (*see* **Subheading 3.3** and **Fig. 4A**) are transformed into yeast strains expressing the Bait of interest. Ost1p is an ER resident protein whereas Fur4p is plasma membrane localized. NubI constructs (positive controls) spontaneously interact with Cub independent of an association between Bait and Prey whereas NubG constructs (negative controls) do not *(11)*. This test is also used to select the optimum concentration of 3-amino-1,2,4-triazole (3-AT) to suppress background growth (*see* **Subheading 3.4**). Typical concentrations of 3-AT tested include 10 mM, 25 mM, 50 mM and 100 mM. Lack of growth of Bait strains containing NubI Prey indicates a problem of Bait expression/stability. Alternatively, growth of Bait strains containing NubG Prey indicates spontaneous activation of the reporter system and may be addressed by increasing the concentration of 3-AT in the medium or expressing the Bait from a different vector.

9. Transformants should appear after 2-3 days on efficiency plates and after 3-4 days on selective media. Calculate the total number of transformants and the transformation efficiency from the efficiency plates (the total number of transformants should be greater than 2×10^6 in order to cover the library effectively) *(14)*:

"Total number of transformants" = number of colonies on efficiency plate * dilution factor * 10 * 4.8
Transformation efficiency (clones/µg DNA) = "total number transformants"/28 µg

References

1. Fields, S. and Song, O. (1989) A novel genetic system to detect protein-protein interactions. *Nature* 340, 245–246
2. Uetz, P. and Hughes, R. E. (2000) Systematic and large-scale two-hybrid screens. *Curr Opin Microbiol* 3, 303–308
3. Stagljar, I., Korostensky, C., Johnsson, N. and te Heesen, S. (1998) A genetic system based on split-ubiquitin for the analysis of interactions between membrane proteins in vivo. *Proc Natl Acad Sci U S A* 95, 5187–5192
4. Fetchko, M., Auerbach, D. and Stagljar, I. (2003) Yeast genetic methods for the detection of membrane protein interactions: potential use in drug discovery. *BioDrugs* 17, 413–424
5. Paumi, C. M., Menendez, J., Arnoldo, A., Engels, K., Iyer, K. R., Thaminy, S., Georgiev, O., Barral, Y., Michaelis, S. and Stagljar, I. (2007) Mapping protein-protein interactions for the yeast ABC transporter Ycf1p by integrated split-ubiquitin membrane yeast two-hybrid analysis. *Mol Cell* 26, 15–25
6. Auerbach, D., Galeuchet-Schenk, B., Hottiger, M. O. and Stagljar, I. (2002) Genetic approaches to the identification of interactions between membrane proteins in yeast. *J Recept Signal Transduct Res* 22, 471–481
7. Thaminy, S., Miller, J. and Stagljar, I. (2004) The split-ubiquitin membrane-based yeast two-hybrid system. *Methods Mol Biol* 261, 297–312
8. Miller, J. and Stagljar, I. (2004) Using the yeast two-hybrid system to identify interacting proteins. *Methods Mol Biol* 261, 247–262
9. Fetchko, M. and Stagljar, I. (2004) Application of the split-ubiquitin membrane yeast two-hybrid system to investigate membrane protein interactions. *Methods* 32, 349–362
10. O'Brien, T. D., Butler, A. E., Roche, P. C., Johnson, K. H. and Butler, P. C. (1994) Islet amyloid polypeptide in human insulinomas. Evidence for intracellular amyloidogenesis. *Diabetes* 43, 329–336
11. Iyer, K., Burkle, L., Auerbach, D., Thaminy, S., Dinkel, M., Engels, K. and Stagljar, I. (2005) Utilizing the Split-Ubiquitin Membrane Yeast Two-Hybrid System to Identify Protein-Protein Interactions of Integral Membrane Proteins. *Sci. STKE* 275, pl3
12. OriGene Technologies, I. (1998) DupLEX-A™ Yeast Two-Hybrid System. Maryland.
13. Russell, S. A. (2001) Molecular Cloning A Laboratory Manual. Cold Spring Harbor Laboratory Press, Cold Spring Harbor, New York
14. Biotech, D. (2006) DUALmembrane kit 3 P01001. Zurich, Switzerland
15. Ge, H., Liu, Z., Church, G. M. and Vidal, M. (2001) Correlation between transcriptome and interactome mapping data from Saccharomyces cerevisiae. *Nat Genet* 29, 482–486
16. Lee, H. K., Hsu, A. K., Sajdak, J., Qin, J. and Pavlidis, P. (2004) Coexpression analysis of human genes across many microarray data sets. *Genome Res* 14, 1085–1094
17. Ashburner, M., Ball, C. A., Blake, J. A., Botstein, D., Butler, H., Cherry, J. M., Davis, A. P., Dolinski, K., Dwight, S. S., Eppig, J. T., Harris, M. A., Hill, D. P., Issel-Tarver, L., Kasarskis, A., Lewis, S., Matese, J. C., Richardson, J. E., Ringwald, M., Rubin, G. M. and Sherlock, G. (2000) Gene ontology: tool for the unification of biology. The Gene Ontology Consortium. *Nat Genet* 25, 25–29
18. Kim, H., Melen, K., Osterberg, M. and von Heijne, G. (2006) A global topology map of the Saccharomyces cerevisiae membrane proteome. *Proc Natl Acad Sci U S A* 103, 11142–11147
19. Krogh, A., Larsson, B., von Heijne, G. and Sonnhammer, E. L. (2001) Predicting transmembrane protein topology with a hidden Markov model: application to complete genomes. *J Mol Biol* 305, 567–580

20. Miller, J. P., Lo, R. S., Ben-Hur, A., Desmarais, C., Stagljar, I., Noble, W. S. and Fields, S. (2005) Large-scale identification of yeast integral membrane protein interactions. *Proc Natl Acad Sci U S A* 102, 12123–12128

21. Jansen, R., Greenbaum, D. and Gerstein, M. (2002) Relating whole-genome expression data with protein-protein interactions. *Genome Res* 12, 37–46

22. Drawid, A., Jansen, R. and Gerstein, M. (2000) Genome-wide analysis relating expression level with protein subcellular localization. *Trends Genet* 16, 426–430

23. Agatep, R., Kirkpatrick, R. D., Parchaliuk, D. L., Woods, R. A. and Gietz, R. D. (1998) Transformation of Saccharomyces cerevisiae by the lithium acetate/ single-stranded carrier DNA/ polyethylene glycol (LiAc/ ss-DNA/ PEG) protocol.

Chapter 15

Computational Analysis of the Yeast Proteome: Understanding and Exploiting Functional Specificity in Genomic Data

Curtis Huttenhower, Chad L. Myers, Matthew A. Hibbs, and Olga G. Troyanskaya

Summary

Modern experimental techniques have produced a wealth of high-throughput data that has enabled the ongoing genomic revolution. As the field continues to integrate experimental and computational analyzes of this data, it is essential that performance evaluations of high-throughput results be carried out in a consistent and biologically informative manner. Here, we present an overview of evaluation techniques for high-throughput experimental data and computational methods, and we discuss a number of potential pitfalls in this process. These primarily involve the biological diversity of genomic data, which can be masked or misrepresented in overly simplified global evaluations. We describe systems for preserving information about biological context during dataset evaluation, which can help to ensure that multiple different evaluations are more directly comparable. This biological variety in high-throughput data can also be taken advantage of computationally through data integration and process specificity to produce richer systems-level predictions of cellular function. An awareness of these considerations can greatly improve the evaluation and analysis of any high-throughput experimental dataset.

Key words: Systems biology, High-throughput data, Genomic data, Functional relationships, Data integration, Evaluation, Context specific

1. Introduction

The explosion of genomic sequencing and the subsequent rapid generation of functional genomic data, including proteomics techniques, provide an unprecedented view of whole networks of interacting proteins and small molecules. Such high-throughput studies, combined with traditional experimentation, have the potential to provide us with a truly mechanistic understanding of how a cell functions. The first steps in gaining such an understanding are to decipher individual protein functions and protein–protein interactions – these two problems are often considered the next key challenges in systems biology *(1–3)*.

While classical genetic and cell biology techniques continue to play an important role in the detailed understanding of cellular mechanisms, they are far too slow to provide a comprehensive genome-level understanding of protein function. Even in yeast, the most well-studied eukaryote and a tractable unicellular organism, nearly a fifth of the predicted genes have no known function. Cataloging all proteins in the genome by traditional methods could take decades. In an attempt to rapidly achieve the goals of gene function annotation and understanding of protein–protein interactions, several high-throughput (large-scale) experimental methods such as proteomic microarrays *(4–8)*, yeast two-hybrid assays *(9)*, and tandem mass spectrometry *(2)* have been developed. Each of these techniques is designed to measure protein levels or protein–protein interactions in a highly parallel fashion, enabling practical whole-genome analysis.

High-throughput functional data are important for rapid functional annotation of unknown genes, but this increase in throughput sacrifices accuracy for scale, potentially increasing false positive and negative detection rates *(2, 10–14)*. Recent work has highlighted this problem, showing that supposedly identical yeast two-hybrid datasets share few overlaps *(15)* and that different cDNA microarrays exhibit 10–30% variation among corresponding microarray probes *(16)*. For accurate gene function annotation, prediction of protein–protein interactions, and later pathway and network analysis, it becomes necessary to mine high-throughput data for only the most reliable and specific portions, even if this comes at the cost of some sensitivity *(12)*.

Precise and flexible algorithms and bioinformatics tools are necessary to extract reliable functional information from high-throughput genomics and proteomics data *(2, 3)*. Since high-throughput data generally represent a speed–quality compromise, substantial benefits can be realized by discovering the most accurate components of each dataset. An integrated analysis of these diverse data can then present only the most informative results drawn from a variety of genome-scale sources. Thus, a key challenge in interpreting such data is separating accurate, functionally relevant information from noise. This task can be complicated by the fact that some high-throughput experiments will capture certain areas of biology better than others, leading to a variety of functional specificities from dataset to dataset. Conversely, integrated analyses can take advantage of this diversity to construct a more unified biological picture from a collection of functional genomic data. This chapter will provide a brief overview of some methodologies for evaluation and analysis of proteomic and other functional genomic datasets in yeast.

1.1. Functional Diversity

When using high-throughput data or methods to direct experiments, scientists typically have a domain of interest in mind. A particular set of results may elucidate one or more biological processes while presenting little information about others. For example, *S. cerevisiae* sporulates only under fairly specific conditions *(17)*, so most datasets would not be expected to indicate which proteins interact during the process of sporulation. Conversely, many laboratory environments will provoke a cellular stress response or changes in growth rate, so information about these functions may be available in a variety of data sources *(18)*. We will refer to this variability in biological coverage as (equivalently) the function, process, or context specificity of a dataset.

Particularly with genome-scale data, functional content can vary widely by experimental platform, conditions, and analysis methods – and it may or may not correspond to the expectations of the experimenter. As with traditional bench experiments, high-throughput results may often contain information in biological areas beyond the intent of the original design, and it is the responsibility of thorough analysis techniques to take advantage of this variety. Similarly, individual datasets may provide moderately accurate information regarding many different biological processes, or they may provide more detailed information in a smaller number of areas. This heterogeneity of functional specificity means that it is crucial to ask not only, "How accurately does this data capture real biology," but also, "How much does this data tell me about each specific biological area?"

This variation can have a large impact on real-world analyzes; methods that take advantage of functional diversity can be more informative, and methods that ignore it can produce misleading results. **Figure 1** contrasts two evaluations of a collection of experimental data types drawn from a recent publication *(19)*. Briefly, experimental results are scored on the basis of the extent to which they include known protein–protein functional relationships; data that are more accurate appear towards the top, and data covering a larger portion of the genome appears towards the right. Evaluations of this type will be discussed in more detail below.

The striking difference between **Fig. 1a** and is due to a single seemingly minor change in the biological processes under consideration. While **Fig. 1a** examines performance in all biological functions equivalently, **Fig. 1b** excludes ribosomal functions (while still including nearly 100 other biological processes). Coexpression data performs so well in this single area that it dominates the global evaluation in **Fig. 1a**, making microarrays appear vastly more powerful than any other type of experiment. **Figure 1b** shows that if one is investigating any biological

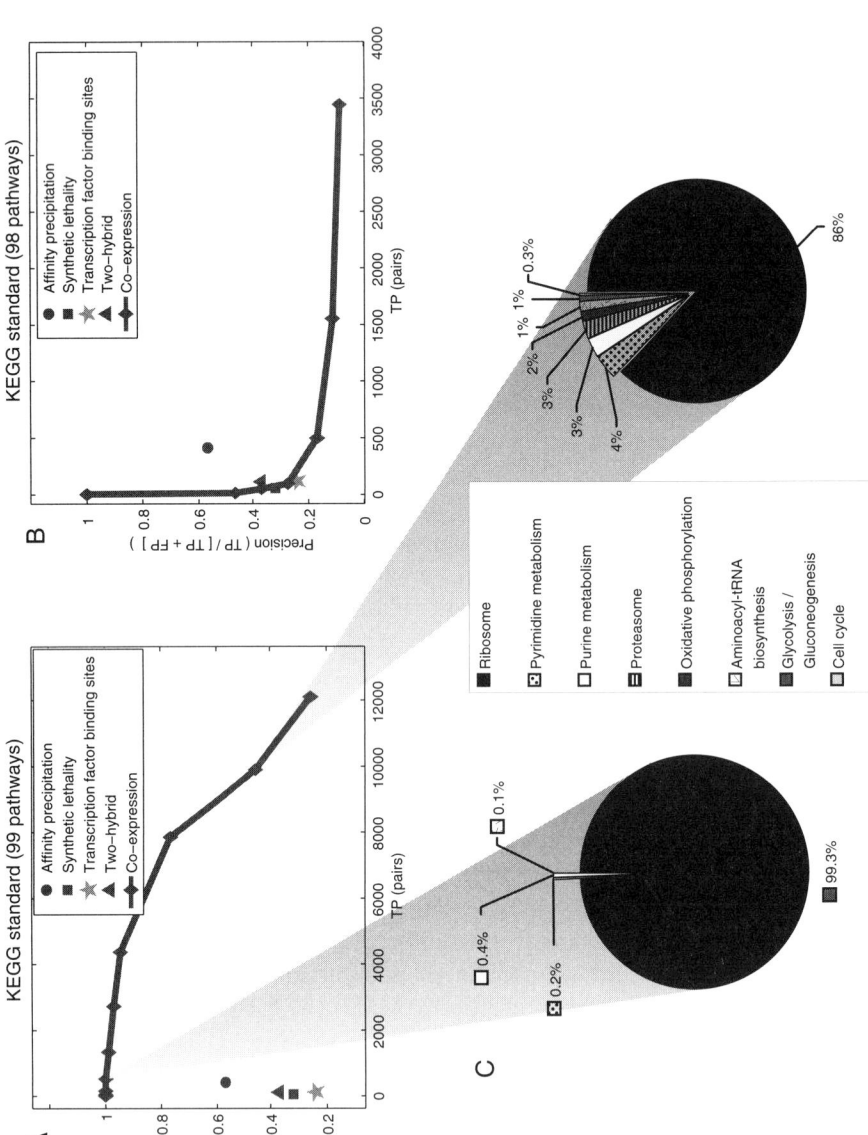

Fig. 1. Inconsistencies in evaluation due to process-specific variation in data. (a and b) Comparative functional evaluation of several high-throughput datasets based on a KEGG-derived gold standard. The evaluation pictured in (b) is identical to that in (a) except that 1 of 99 KEGG pathways was excluded from the analysis ("Ribosome," sce03010). Gold standard positive pairs were obtained by considering all protein pairs sharing a KEGG pathway annotation to be related; pairs of proteins occurring in at least one KEGG pathway but with no coannotation were considered to be unrelated. Performance is measured as the trade-off between precision and true positive pairs (a measure of recall). For the evaluation in (b), both precision and sensitivity drop dramatically for coexpression data. (c) Composition of correctly predicted functional relationships at two different precision levels. Of the 0.1% most coexpressed pairs, 99.3% of the true positive pairs (842 of 848) are due to coannotation to the ribosome pathway (left pie chart). This bias is less pronounced (but still present) at lower precision. Of the 1% most coexpressed pairs, 86% of the true positive pairs (8,500 of 9,900) are due to coannotation to the ribosome pathway (right pie chart). Reproduced from ref. 18.

process *other* than ribosomal function, coexpression data is neither more nor less informative than most other data types. When considering the results of any experiment or analysis, whether to evaluate accuracy or to generate biological predictions, it is critical to consider the biological functions for which the results are intended to be applicable.

1.2. Evaluation of High-Throughput Datasets: Standards

Performance evaluations of high-throughput experimental results and computational methods are generally done using one of a number of techniques drawn from signal detection theory and information retrieval, most notably precision–recall (PR) curves *(20)* and receiver operating characteristic (ROC) curves *(21)*. Both these curves place a measure of accuracy on one axis (precision in the former, specificity in the latter) and a measure of coverage on the other (recall in both cases, also called sensitivity). Precision is a measure of true positive rate, specificity a measure of true negative retrieval, and sensitivity a measure of true positive retrieval; for more information, *see* **ref**. *22*. These curves quantify the throughput versus accuracy trade-off mentioned previously: generally, data will describe either a few genes very accurately, many genes without much accuracy, or somewhere in between. An ideal dataset would lie in the upper corner of such a graph, describing every gene perfectly.

In order to generate such a curve, one needs a reference indicating the "correct" answer for each data point, be it known functions for individual genes or known gene pair interactions. Such a reference is generally referred to as a "gold standard" or "answer set." Several gold standards and approaches to evaluation of functional genomic data have been employed in the area of biological function prediction. Genomic data are often in the form of associations of genes or gene products (e.g., protein–protein physical interactions, synthetic lethality interactions, mRNA coexpression), and thus many of the proposed standards focus on functional relationships between pairs of genes or proteins. Such standards are often derived from curated databases such as the Munich Information Center for Protein Sequences (MIPS) *(23)*, Saccharomyces Genome Database (SGD) *(24)*, Kyoto Encyclopedia of Genes and Genomes (KEGG) *(25)*, and the Gene Ontology (GO) *(26)*, all of which define vocabularies of biological functions/processes and annotate proteins to specific terms in such vocabularies based on biological literature.

On the basis of these annotations, one can define a gold standard, or sets of gene pairs marked as either "positive" or "negative" examples, i.e., pairs of genes or proteins known to share or not share a common function. While the common function catalogs listed above share many invaluable features, such as controlled vocabularies and generality across organisms, special care must be taken in producing gold standards from these resources. There is no one "right" way to simplify a complex, hierarchical functional catalog into sets of related and unrelated gene pairs. Furthermore, given such a gold standard, the manner in which it is employed in a performance evaluation is critical to an accurate understanding of the evaluation's results. This becomes of immediate import in the laboratory when evaluating, for example, how computational approaches will perform when used to predict gene function or to drive experimental efforts.

There are a number of important issues in existing standards and evaluation approaches: (1) significant differences exist between standards, making published evaluations of data or methods incomparable; (2) process-specific performance (as discussed above) is often not considered in such evaluations, making the reported results potentially misleading; (3) negative examples – that is, gene pairs that are definitely unrelated – can be difficult to find consistently; (4) the relative proportions of positive and negative examples in such standards can fail to mirror biological reality (e.g., it is improbable that 50% of the gene pairs in a cell interact and 50% do not). In short, many published evaluations of functional genomic datasets and computational methods are at best nonstandard, making reported results incomparable between publications, and at worst biased enough to be completely biologically misleading; this chapter discusses guidelines for avoiding these pitfalls and performing accurate evaluations.

1.3. Function Prediction Versus Functional Relationship Prediction

There are at least two broad goals that can be the desired result of a functional genomic analysis: function prediction, the task of assigning one or more functions to individual genes, and functional relationship prediction, the task of determining whether pairs of genes participate in the same biological processes. Functional relationships are thus similar to protein–protein interactions, but more general in that a functional relationship may entail direct binding, participation in the same or related pathways, transcriptional control, or any shared cellular responsibilities.

Evaluation techniques are basically the same for these two tasks, in that both require gold standards and can be visualized using PR or ROC curves. However, results from function prediction are by no means directly comparable to results from functional relationship prediction. In the case of function prediction, evaluations are generally performed on a per-function basis; the gold standard is a list of positive genes known to participate in the function and negative genes known not to. In the case of functional relationship prediction, gold standards consist of positive pairs of genes known to be related and negative pairs known to be unrelated, and evaluations can be either global (gene pairs related in at least one of many functions) or process specific. Most of the work in this chapter refers to functional relationship prediction, but the techniques and pitfalls apply equally to function prediction, and we provide a few examples of the latter (e.g., **Fig. 6**).

2. Methods

2.1. GRIFn: A System for Evaluating Genomic Methods and Data

The GRIFn system (*(18)*, http://function.princeton.edu/grifn) represents one example of an end-to-end data and method evaluation framework designed to abrogate these difficulties. Specifically, GRIFn employs three main features to avoid the pitfalls of biological evaluation described above. First, information regarding functional coverage is always provided alongside global evaluations, allowing an investigator to see whether a dataset is performing well in many biological contexts or in just a few (**Fig. 2**). Second, an evaluation mode exists to exclusively detail any dataset's coverage of any biological area, providing a high-level view of which experimental or computational methods might best elucidate any particular process (**Fig. 3**). Third, the gold standard GRIFn used in these evaluations is derived from the Gene Ontology in such a way as to specifically avoid inconsistencies in coverage and to provide information that is maximally relevant in a laboratory setting.

2.1.1. Context-Aware Global Evaluations

GRIFn's primary purpose is to provide global evaluations of experimental datasets and computational predictions as they pertain to gene pair functional relationships (**Fig. 2**). In addition to the usual estimation of precision–recall characteristics, it computes the distribution of biological processes represented in the set of correctly classified positives (true positives) at any point along the precision–recall curve. This distribution allows the user to identify and measure any biases toward a specific biological process in the set of positive results and to interpret evaluation results accordingly. This information is summarized and presented in a dynamic and interactive visualization framework that facilitates quick but complete understanding of the biological information present in the data.

Figure 2 illustrates an example of a genome-wide evaluation of several different high-throughput datasets using this framework. These datasets comprise five protein–protein interaction datasets, including yeast two-hybrid *(7, 24, 25)* and affinity precipitation data *(5, 26)*, and two gene expression microarray studies *(27, 28)*. At first glance, this general evaluation indicates that the Gavin et al. data is perhaps the most precise and sensitive of these studies, but the Gasch et al.microarray data is a close second (**Fig. 2a**). The Gasch et al. data appears to offer more reliable information than four of the five protein–protein interaction datasets.

280 Huttenhower et al.

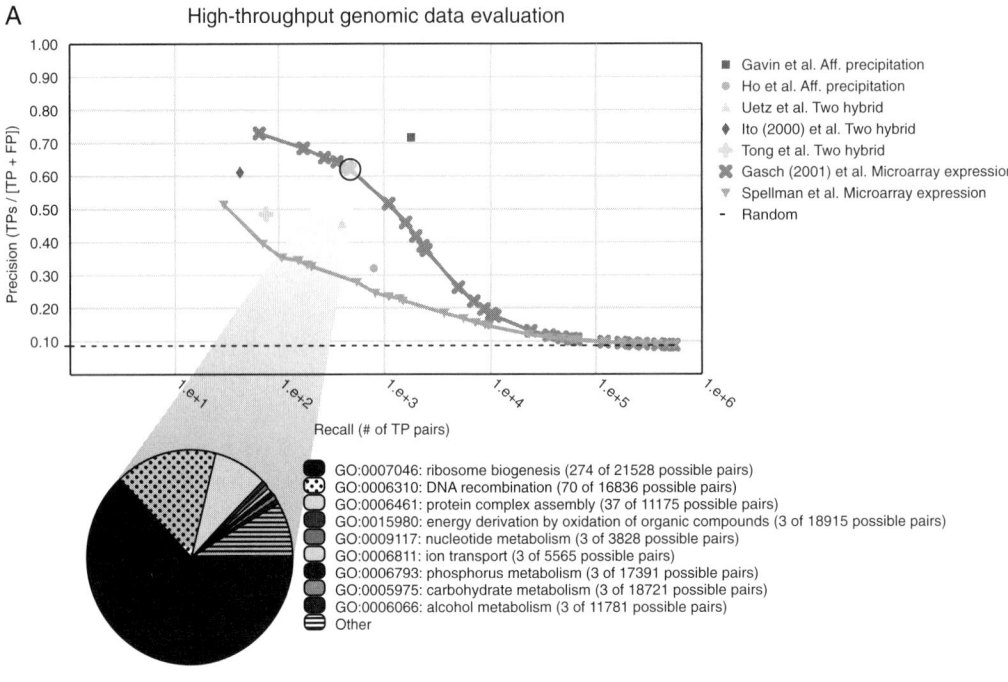

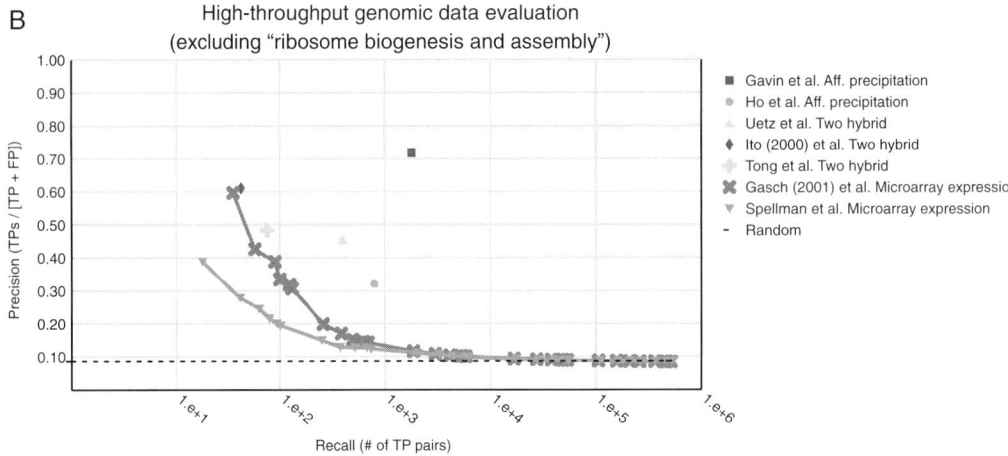

Fig. 2. Global functional evaluation. (**a**) A genome-wide evaluation of several high-throughput datasets, including five protein–protein interaction datasets (yeast two-hybrid *(16,34,35)* and affinity precipitation data *(14,36)*) and two gene expression microarray studies *(37,38)*. This analysis reveals that a large fraction of the true positive predictions made by the coexpression datasets are associations of proteins involved in ribosome biogenesis. (**b**) The same form of evaluation as in (**a**), but with a single biological process (*ribosome biogenesis and assembly*) excluded from the analysis. With this term excluded, the evaluation shows that neither of the coexpression datasets is as generally reliable as the physical binding datasets. Reproduced from ref. *18*.

However, an analysis of the processes represented in the set of correctly classified pairs reveals that this may not be an accurate generalization. In fact, approximately 60% of the true relationships predicted by the coexpression data are related to the process of ribosome formation (**Fig. 2a**, bottom). This type of analysis is included for any evaluation done with

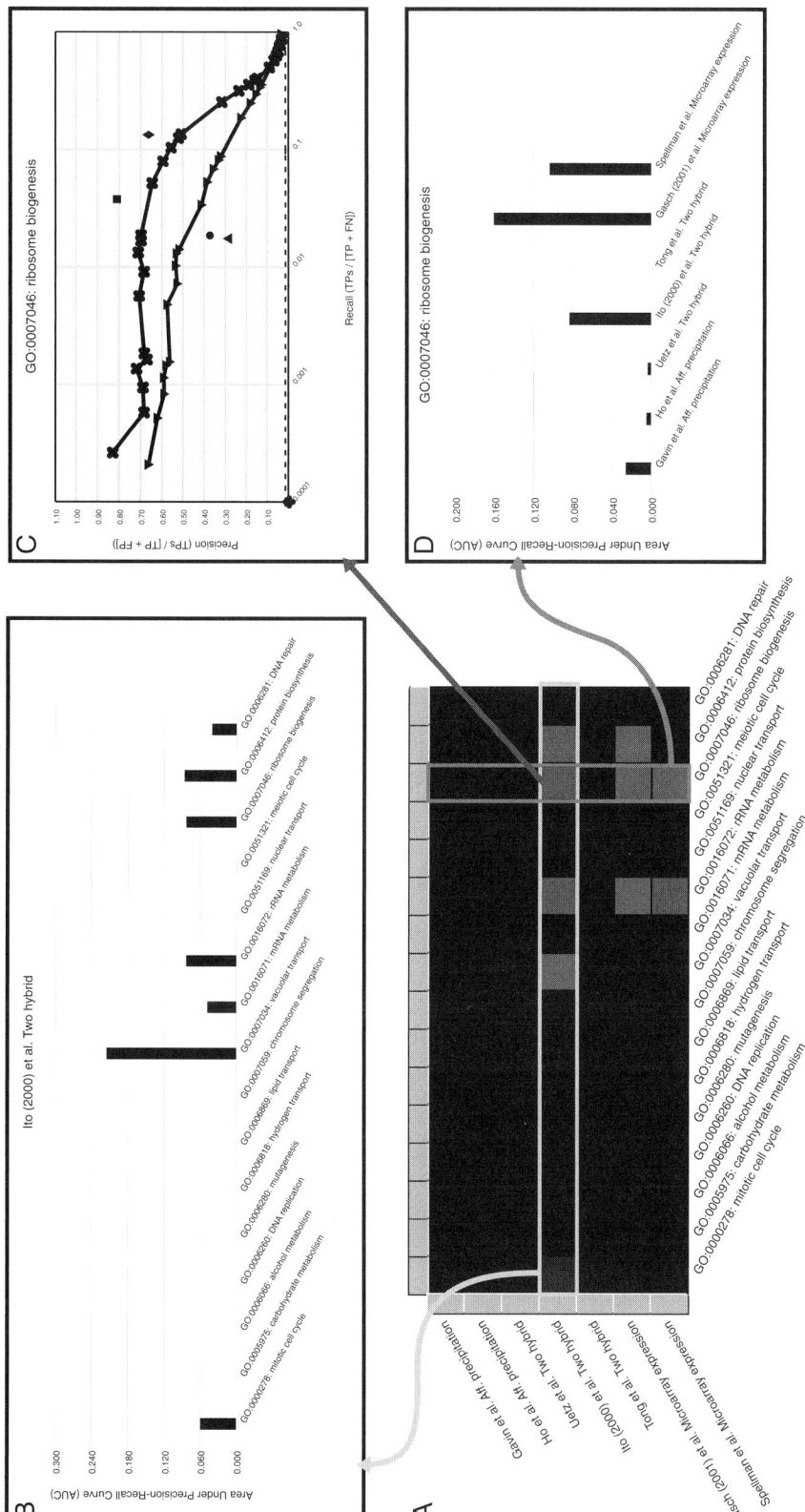

Fig. 3. Process-specific evaluation. (**a**) Example of an evaluation of seven high-throughput datasets over a set of 16 biological processes. The precision–recall characteristics of each dataset/process combination were computed independently and the intensity of the corresponding square in the matrix is scaled according to the area under the precision–recall curve (AUPRC). (**b**) Detailed comparison of results for a single dataset. The AUPRC statistic of a particular dataset (e.g., Ito et al. two-hybrid) for each process is plotted to allow for comparison across a single dataset. (**c**) The PR curve from which the AUPRC was computed. (**d**) The AUPRC values for a single biological process across all datasets can also be obtained from an evaluation result. This allows for a direct measure of which datasets are the most informative for a process of interest.

GRIFn, and interactive visualization allows for dynamic exploration of any biases that might be present.

Once such biases are identified within an evaluation, they can be temporarily hidden in order to provide a more normalized view of the functional content of a dataset. In GRIFn, a user can choose to exclude all examples related to specific biological processes from an evaluation, producing a precision–recall plot that is "neutral" with respect to the selected processes. **Figure 2b** illustrates an example of this functionality for the evaluation discussed above. On the basis of the apparent ribosomal bias, the same datasets were re-evaluated excluding all proteins involved in *ribosome biogenesis and assembly* (GO:0042254). While none of the interaction datasets changes significantly with this function excluded, both gene expression datasets show substantial decay in their precision–recall characteristics, suggesting that they are generally less reliable at predicting functional relationships over a broad range of processes (but probably quite accurate when predicting ribosomal function). This conclusion is quite different from what one might have derived from a traditional performance evaluation insensitive to functional diversity.

2.1.2. Process-Specific Evaluations

When using computational methods to direct experiments, or when assessing experimental results, scientists typically have a domain of interest in mind. In such a situation, a focused, process-specific evaluation is often more appropriate than a genome-wide evaluation. In an exploratory setting, this can more effectively identify a set of methods or data to generate experimental hypotheses relevant to the area of interest; in an analysis setting, this can provide a more specific indication of experimental performance in the appropriate domain. Even when examining high-throughput data or methods that might cover a wide variety of functional areas, a researcher with specific experimental goals may gain more focused insights from a performance evaluation dealing only with the biological process of interest.

GRIFn facilitates process-specific evaluations by providing a way to address the question, "How effective is the dataset or computational result X in describing genes in the biological process Y?" One way to answer this question is to perform independent precision–recall analyzes for each process of interest. Since this leads to a separate performance curve for each dataset/process pair, these context-sensitive results are summarized in an interactive heat map (**Fig. 3**).

To allow rapid visualization and comparison of many datasets' accuracies within many biological processes, the heat map in **Fig. 3** summarizes each performance curve with a single statistic, the area under the precision–recall curve (AUPRC). This scale normalizes each result into the 0–1 range (0, in black, corresponding to low performance and 1, in red, corresponding to optimal performance) and guarantees equitable inter-dataset and inter-process comparisons. The results for any one dataset (row) can be displayed as a bar chart containing more explicit numerical results, as can the results for any one biological process (column). In addition to this summarization, each cell in the matrix can be expanded into its fully detailed precision–recall curve. This combination of dynamic visualization, high-level summarization, low-level detail, and (most importantly) context sensitivity allows an experimenter to quickly determine the biological characteristics of datasets or, conversely, which datasets are appropriate for investigating a biological area of interest.

2.1.3. A Consistent Gold Standard

There has been little consensus in the literature regarding the derivation of gold standards from functional catalogs such as GO, MIPS, and KEGG *(18)*. The process of reducing a rich, hierarchical structure such as the GO to sets of related and unrelated gene pairs must of necessity discard information, and if this is done carelessly, the result can be a gold standard that is disconnected from biological reality. Perhaps even more troubling is the fact that standards derived in similar manners from different functional catalogs can produce surprisingly dissimilar evaluation results (**Fig. 4**). These issues imply that it is essential to consider not only the potential functional biases of an evaluation, but also the source and characteristics of the evaluation's standard.

Experimental validation is perhaps closest to a ground truth in biology, and so when deriving an evaluation standard from functional annotations, one should do so in a way that is maximally experimentally informative. Particularly for *S. cerevisiae*, one can also argue that the GO represents the most comprehensive functional catalog, since the GO structure is assiduously maintained in a consistent manner and yeast protein annotations are actively curated by SGD *(24)*. Combining these two observations, one way in which a consistent and biologically relevant evaluation standard can be produced is by selecting an experimentally relevant subset of GO based on expert knowledge.

When deriving positive examples (functionally related gene pairs) from GO, it is fairly intuitive to consider genes coannotated to the same biological process to be functionally related. However, because of GO's hierarchical structure, genes may be coannotated to a low, specific term (e.g., *establishment of mitotic spindle orientation*) or to a high, general term (e.g., *cellular process*). Selecting positive pairs from overly general coannotations will render them biologically uninformative, and selecting them from overly specific coannotations will discard large amounts of relevant data. Positive pairs for an evaluation standard should thus be selected only from coannotations at an appropriate level of specificity within GO.

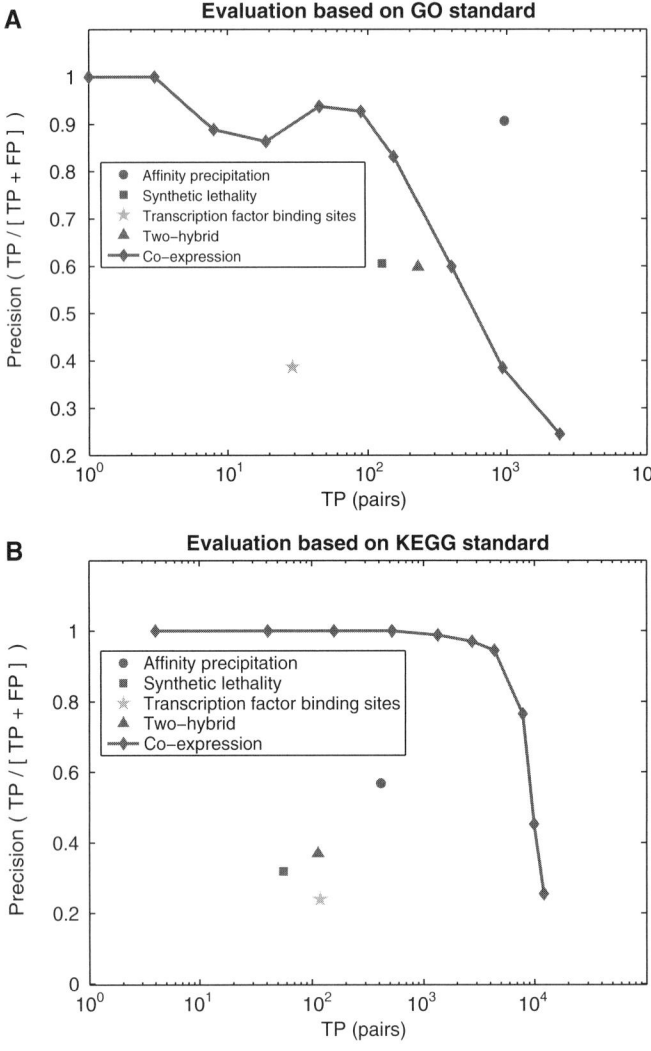

Fig. 4. Comparison of functional genomic data evaluation on gold standards derived from different functional catalogs. (a) Comparative functional evaluation of several high-throughput evidence types based on a typical Gene Ontology (GO) gold standard. Positive pairs were obtained by finding all protein pairs with coannotations to terms at depth eight or lower in the biological process ontology. Negative pairs were generated from protein pairs whose most specific coannotation occurred in terms with more than 1,000 total annotations. (b) Evaluation of the same data against a KEGG-based gold standard. Gold standard positives were obtained by considering all protein pairs sharing a KEGG pathway annotation to be related, while gold standard negatives were drawn from pairs of proteins occurring in at least one KEGG pathway but with no coannotation. There are clear inconsistencies between the two evaluations. In addition to vastly different estimates of the reliability of coexpression data, other evidence types change relative positions. For instance, transcription factor binding site predictions appear competitive with both two-hybrid and synthetic lethality in the KEGG evaluation but are substantially outperformed in the GO evaluation.

Unfortunately, GO terms contain no inherent indication of their biological specificity. Two common solutions are to use either term size (number of annotated genes) or hierarchical depth as proxies for specificity. The former is inappropriately sensitive to how well studied an area is; *multi-organism process*, for example, is quite general but has only 142 annotations in yeast. The latter can be misleading in that, while child terms in GO are always more specific than their parents, one term at a particular level is not necessarily as specific as another. *Regulation of cellular metabolic process*, for instance, lies three levels deep in the GO hierarchy with 775 gene annotations; *intracellular sequestering of iron ion*, a much more specific process, lies at the same depth with only three annotations. Neither the number of annotations nor term depth is a consistent indicator of the biological specificity of a GO term.

To solve this problem, the standard employed by GRIFn was derived by polling a panel of yeast domain experts to determine which terms would be informative in an experimental context. This curation was performed by having each expert indicate for all GO terms, without information about their annotations or hierarchical relationships, whether or not an annotation to that term would aid in designing experiments to probe a gene's function. Terms receiving votes from half the panel or less were deemed insufficiently informative for a gold standard, and gene pairs coannotated to the remaining terms formed the positive examples in the GRIFn standard.

Since most gene pairs in the genome are unlikely to be functionally related, randomly selected pairs can form a reasonable negative example set. The GRIFn standard improves upon this by considering only pairs of genes with known but dissimilar functions to be unrelated. Specifically, a pair of genes is a negative in the GRIFn standard if both genes are annotated to some term deemed experimentally informative but not coannotated to any term containing fewer than 1,000 genes. The resulting negative set is thus more accurate than random pairs of proteins but is still large enough to accurately reflect reasonable biological proportions of related and unrelated gene products.

The resulting gold standard is quite different from previous GO-based standards using term size or depth as a measure of biological specificity. Because this gold standard is based on direct re-evaluation of the GO with respect to functional genomics, there are a number of nonspecific but hierarchically deep GO terms excluded on the basis of the voting results. Conversely, a number of relevant GO terms are included that appear near the ontology root. A similar trend is true regarding the GO term sizes of the selected and excluded sets: Many GO terms excluded on the basis of expert voting have relatively few annotations. Thus, it is clear that neither size nor depth in the ontology serve as good measures of biological specificity. Basing the criteria for generating an ontology-based gold standard on expert knowledge ensures that the standard is consistent in terms of the functional relevance of the relationships it captures and can therefore provide a meaningful basis for evaluation and experimental direction. This expert-curated set of GO terms is available for download from http://function.princeton.edu/grifn.

2.2. The Benefits of Breadth: Integrating Diverse Data

Once one is aware of the functional diversity of high-throughput data and the potential pitfalls in comparative evaluations, it is possible to leverage datasets' varying biological content in order to provide a more unified, systems-level view of an organism. When performing a computational analysis or generating function predictions, the benefits of integration are clear: The more data you have, the more you can learn, and the more sensitive you are to the functional specificity of your input, the more precise your answers can be. When analyzing experimental results, consideration of new data in the context of existing results can provide important information about functional content and amplify biological signals too weak to be detected in individual datasets.

Just as comparative evaluation between datasets can be difficult (as discussed above), combining information from multiple datasets can be equally complex. Different data types (e.g., synthetic lethality versus coexpression) can have vastly different biological meanings, different experimental platforms can have different noise characteristics, and even nominally similar datasets can have surprisingly different functional biases *(18)*. The more optimistic corollary to this is that successful integration can provide substantial benefits: By selecting only the "best" parts of each dataset, functional predictions based on many datasets can be much more accurate than those from any one dataset alone (**Fig. 5**).

A wide variety of meta-analytic *(27–29)*, probabilistic *(19, 30, 31)*, and machine-learning techniques *(32–34)* have been

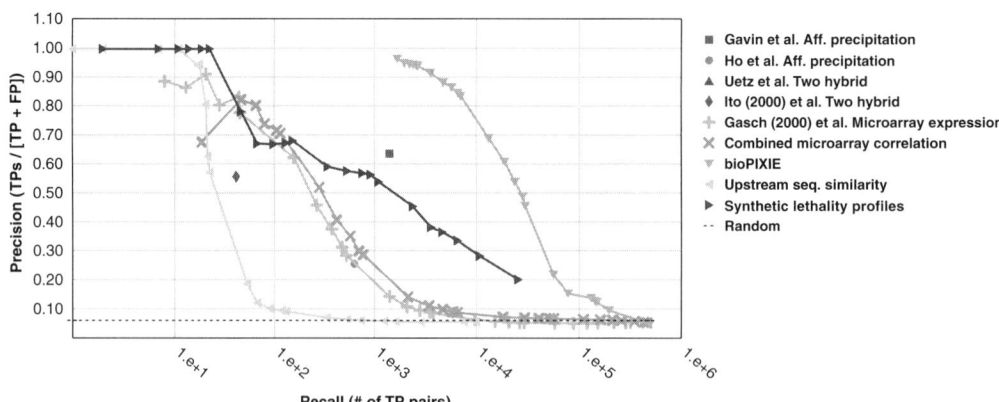

Fig. 5. Integration of diverse data sources improves functional analysis. The predicted probabilities of functional relationship produced by bioPIXIE outperform those from any one input dataset in both precision and recall characteristics. By taking advantage of the process specificity of individual datasets, computational methods can predict gene functions and functional relationships with greater overall accuracy.

employed to solve this problem, both in computational biology and in other fields. Three systems that have demonstrated significant benefits specifically in the area of predicting gene function and functional relationships are bioPIXIE *(35, 36)*, MEFIT *(18)*, and SPELL *(37)*. bioPIXIE predicts pairwise functional relationships from heterogeneous data sources using a Bayesian network. MEFIT and SPELL both predict functional relationships using a large collection of microarray data (~2,500 conditions), but while MEFIT uses Bayesian classification, SPELL uses a query-driven system combined with singular value decomposition (SVD, *(38)*) and dataset weighting. All three systems have been evaluated using GRIFn's GO-derived gold standard, which bioPIXIE and MEFIT also use during training, and all three have been shown to produce more accurate predictions than can be generated from their individual inputs. Most importantly, though, these three systems produce predictions that are equally accurate yet explore different areas of biology – that is, they represent computational methods with different functional specificities.

bioPIXIE addresses one important aspect of diverse data integration: How can genomic datasets from very different experimental systems be made to look "similar" enough to allow unified function prediction while still taking advantage of their informative differences? This problem is solved by mapping each dataset into pairwise gene scores in a manner specific to the individual datasets. Direct binding data from yeast two-hybrid or coimmunoprecipitation assays, for example, may produce simply "yes" or "no" scores for each gene pair, whereas synthetic genetic effects might be scored continuously by the degree to which an interaction is aggravating or alleviating. These per-dataset scores are then integrated in a Bayesian framework, allowing bioPIXIE to answer questions such as, "If dataset X gives a gene pair some score Y, how likely are the genes to be functionally related," and, "Given two genes' scores in many datasets, how likely are the genes to be functionally related overall?"

MEFIT extends this Bayesian framework to include microarray data, which introduces a number of new challenges. Microarray data are not inherently pairwise, and the choices of similarity function (Pearson correlation, Euclidean distance, etc.), intra-dataset normalization, and inter-dataset normalization can all greatly influence predictive accuracy. Additionally, coexpression data tends to be noisier yet higher throughput than even other genome-scale techniques; a single microarray will generally produce a measurement for each gene in the genome, and there are thousands of microarray conditions already publicly available for yeast. This implies that microarray data can best be analyzed by looking for functional relationships that might be weak but appear consistently in many datasets. Thus, MEFIT can answer questions such as, "If two genes are correlated in many coexpression datasets, how likely are they to be functionally related?"

The SPELL system is a query-driven framework that also analyzes microarray conditions. When provided with a set of query genes, SPELL processes each microarray dataset using SVD to normalize the presentation of information across datasets. Each dataset is then weighted on the basis of the SVD-normalized correlation of the query genes within that dataset; that is, datasets in which the query genes are coexpressed are upweighted, and datasets in which the query is uncorrelated are downweighted. Finally, other genes that correlate well with the query in the upweighted datasets (and are thus likely to be functionally related) are presented in a ranked manner. This means that SPELL can answer questions like, "Given several query genes, what other genes are likely to be functionally related to them based on relevant microarrays?"

In global evaluations, these three computational methods predict gene function with similar accuracies. However, in process-specific evaluations, substantial differences are observed in the biological areas predicted well by the three methods (**Fig. 6**). These differences are even more striking given the ostensible similarities between the three techniques; bioPIXIE and MEFIT, for example, both rely on probabilistic Bayesian frameworks, and MEFIT and SPELL both examine large collections of microarray data. Even with these similarities, though, different integration and analysis techniques can produce different biologically relevant conclusions as a result of the functional variety of high-throughput data.

2.3. The Benefits of Depth: Context Specificity

The benefits that process specificity brings to functional evaluation raise the question of how such process specificity can be also be applied to data integration and computational methods prior to evaluation. In other words, if you want to learn about a dataset or to predict functional relationships, how can an awareness of varying biological processes be beneficial? The three methods described above, in addition to evidencing varied functional specificity during evaluation, also all incorporate process specificity into their analyzes in order to deliver more accurate biological predictions.

Both bioPIXIE and MEFIT rely on probabilistic frameworks that learn which datasets to "trust" when predicting functional relationships. These frameworks can thus incorporate process specificity by learning different trusts (i.e., probabilities) not only for each dataset, but also for each biological context. For example, if a context-specific evaluation shows that a two-hybrid dataset is particularly good at predicting genes functioning together in *vacuolar transport*, it makes sense to upweight its predictive contribution in that term and downweight it in others. Incorporating contextual information in this manner nearly always improves the accuracy of predicted functional relationships (**Fig. 7**).

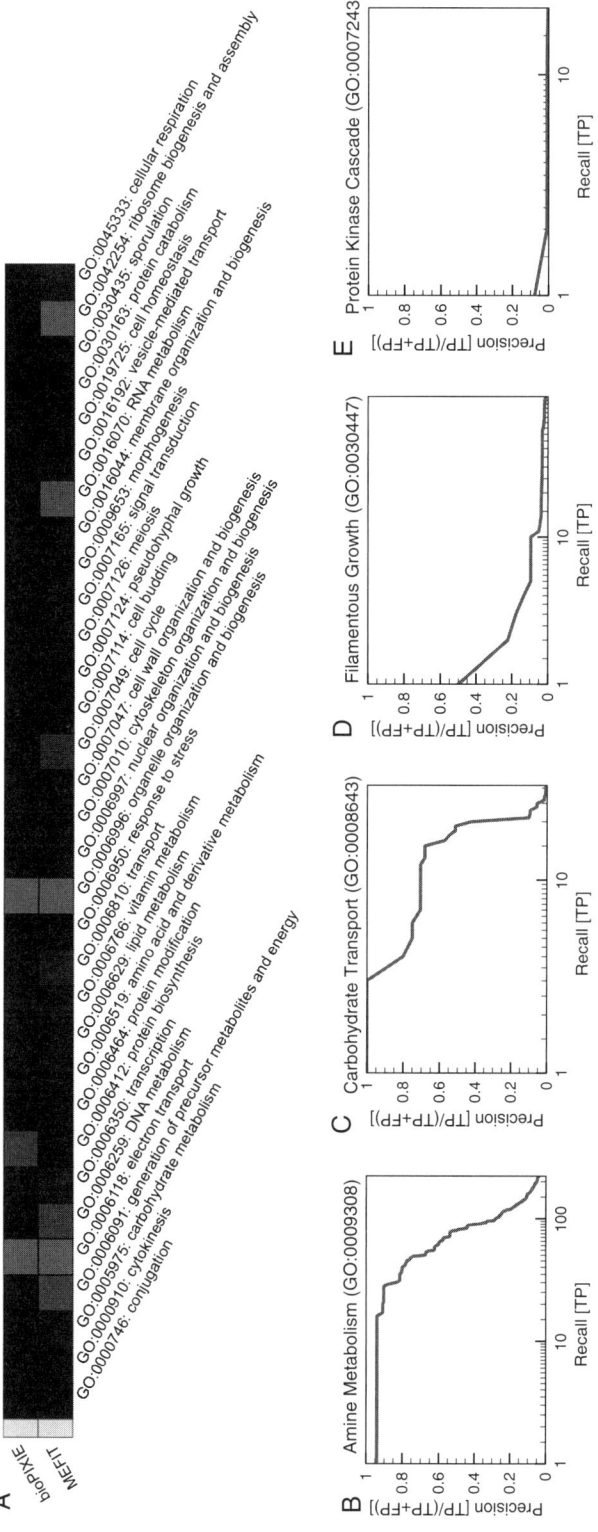

Fig. 6. Variation in process-specific performance between computational methods. (**a**) A selection of biological processes in which bioPIXIE and MEFIT demonstrate different accuracies in predicting functional relationships because of their different input data. Note that, while overall performance is similar, one method can substantially outperform the other in any one biological context. (**b–e**) SPELL performance when predicting gene function in a selection of GO terms. (**b** and **c**) Large and small terms that are both predicted well. (**d**) A term that performs poorly due to lack of appropriate data (none of the SPELL compendium's datasets study filamentous growth). (**e**) A term not captured well by coexpression data (since kinase cascades operate post-transcriptionally).

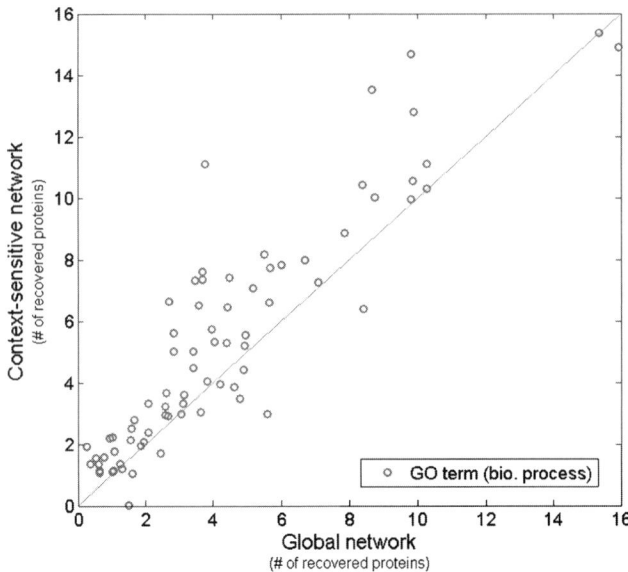

Fig. 7. Context specificity during computational analysis improves prediction accuracy. A comparison of the accuracy of process-specific and global approaches to function prediction in which the function-specific approach improves recovery by more than 2 standard deviations for 51% of the terms evaluated and only causes deterioration by this amount on 8% of the terms. Many cases where performance deteriorated are indicative of small biological processes with insufficient data available for such specific learning. In the majority of cases, incorporating process-specific knowledge into the analysis process results in greater overall accuracy. Reproduced from ref. *36*.

In addition to improving overall performance, this probabilistic weighting conveys additional information about the functional content of each dataset. This is particularly evident in MEFIT, since the number of microarray datasets under analysis is so large and each represents a continuous distribution of pairwise correlation scores. Function-specific learning allows one to examine the specific distribution of correlations in each dataset, which can vary surprisingly by biological context. **Figure 8** demonstrates this effect for the Primig et al. dataset *(17)*, in which genes related in the process of meiosis and sporulation are much more highly correlated than expected, but genes related in *carbohydrate metabolism* are not informatively correlated.

SPELL provides an even finer level of contextual control by providing all analysis in response to a specific user query. This query can contain one or many genes, and it becomes the biological context for all subsequent data weightings and functional predictions. Conversely, in bioPIXIE and MEFIT, contexts are limited to GO terms selected as described above in the GRIFn gold standard. By combining this fine-grained contextual control with iterative query refinement, SPELL allows a biologist

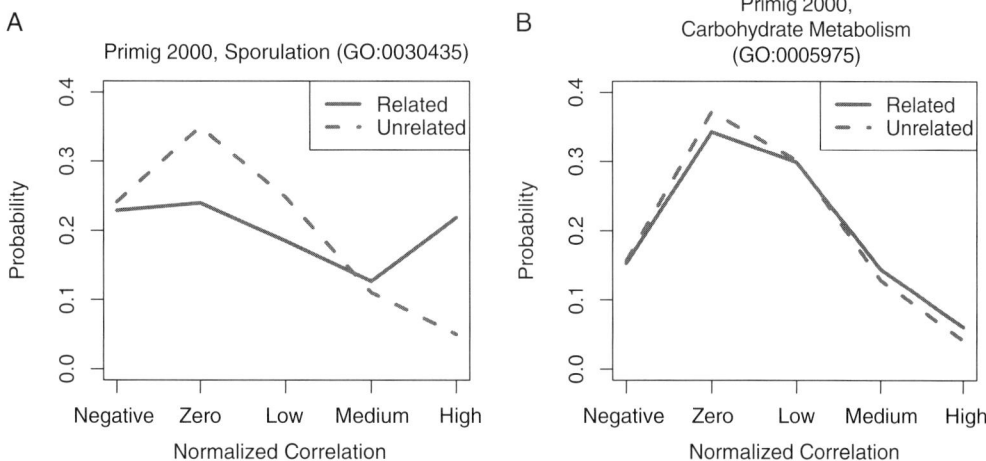

Fig. 8. Distributions of coexpression values within one dataset for genes in different biological processes. Another way of viewing the process specificity of a dataset is to consider the distribution of data for genes functionally related or unrelated within different biological contexts. (**a**) The distribution of correlations between gene pairs functionally related or unrelated within the process of *sporulation*. (**b**) The distribution of correlations for gene pairs within *carbohydrate metabolism*. Note that high correlation within the Primig et al. dataset *(17)* is quite predictive of sporulation relationships, but it is only very slightly informative of relationships in carbohydrate metabolism.

to obtain functional predictions that are both tailored to and evaluated in his or her specific area of interest.

Context specificity provides additional information to biologists when used to more thoroughly describe functional evaluations, and it provides additional power to computational methods when predicting gene function and functional relationships. An interesting added benefit of this integration is that it allows one to detect biological processes not described well by any current datasets or prediction techniques. This can imply that there is simply insufficient data to detail the process (e.g., functions based on post-transcriptional regulation will rarely be detectable in coexpression data) or that the process is not sufficiently cohesive to be captured well. By examining function-specific information in this manner, an experimenter can determine which biological areas are described well by current data and which areas remain to be explored.

2.4. Conclusions

Proteomic data, in the form of either experimental results or computational analyzes, must be evaluated in the context of specific biological processes in order to convey accurate biological information. High-throughput experimental results can provide a wealth of predictive data, and any one dataset may be useful in a few detailed biological contexts or in many broader contexts. When determining the characteristics of a dataset or computational method, any evaluation is sensitive to the gold standard being employed, which should be selected in such a way as to most accurately reflect biological reality. An awareness of both of these effects – functional variety and gold standard selection – can greatly simplify the process of evaluating diverse genome-scale data.

If a method or evaluator is aware of this functional diversity, it can be taken advantage of to provide richer results. In a laboratory context, this might mean choosing an experimental platform known to describe some pathway of interest particularly well. In a computational setting, this variety can provide additional information when learning to predict new biological relationships. Particularly when combining these and using computational predictions to guide laboratory experiments (hopefully the best of both worlds), performing analyzes in the context of specific biological processes can help to realize the full potential of high-throughput genomics.

References

1. Kitano H. (2002). Looking beyond the details: a rise in system-oriented approaches in genetics and molecular biology. Curr Genet;41(1):1–10.
2. Steinmetz LM, Deutschbauer AM. (2002). Gene function on a genomic scale. J Chromatogr B Analyt Technol Biomed Life Sci;782(1–2):151–63.
3. Ideker T, Galitski T, Hood L. (2001). A new approach to decoding life: systems biology. Annu Rev Genomics Hum Genet;2:343–72.
4. Cahill DJ, Nordhoff E. (2003). Protein arrays and their role in proteomics. Adv Biochem Eng Biotechnol;83:177–87.
5. Sydor JR, Nock S. (2003). Protein expression profiling arrays: tools for the multiplexed high-throughput analysis of proteins. Proteome Sci;1(1):3.
6. Oleinikov AV, Gray MD, Zhao J, Montgomery DD, Ghindilis AL, Dill K. (2003). Self-assembling protein arrays using electronic semiconductor microchips and in vitro translation. J Proteome Res;2(3):313–9.
7. Huang RP. (2003). Protein arrays, an excellent tool in biomedical research. Front Biosci;8:d559–76.
8. Cutler P. (2003) Protein arrays: the current state-of-the-art. Proteomics;3(1):3–18.
9. Bartel PL, Fields S. (1995). Analyzing protein-protein interactions using two-hybrid system. Methods Enzymol;254:241–63.
10. Grunenfelder B, Winzeler EA. (2002). Treasures and traps in genome-wide data sets: case examples from yeast. Nat Rev Genet;3(9):653–61.
11. Chen Y, Xu D. (2003). Computational analyses of high-throughput protein-protein interaction data. Curr Protein Pept Sci;4(3):159–81.
12. Bader GD, Heilbut A, Andrews B, Tyers M, Hughes T, Boone C. (2003). Functional genomics and proteomics: charting a multidimensional map of the yeast cell. Trends Cell Biol;13(7):344–56.
13. von Mering C, Krause R, Snel B, et al. (2002). Comparative assessment of large-scale data sets of protein-protein interactions. Nature;417(6887):399–403.
14. Deane CM, Salwinski L, Xenarios I, Eisenberg D. (2002). Protein interactions: two methods for assessment of the reliability of high throughput observations. Mol Cell Proteomics;1(5):349–56.
15. Ito T, Chiba T, Ozawa R, Yoshida M, Hattori M, Sakaki Y. (2001). A comprehensive two-hybrid analysis to explore the yeast protein interactome. Proc Natl Acad Sci USA;98(8):4569–74.
16. Yue H, Eastman PS, Wang BB, et al. (2001). An evaluation of the performance of cDNA microarrays for detecting changes in global mRNA expression. Nucleic Acids Res;29(8):E41-1.

17. Primig M, Williams RM, Winzeler EA, et al. (2000). The core meiotic transcriptome in budding yeasts. Nat Genet;26(4):415–23.
18. Myers CL, Barrett DR, Hibbs MA, Huttenhower C, Troyanskaya OG. (2006). Finding function: evaluation methods for functional genomic data. BMC Genomics;7:187.
19. Lee I, Date SV, Adai AT, Marcotte EM. (2004). A probabilistic functional network of yeast genes. Science;306(5701):1555–8.
20. van Rijsbergen CJ. (1979). Information retrieval. London, Boston: Butterworth.
21. Egan JP. (1975). Signal detection theory and ROC-analysis. New York: Academic.
22. Davis J, Goadrich M. (2006). The relationship between precision-recall and ROC curves. 23rd international Conference on Machine Learning, 2006, Pittsburgh, PA: ACM. pp233–40.
23. Mewes HW, Frishman D, Guldener U, et al. (2002). MIPS: a database for genomes and protein sequences. Nucleic Acids Res;30(1):31–4.
24. Ball CA, Dolinski K, Dwight SS, et al. (2000). Integrating functional genomic information into the *Saccharomyces* genome database. Nucleic Acids Res;28(1):77–80.
25. Kanehisa M, Goto S. (2000). KEGG: kyoto encyclopedia of genes and genomes. Nucleic Acids Res;28(1):27–30.
26. Ashburner M, Ball CA, Blake JA, et al. (2000). Gene ontology: tool for the unification of biology. The Gene Ontology Consortium. Nat Genet;25(1):25–9.
27. Choi JK, Yu U, Kim S, Yoo OJ. (2003). Combining multiple microarray studies and modeling interstudy variation. Bioinformatics (Oxford, England);19(Suppl 1):i84–90.
28. Moreau Y, Aerts S, De Moor B, De Strooper B, Dabrowski M. (2003). Comparison and meta-analysis of microarray data: from the bench to the computer desk. Trends Genet;19(10):570–7.
29. Hu P, Greenwood CM, Beyene J. (2005). Integrative analysis of multiple gene expression profiles with quality-adjusted effect size models. BMC Bioinformatics;6:128.
30. Troyanskaya OG, Dolinski K, Owen AB, Altman RB, Botstein D. (2003). A Bayesian framework for combining heterogeneous data sources for gene function prediction (in *Saccharomyces cerevisiae*). Proc Natl Acad Sci USA;100(14):8348–53.
31. Jaimovich A, Elidan G, Margalit H, Friedman N. (2006). Towards an integrated protein-protein interaction network: a relational Markov network approach. J Comput Biol;13(2):145–64.
32. Deng M, Chen T, Sun F. (2004). An integrated probabilistic model for functional prediction of proteins. J Comput Biol;11(2–3):463–75.
33. Karaoz U, Murali TM, Letovsky S, et al. (2004). Whole-genome annotation by using evidence integration in functional-linkage networks. Proc Natl Acad Sci USA;101(9):2888–93.
34. Barutcuoglu Z, Schapire RE, Troyanskaya OG. (2006). Hierarchical multi-label prediction of gene function. Bioinformatics (Oxford, England);22(7):830–6.
35. Myers CL, Robson D, Wible A, et al. (2005). Discovery of biological networks from diverse functional genomic data. Genome Biol;6(13):R114.
36. Myers CL, Troyanskaya OG. (2007). Context-sensitive data integration and prediction of biological networks. Bioinformatics (Oxford, England);23(17):2322–30.
37. Hibbs MA, Hess DC, Myers CL, Huttenhower C, Li K, Troyanskaya OG. (2007). Exploring the functional landscape of gene expression: directed search of large microarray compendia. Bioinformatics (Oxford, England);23(20):2692–9.
38. Alter O, Brown PO, Botstein D. (2000). Singular value decomposition for genome-wide expression data processing and modeling. Proc Natl Acad Sci USA;97(18):10101–6.

INDEX

A

Aging... 101–104
AH109 236
Alpha-factor (α-factor) 147
Alzheimer's disease (AD) 164, 165
Aminoallyl labelling 27
3-Aminotriazole (3-AT). *See HIS3* reporter activity assay
Amyotrophic lateral sclerosis (ALS).......... 164, 165
Antisilencing drugs............................... 146, 150
    specificity of................................... 150–153
    splitomicin 146
Area under precision-recall curve 281–282
Array..10, 11, 15
    analysis (*see* Data analysis)
    hybridization................................... 10, 15
        air bubbles........................... 15
    scanning... 11
    washing.. 11
Array platforms
    fast slides 218
    path slides 218
A-synuclein 164
AUPRC. *See* Area under precision-recall curve
Axon GenePix scanner 24

B

Bait.. 248
    construction/generation of baits 250–254, 258–259
        iMYTH 253, 259
        MYTH 250, 258
    verification of bait.............254, 259–260, 270
Bait-dependency testing 255, 263
Barcode arrays, DNA, *See* Pooled analysis of yeast deletion strains using DNA barcodes
Barcodes, DNA, 115–116. *See also* Pooled analysis of yeast deletion strains using DNA barcodes
Beta-galactosidase activity assay (array format).........59, 63–66
    confirmation 59, 64–66
    considerations 70
    lacZ.................................56, 60, 63–66, 70
    media 59
    protocol... 62–63
    reagents 59

Beta-galactosidase filter lift assay 233, 241
Biological context 279, 288–290
BioMatrix 41–42, 44, 50
BioPIXIE 287–289
BioProspector 34
Bio-Rad Chipwriter Pro............................ 213, 218
Bioscreen 104, 106, 108–109, 112–113
    F1.b2 error... 112
    honeycomb plate 106–108
    maintenance... 112
BsrG1 209–221
Buffers
    blocking buffer................................ 214, 220
    elution buffer 213, 218, 219
    kinase buffer 214, 221
    lysis buffer........................ 150, 212, 216, 217
    PBS lysis buffer 213, 219
    wash buffer 212, 213, 217, 219

C

C2-8 .... 170, 171
Cell culturing... 24
    media 24
    strains 24
Cell cycle 55, 56
Cell cycle-regulated transcription................ 55, 56
Characterization of hits 151
    histone modifications........................... 152
    identifying molecular targets 151–152
    transcriptional profiling 152
    in vitro assays....................................... 152
Chemical genetics...........................145–147, 151
Chromatin 145–148
    structure 145
Chromatin modifications 176–184
Chronological life span..........................101–104, 106–107
Cloning 234
    gateway 234–235
    homologous recombination 234
Comparative genome hybridization (CGH) 2, 3
    array-CGH.. 2, 3
        mammalian 2, 3
        resolution 2, 3
        Saccharomyces cerevisiae 3
COMPASS 176, 182

Competent cells. *See* Yeast
Context-specificity. *See* Biological context
3C Protease 210, 213, 218, 221
C-terminal moiety of ubiquitin. *See* Cub
Cub 248
Cy dyes 23, 28
    fluor reversal 27
    labeling efficiency 29
    preparation 28
    quenching 29

D

Data analysis 12
    GeneChip operating software (GCOS) 12
        .CEL file 12, 15
        grid alignment 15, 16
    integrated genome browser (IGB) 13, 16
        aneuplodies 16
        .BAR file 13, 16
        log2 ratio 13, 16
    tiling analysis software (TAS), 12
        parameters 12, 13
        .TAG file 12
Data integration 273–274, 286
DH5α competent cells 211, 215
Dietary restriction 102
Differential gene expression 31–34
Differential interference contrast (DIC) 80, 85
Disease-associated genes 39
    Shwachmann–Diamond syndrome 50
DNA 8–12, 14, 15
    amplification (*see* Whole genome amplification)
    fragmentation 9, 15
    labeling 10
    precipitation 12
    preparation 8
        midi-prep 8
        mini-prep 9
    purification 12, 15
    quantification 11, 14
        absorbance 14
        fluorometry 11, 14
Drosophila 162, 164, 168, 169, 171
Drug function identification. *See* Pooled analysis of yeast deletion strains using DNA barcodes
Drug target identification. *See* Pooled analysis of yeast deletion strains using DNA barcodes

E

Easy Bioscreen Experiment (software) 108
EGCG 171
Epigenetics 145–146

F

False negatives. *See* Yeast two-hybrid system
False positives. *See* Yeast two-hybrid system
Flies. *See* Drosophila
FLIP. *See* Fluorescence loss in photobleaching
Fluorescence labeling
    DASPMI 79
    FM4-64 79
    MitoTracker 79
    Nile Red 79
Fluorescence loss in photobleaching 85
Fluorescence recovery after photobleaching 85
    microscope setup 86–87
FRAP. *See* Fluorescence recovery after photobleaching
Friedrich's ataxia 164, 165
Fruit flies. *See* Drosophila
Functional evaluation 276, 280
Functional interaction. *See* Functional relationship
Functional relationship 278
Funspec 31

G

GAL1/10 promoter 20, 24, 33
Gateway cloning sequences
    att B1 .... 210, 215
    att B4 .... 210, 215
    F5 210
    R3 210
Gateway vectors
    pBG1805 .... 210, 215
    pDONR221 .... 210, 215, 221
    pYES-DEST52 .... 210, 215, 219, 221
Gel mobility shift assay 32
Gene function identification. *See* Pooled analysis of yeast deletion strains using DNA barcodes
Gene ontology 277
Gene ontology functional enrichment 31, 33–34
Genetic suppression
    extra-genetic suppressors 38–39, 43, 49
Genomic instability 3
Global proteomics in saccharomyces cerevisiae (GPS) 175–186
GO. *See* Gene ontology
Gold standard 277
GRIFn 279–282

H

Haploinsufficiency profiling (HIP) assay. *See* Pooled analysis of yeast deletion strains using DNA barcodes
Hardware
    DIC (*see* Differential interference contrast)
    microscope 80

objectives 80
temperature control 80
Herpes simplex VP16 transactivator protein 248
Heterochromatin 147
  mating type loci 147, 149
  rDNA 147, 150
  telomeres... 147–149
High-content imaging
  cell preparation 84
High-throughput plasmid transfer 55–63
High throughput screen. *See* Large-scale library transformation
HIS3... 224
HIS3 reporter activity assay 60, 66–68
  3-aminotriazole 60, 66–68, 71
  array-format 60, 66–67
  confirmation (*see* Serial spot dilutions)
  media 60
  protocol.. 66
  serial spot dilutions 60, 66–68
Histone corsstalk 175–177
Histone H3K4 trimethylation 182
Histone modifications 175–176
His6X affinity tag 210
H3K56 acetylation 182, 184
HML silencing... 149, 156
HMR silencing... 149, 156
Homozygous profiling (HOP) assay. *See* Pooled analysis of yeast deletion strains using DNA barcodes
Huntingtin (htt) 161–171
  aggregation (*see* inclusions)
  chaperone co-expression 166, 170
  cloning.. 161
  HEAT domains... 163
  Htt25Q and Htt103Q in yeast....................... 166–167
  inclusions.......................... 161, 163, 166, 171
  misfolding (*see* inclusions)
  polyproline regions 164
  [RNQ+] and htt 166
  structure... 163–164
  toxicity in yeast 167
Huntington's disease (HD)............................. 161, 163, 164
  age of onset.. 163
  CAG repeat 163
  neuropathology... 163
  polyglutamine (polyQ) tract 163
  prevalence 163
  symptoms... 163
  vesicle trafficking....................................... 164

I

iMYTH. *See* Integrated MYTH
Integrated MYTH 250, 259

K

KEGG. *See* Kyoto encyclopedia of genes and genomes
Kynurenine pathway................................. 167–168
  BNA4 (*see* kynurenine 3-monooxygenase (KMO))
  kynurenine 3-monooxygenase (KMO)............ 167–168
  reactive oxygen species (ROS) 168
  Ro 61-8048.. 168
Kyoto encyclopedia of genes and genomes 277

L

lacZ. *See* Beta-galactosidase activity assay (array format)
Large-scale library transformation 254, 260
LexA-DNA binding domain.......................... 248
Life span analysis................................. 103, 109–110, 112
  growth curves 110
  survival... 103, 110
  survival integral (SI) 112
  time shift 110
Linkage group 48, 52

M

Mating. *See* Yeast
MATLAB 109
Media.. 7, 153–155, 231, 256
  basic... 104–105
  C+5-FOA... 153, 154
  complete synthetic media (C)................. 153, 154
  C–TRP.. 154
  C–URA... 153, 154
  dropout mix 79, 232
  LB media.. 211, 215
  minimal 231
  Sc-Ura/dextrose media 212, 216
  Sc-Ura/raffinose Media 216, 217, 219
  SD 104–105
  synthetic drop-out 256
  10x amino acid drop-out mix................... 256
  X-gal.. 256
  yeast.. 231
  yeast extract/peptone/dextrose (YPD)........... 79
  YEPD 153, 231
  YEPD agar.. 155
  YEPD agar plates 104
  YEPD liquid... 104
  YPAD 211, 215, 256
MEFIT 287–289
Membrane proteins 247
Membrane yeast two-hybrid 248
Microarray... 20, 29–31
  hybridization... 30
  noise 20, 31, 33

Microarray (*Continued*)
  prehybridization 29
  wash 30
Mini-prep 255–256, 264–265
  E. coli 256, 265
  yeast 255, 264
MIPS. *See* Munich information center for protein sequences
Mouse models 171
Munich information center for protein sequences 277
MYTH. *See* Membrane yeast two-hybrid

N

Negative selection screen 155
Nematodes 164
N-terminal moiety of ubiquitin. *See* Nub
Nub 248
  NubG 248
  NubI 248
NubG/NubI Test. *See* Bait, verification of bait

O

Oligo-dT cellulose 25
  preparation 25
  recycling 27
OREeome 227, 235

P

Parkinson's disease (PD) 164–165
pAS1. *See* Yeast two-hybrid vectors
Pathogen Functional Genomics Resource Center 235
PC12 cells 164
pDEST22. *See* Yeast two-hybrid vectors
pDEST32. *See* Yeast two-hybrid vectors
Pearson correlation 20, 31
96PEG 232
PFGRC. *See* Pathogen Functional Genomics Resource Center
pGADT7g. *See* Yeast two-hybrid vectors
pGBKT7g. *See* Yeast two-hybrid vectors
Phenotypic activation 19–20, 33
Pitfalls
  fluorescence activation 87
  labeling artifacts 87
  morphological alterations 87
  preparation-dependent localization 90
pOBD2. *See* Yeast two-hybrid vectors
Poly-A mRNA isolation 25–27
  precipitation 27
  yield 27
Pooled analysis of yeast deletion strains using DNA barcodes 115–120
  array hybridization and scanning 122–123, 126–127
  cell samples, purification and amplification of barcodes from 122, 126
  drug function identification using 116–117
  drug target identification using 118
  gene function identification using 116
  pooled cell aliquots, preparing 120–121, 124–125
  pooled cultures, growing 121, 125–126
  tag array results, analysis of 128–131
Pooling. *See* Yeast two-hybrid system
Positive selection screen 155
Precision 277
Prey protein(s) 248
Primary screen 150, 155
  negative selection protocol 155
  positive selection protocol 155
Protein kinase A (PKA) 102
Protein–protein interaction (PPI) 188, 190
  protein complexes 176, 187–190, 203, 223, 247
Purification of *S. Cerevisiae* TAP-tagged fusion proteins 194
  calmodulin 187–189, 191–192, 197, 204–205
  IgG 6, 8, 11, 165, 188–189, 196–197, 212–213, 215, 217
  LC-MS 187–189, 194, 198, 204
  MALDI-TOF MS 188–189
  mass spectrometry 187–190, 192–194, 198–199, 203–206, 274
  protein identification 187, 189, 192, 194, 198–199, 202, 204
  protein purification, 187, 196
  sequest 189, 204, 206
  solvent 125, 194, 204
  statquest 204, 206
  strains 3, 13, 16, 19–21, 24, 31, 38, 42–45, 47–50, 52, 55–56, 62–66, 68–72, 78, 80, 82, 104, 106–112, 115–119, 124–125, 128, 130–142, 146, 148–149, 151, 162, 169–170, 181, 188, 190, 194, 210, 225, 227, 229, 232–233, 235–236, 238, 241–243, 262–263, 266–270
  tandem affinity 187–189
  tandem affinity purification (TAP) 187–192, 194, 203, 205
  tobacco etch virus (TEV) 187–188
  trypsin 189, 193–194, 198, 200–201, 203–204
PVDF filter plate 217

R

Rad6/Bre1 180
Rank Motif 32, 34
rDNA silencing 147–148, 156
Recall 277
Receiver operating characteristic 277
Replicative life span 101

Reporter-based synthetic genetic array
(R-SGA) 55–73
    cloning reporter constructs 60, 69
    control promoters 60, 69
    media 57–59
    protocol 61–63
    reporters 60
    sensitivity 71
    specificity 71
    starting vectors 57, 60
    strains 57
Reverse transcription 27–28
Robotics 41–42, 44, 50
ROC. *See* Receiver operating characteristic
RoTor 42, 50
R-SGA. *See* Reporter-based synthetic
genetic array
RTT109 .... 182, 184

S

Saccharomyces cerevisiae 3, 20, 37, 39, 55–56, 75,
78, 161–162, 175–176, 187–188, 253–254, 268.
See also Yeast
Saccharomyces genome database 277
SCH9 102
Screening strategy 150–151
Screening with yeast models
    compound screens 170, 171
    enhancer screen (*see* synthetic lethal screen)
    suppressor screen 167, 170
    synthetic lethal screen 169–171
SDS. *See* Shwachmann–Diamond
syndrome
Secondary assays 150
    alpha-factor resistance assay
      (HML silencing) 156
    dose response protocol 155–156
    HMR silencing assay 156
    rDNA silencing assay 156
    red/white colony assay 156
Self-activation test 232
Serial spot dilutions .... 60, 66–68.
    See also HIS3 reporter activity assay
SGA. *See* Synthetic genetic array
SGAM. *See* SGA mapping
SGA mapping 38–53
    analysis 46–47
    applications 49–50
    confirmation 48–49
    considerations 51
    isolating suppressors 43, 46, 48
    media 39
    pin tools 41–44
    plates 41, 50

SGD. *See* Saccharomyces genome database
Shwachmann–Diamond syndrome 50
Silencing assays in yeast 146–156
SIR2 102
Sirtuins 146–147
Smart pool-array (SPA) 229
Soluble proteins 188, 263
Solutions 7
SPA. *See* Smart pool-array
SPELL 287–289
Splitomicin 146
Split-ubiquitin membrane yeast two-hybrid.
    See Membrane yeast two-hybrid
Strains. *See* Yeast
Superblock (pierce) 214, 221
Synthetic genetic array (SGA) 38–53, 55, 56
    cell preparation 80, 84
    C-terminal GFP fusions 82
    haploid selection 82
    mating 82
    media 39
    N-terminal GFP fusions 82
    plasmids 42
    protocol 44–46
    query strain construction 42–43, 48, 80
    query strain selection 82
    sporulation 82
    strains 42
Synthetic lethality 38
Synthetic lethal screen 169, 170

T

Tag arrays, DNA. *See* Pooled analysis of yeast deletion
    strains using DNA barcodes
Tags, DNA. *See* Barcodes, DNA
Telomeric silencing 147–148, 150, 155
The Paf1 complex 181
Three-dimensional time-lapse imaging
    setup 85
TOR 102
Total RNA isolation 24–25
    precipitation 25
    yield 25
Transcription elongation factors 181–182
Transcription factor 19–20, 31–34
    dimerization 33
    motif searching 32–34
    nontoxic overexpressors 33
    overexpression 20, 31, 33
    target identification 31–34
Transformation .... 255–256, 264–265.
    See also Yeast
    E. coli 256, 265, 269
    yeast 255, 264

U

Ubiquitin 248
    pseudo-ubiquitin 248
    split-ubiquitin 248
Ubiquitin specific proteases 248
UBPs. *See* Ubiquitin specific proteases

V

V5-alexafluor antibody 220
Validation 263, 265–267
    bioinformatics 263
    co-immunoprecipitation 265
    fluorescence microscopy 267
    in-vitro radiolabeled substrated
        transport assay 267
V5-epitope 5, 210, 215, 219–221

W

Whole genome amplification (WGA) 9, 13, 14
Wilcoxon-Mann-Whitney 34

X

3XYEP-Gal 212, 216, 217, 219

Y

Y187 236
Yeast .... 101–113, 162, 164–172, 223 ff.
    See also Saccharomyces
    cell cycle 162
    colony forming units (CFUs) 104, 113
    competent cells 236
    doubling time 109
    fixation 83
    gasping 105
    genome 162
    Huntington's disease model 165–168
    immobilization 80, 83–84
    mating 162, 240
    media (*see* Media)
    neurodegenerative disease models 164–165
    optical density (OD) 103, 104, 108–110, 112
    stationary phase 101
    strains 78, 236
    transformation 232, 236
Yeast deletion collection .... 115–116.
    See also Pooled analysis of yeast deletion strains
        using DNA barcodes
Yeast gene deletion collection 38, 42, 44, 46, 51, 55–57
Yeast gene deletion strains (YGDS) 169
Yeast ORF collection (YOC) 171, 172
Yeast ORF deletion collection 107
Yeast overexpression array 20, 33
Yeast strains 148
    BY4700 210
    Y258 .... 211, 215
Yeast two-hybrid 248
Yeast two-hybrid screens 223
    array-based 223, 227
    protocol 232
    self-activation test 232, 238
Yeast two-hybrid system 224
    applications 224
    false negatives 226
    false positives 226
    limitations 226
    pooled array screening 228
    pooling strategy 228
    requirements 230
Yeast two-hybrid vectors 230
    pAS1 230
    pDEST22 230
    pDEST32 230
    pGADT7g 230
    pGBKT7g 230
    pOBD2 230
YEB 191, 195–196, 205
YEPD 39–40, 42–44, 46, 58–62, 64, 66,
    68–69, 153, 155–156, 190, 194, 225, 230–233,
    236, 239–243
Y2H. *See* Yeast two-hybrid

Printed in the United States of America